Android App开发从入门到精通

· 安辉 编著 ·

清华大学出版社
北京

内 容 简 介

本书使用Android Studio 3.0开发环境，同时适配新版的Android 8.0操作系统，由浅入深地学习Android App的开发。全文共分为10章，涵盖Android Studio的开发环境搭建、Android控件的使用、四大组件的使用、Fragment（碎片）、多线程开发、网络编程与数据存储等内容。最后通过项目实战，对所学知识点融会贯通，进一步增强开发能力。

本书内容通俗易懂，案例丰富，不仅适用于Android开发的广大从业人员、App开发的业余爱好者，也可作为大中专院校与培训机构的培训教程。

图书在版编目（CIP）数据

Android App开发从入门到精通 / 安辉编著. —北京：清华大学出版社，2018（2022.1重印）
ISBN 978-7-302-51358-2

Ⅰ. ①A… Ⅱ. ①安… Ⅲ. ①移动终端－应用程序－程序设计 Ⅳ. ①TN929.53

中国版本图书馆CIP数据核字（2018）第229155号

责任编辑： 王金柱
封面设计： 王　翔
责任校对： 闫秀华
责任印制： 沈　露

出版发行： 清华大学出版社
网　　址： http://www.tup.com.cn，http://www.wqbook.com
地　　址： 北京清华大学学研大厦A座　　**邮　　编：** 100084
社 总 机： 010-62770175　　**邮　　购：** 010-62786544
投稿与读者服务： 010-62776969，c-service@tup.tsinghua.edu.cn
质 量 反 馈： 010-62772015，zhiliang@tup.tsinghua.edu.cn

印 装 者： 北京富博印刷有限公司
经　　销： 全国新华书店
开　　本： 190mm×260mm　　**印　　张：** 26　　**字　　数：** 666千字
版　　次： 2018年12月第1版　　**印　　次：** 2022年1月第6次印刷
定　　价： 79.00元

产品编号：072653-01

前　　言

我在写书之前一直在 CSDN 上发表文章，同时在“知乎”等网站回答一些 Android 相关的问题，后来有幸收到了清华大学出版社编辑的邀请，于是产生了编写本书的想法。

我最早是从事 Java Web 开发的，出于对 Android 的浓厚兴趣，后来又开始从事 Android 的开发。在开发过程中，我走了很多弯路，阅读了很多 Android 方面的书，从入门类到高级开发类都有，美中不足的是这些书要么篇幅过长要么技术过时，浪费了很多时间。鉴于此，本书将结合我多年的 Android 开发经验，总结企业中常用的开发技术，使用前沿技术兼容最新的 Android 操作系统，使初学者快速加入 Android 开发阵营。即使是中、高级开发者，阅读本书后也能从中获益。

Android 操作系统经过将近 10 年的发展。随着移动 App 的热潮，越来越多的人加入移动开发的大军，企业对 Android 招聘的需求也越来越高。本书内容从基础入门到高级开发，涵盖企业开发中常用的技术点，能让读者对 Android 开发有一个学习框架。最后一章通过模仿商业 App 开发，融会贯通前面的知识点，以提高读者项目开发的实战能力。

本书内容

本书共有 10 章，主要内容如下：

- 第 1 章学习开发工具 Android Studio 的使用，一个好的开发工具可以大大提高开发人员的工作效率。
- 第 2 章讲解 Android 控件相关知识，一个 UI 界面由多个控件组成，只有熟练使用各种控件才能设计出好看的 App，达到 UI 设计师想要的效果。
- 第 3 章学习 Android 中四大组件的使用。在企业的项目开发中，四大组件中的 Activity（活动）、Service（服务）、Broadcast Receiver（广播接收器）使用很频繁，ContentProvider（内容提供者）使用频率相对少一些，只有某些特定需求时才会用到。
- 第 4 章学习 Fragment（碎片）的使用方法，从 Fragment 简单使用到最后的案例开发，一步步深入地学习 Fragment。使用 Fragment 会让 App 模块化，还能解决手机与平板电脑的适配问题。
- 第 5 章学习多线程开发。从多线程的创建，到子线程如何更新 UI，通过阅读源码分析 Handle 的实现原理，最后介绍线程池的使用方法。
- 第 6 章首先学习 Android 的网络编程，通过 Get/Post 方式向服务器发送 HTTP 请求。现在市面上大部分 App 与服务器交互都是返回 Json 数据，所以介绍 Gson 框架，以及 OkHttp 开源项目的使用和封装。最后是数据存储的三种方式。
- 第 7 章学习 Android 的高级应用，主要介绍 Notification 使用、多媒体开发、WebView 使用、定位的三种方式、NDK 和 JNI 开发、Git 管理项目等。

- 第 8 章学习 Android 中各大版本的更新，让我们的 App 解决版本适配问题，完美兼容 5.0 以上的各个版本。
- 第 9 章学习常用功能模板的使用。这些功能是企业开发中可能会碰到的需求，通过模板的学习，知道如何对一个 App 进行功能划分以及如何封装模块。
- 第 10 章通过模仿一个商业 App，从零开始搭建项目，使用前面 9 章所学的内容，将所学知识点融会贯通，并进一步熟练掌握。有了项目开发的经验，你在今后的企业开发中就能快速成为一名合格的开发人员。

本书特色

本书定位为基础类图书，对每一个知识点的讲解都很详细，从基础入门逐步进入高级应用，让读者能系统全面地学习 Android 开发，更深入地了解 Android 开发体系。本书的内容是我多年 Android 开发经验的总结，也是一个合格的 Android 开发者必须掌握的内容，简单来说，就是企业开发中经常用到的技术。

致谢

首先感谢我的好朋友王帅和芮成兵，他们协助我完成了本书的审阅工作，给本书的修订提供了宝贵的意见；然后感谢翼成的伙伴们，在写书的这段时间里他们给予我很大的帮助。

书中案例源码下载

https://github.com/ansen666/book_source_code

资源与勘误

由于技术水平有限，书中难免会有疏漏，欢迎大家通过 androidcoder666@163.com 邮箱提供反馈意见。另外，还可以通过以下方式进行交流：

- CSDN 博客：https://blog.csdn.net/lowprofile_coding
- QQ 群：202928390
- 微信公众号：Android 开发 666

编　者

2018 年 10 月

目　　录

第 1 章

Android Studio 的介绍以及使用

孔子云："工欲善其事，必先利其器"。一个好的开发工具可以让开发人员的工作效率有大幅度的提高，本章主要学习 Android Studio 的使用。以前都是用 Eclipse 开发 Android 程序，自从 2013 年 Google 官方发布了 Android Studio，现在很少有人使用 Eclipse 开发 Android 程序了。本书不对 Eclipse 多做介绍。

Android Studio 是 Google 于 2013 年 I/O 大会针对 Android 开发推出的新开发工具，是基于 IntelliJ IDEA 开发的，IntelliJ 在业界被公认为最好的 Java 开发工具之一。尤其是在智能代码助手、代码自动提示、重构、J2EE 支持、各类版本工具（Git、SVN、GitHub 等）、JUnit、CVS 整合、代码分析、创新的 GUI 设计等方面的功能，可以说是超常的。IDEA 是 JetBrains 公司的产品，它的旗舰版本还支持 HTML、CSS、PHP、MySQL、Python 等，免费版本只支持 Java 等少数语言。

1.1 探索 Android Studio

Android Studio 是基于 IntelliJ IDEA 的官方 Android 应用集成开发环境（IDE）。除了 IntelliJ 强大的代码编辑器和开发者工具，Android Studio 提供了更多可提高 Android 应用构建效率的功能，例如：

- 基于 Gradle 的灵活构建系统。
- 快速且功能丰富的模拟器。
- 可针对所有 Android 设备进行开发的统一环境。
- Instant Run，可将变更推送到正在运行的应用，无须构建新的 APK。
- 帮助构建应用程序和导入示例代码以及 GitHub 集成。
- 丰富的测试工具和框架。
- 可捕捉性能、易用性、版本兼容性以及其他问题的 Lint 工具。

- C++和 NDK 支持。
- 内置对 Google 云端平台的支持，可轻松集成 Google Cloud Messaging 和 App 引擎。

1.1.1 项目结构

Android Studio 中的每个项目包含一个或多个含有源代码文件和资源文件的模块。模块类型包括：

- Android 应用模块。
- 库模块。
- Google App 引擎模块。

默认情况下，Android Studio 会在 Android 项目视图中显示项目文件，如图 1-1 所示。该视图按模块组织结构，便于快速访问项目的关键源文件。

所有构建文件在项目层次结构顶层 Gradle Scripts 下显示，并且每个应用模块都包含以下文件夹：

- manifests：包含 AndroidManifest.xml 文件。
- java：包含 Java 源代码文件，包括 JUnit 测试代码。
- res：包含所有非代码资源，例如 XML 布局、UI 字符串和位图图像。

图 1-1 Android 视图中的项目文件

磁盘上的 Android 项目结构与此扁平项目结构有所不同。要查看实际的项目文件结构，可以从 Project 下拉菜单（在图 1-1 中显示为 Android）选择 Project。

用户还可以自定义项目文件的视图，重点显示应用开发的特定方面。例如，选择项目的 Problems 视图会显示指向包含任何已识别编码和语法错误（如布局文件中缺少一个 XML 元素结束标记）的源文件链接，如图 1-2 所示。

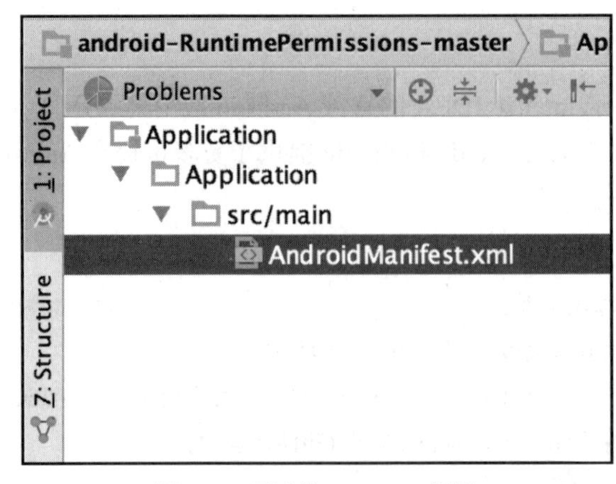

图 1-2 项目的 Problems 视图

1.1.2　Android Studio 主窗口

Android Studio 主窗口由如图 1-3 所示的几个逻辑区域组成。

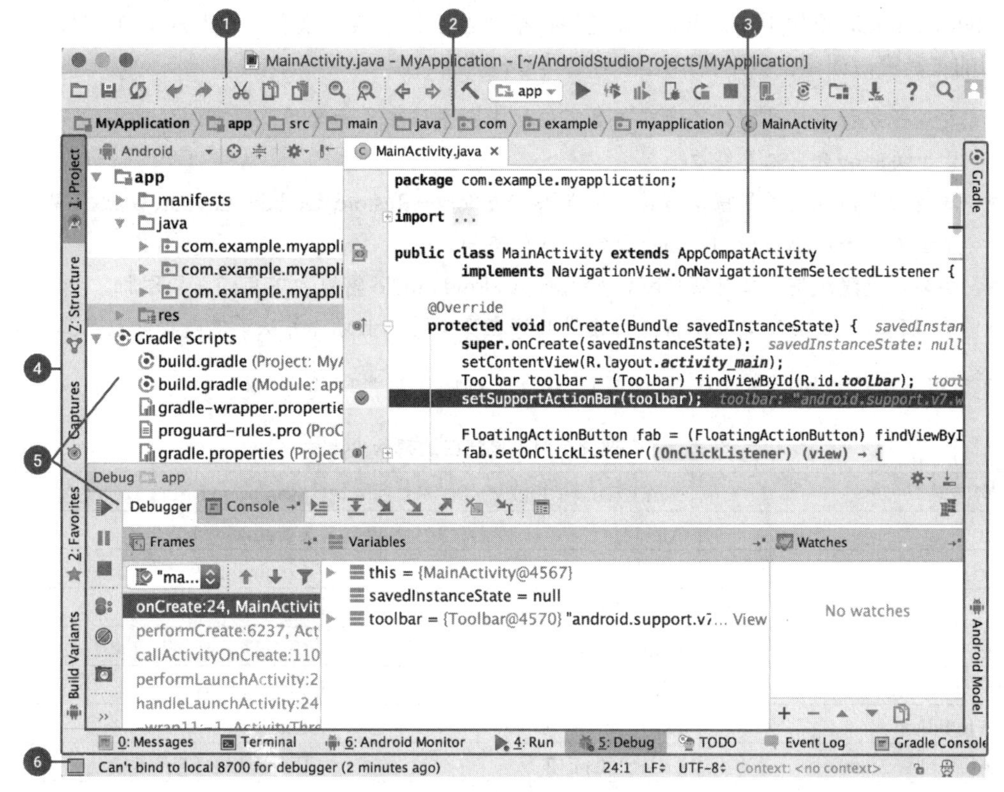

图 1-3　Android Studio 主窗口

① 工具栏，提供执行各种操作的工具，包括运行应用和启动 Android 工具。

② 导航栏，可以帮助在项目中导航，以及打开文件进行编辑。此区域提供 Project 窗口所示结构的精简视图。

③ 编辑器窗口，是创建和修改代码的区域。编辑器可能因当前文件类型的不同而有所差异。例如，在查看布局文件时，编辑器显示布局编辑器。

④ 工具窗口栏，在 IDE 窗口外部运行，并且包含可用于展开或折叠各个工具窗口的按钮。

⑤ 工具窗口，提供对特定任务的访问，例如项目管理、搜索和版本控制等。可以展开和折叠这些窗口。

⑥ 状态栏，显示项目和 IDE 本身的状态以及任何警告或消息。

用户可以通过隐藏或移动工具栏和工具窗口调整主窗口，以便留出更多屏幕空间，还可以使用键盘快捷键访问大多数 IDE 功能。

可以随时通过按两下 Shift 键或点击 Android Studio 窗口右上角的放大镜搜索源代码、数据库、操作和用户界面的元素等。此功能非常实用，例如在忘记如何触发特定 IDE 操作时，可以利用此功能进行查找。

1.1.3 工具窗口

Android Studio 不使用默认窗口，而是根据情境在工作时自动显示相关工具窗口。默认情况下，最常用的工具窗口固定在应用窗口边缘的工具窗口栏上。

- 要展开或折叠工具窗口，请在工具窗口栏中点击该工具的名称，还可以拖动、固定、取消固定、关联和分离工具窗口。
- 要返回当前默认工具窗口布局，请点击 Window→Restore Default Layout 或点击 Window→Store Current Layout as Default 自定义默认布局。
- 要显示或隐藏整个工具窗口栏，请点击 Android Studio 窗口左下角的窗口图标。
- 要找到特定工具窗口，请将鼠标指针悬停在窗口图标上方，并从菜单选择相应的工具窗口。

还可以使用键盘快捷键打开工具窗口。表 1-1 列出了最常用的窗口快捷键。

表 1-1 部分实用工具窗口的键盘快捷键

工具窗口	Windows 和 Linux	Mac
Project	Alt+1	Command+1
Version Control	Alt+9	Command+9
Run	Shift+F10	Ctrl+R
Debug	Shift+F9	Ctrl+D
Android Monitor	Alt+6	Command+6
Return to Editor	Esc	Esc
Hide All Tool Windows	Ctrl+Shift+F12	Command+Shift+F12

如果想要隐藏所有工具栏、工具窗口和编辑器选项卡，请点击 View→Enter Distraction Free Mode。此操作可启用无干扰模式。要退出“无干扰模式”，请点击 View→Exit Distraction Free Mode。

用户可以使用快速搜索在 Android Studio 中的大多数工具窗口中执行搜索和筛选。要使用快速搜索，请选择工具窗口，然后输入搜索查询。

1.1.4 代码自动完成

Android Studio 有三种自动补全代码快捷键，如表 1-2 所示。

表 1-2 代码自动完成的键盘快捷键

类型	说明	Windows 和 Linux	Mac
基本自动完成	显示对变量、类型、方法和表达式等的基本建议。如果连续两次调用基本自动完成，将显示更多结果，包括私有成员和非导入静态成员	Ctrl+空格	Control+空格

（续表）

类型	说明	Windows 和 Linux	Mac
智能自动完成	根据上下文显示相关选项。智能自动完成可识别预期类型和数据流。如果连续两次调用智能自动完成，将显示更多结果，包括链	Ctrl+Shift+空格	Control+Shift+空格
语句自动完成	自动完成当前语句，添加缺失的圆括号、大括号、花括号和格式化等	Ctrl+Shift+Enter	Shift+Command+Enter

还可以按 Alt+Enter 组合键执行快速修复并显示建议的操作。

1.1.5 样式和格式化

在编辑时，Android Studio 将自动应用代码样式设置中指定的格式设置和样式。可以通过编程语言自定义代码样式设置，包括指定选项卡和缩进、空格、换行、花括号以及空白行的约定。要自定义代码样式设置，请点击 File→Settings→Editor→Code Style（在 Mac 上，点击 Android Studio→Preferences→Editor→Code Style）。

IDE 会在你写代码时自动对代码进行格式化，也可以通过按快捷键 Ctrl+Alt+L（在 Mac 上，按 Opt+Command+L）格式化代码、按快捷键 Ctrl+Alt+I（在 Mac 上，按 Alt+Option+I*）自动缩进所有行。图 1-4（a）是格式化之前的代码，图 1-4（b）是格式化之后的代码。

```
public void onCreate(Bundle savedInstanceState) {
    super.onCreate(savedInstanceState);
        setContentView(R.layout.activity_main);
  mActionBar = getSupportActionBar();
        mActionBar.setDisplayHomeAsUpEnabled(true);
```

（a）

```
public void onCreate(Bundle savedInstanceState) {
    super.onCreate(savedInstanceState);
    setContentView(R.layout.activity_main);
    mActionBar = getSupportActionBar();
    mActionBar.setDisplayHomeAsUpEnabled(true);
        Formatted 7 lines
        Show reformat dialog: ⌥⇧⌘L
    // Get reference to the drawer layout and set event listener
```

（b）

图 1-4-2　格式化前后的代码

1.1.6 版本控制基础知识

Android Studio 支持多个版本控制系统（VCS），包括 Git、GitHub、CVS、Mercurial、Subversion

和 Google Cloud Source Repositories。

在将应用导入 Android Studio 后，使用 Android Studio VCS 菜单选项启用对所需版本控制系统的 VCS 支持、创建存储库、导入新文件至版本控制以及执行其他版本控制操作：

- 在 Android Studio VCS 菜单中点击 Enable Version Control Integration。
- 从下拉菜单中选择要与项目根目录关联的版本控制系统，然后点击 OK 按钮。

此时，VCS 菜单将根据选择的系统显示多个版本控制选项。

提 示
还可以使用 File→Settings→Version Control 菜单选项设置和修改版本控制设置。

1.1.7 Gradle 构建系统

Android Studio 基于 Gradle 构建系统，并通过适用于 Gradle 的 Android 插件提供更多面向 Android 的功能。该构建系统可以作为集成工具从 Android Studio 菜单运行，还可以从命令行独立运行。

可以利用构建系统的功能执行以下操作：

- 自定义、配置和扩展构建流程。
- 使用相同的项目和模块为用户的应用创建多个具有不同功能的 APK。
- 在不同源代码集之间重复使用代码和资源。

利用 Gradle 的灵活性，可以在不修改应用核心源文件的情况下实现以上所有目的。Android Studio 构建文件以 build.gradle 命名。

这些文件是纯文本文件，使用适用于 Gradle 的 Android 插件提供的元素以 Groovy 语法配置构建。

每个项目有一个用于整个项目的顶级构建文件，以及用于各模块的单独的模块层级构建文件。在导入现有项目时，Android Studio 会自动生成必要的构建文件。

1.1.8 Debug 调试

使用 Debug 调试功能在调试程序视图中对引用、表达式和变量值进行内联验证，提高代码检查效率，如图 1-5 所示。Debug 调试信息包括：

- 变量值
- 引用某选定对象的引用对象
- 方法返回值
- Lambda 和运算符表达式
- 工具提示值

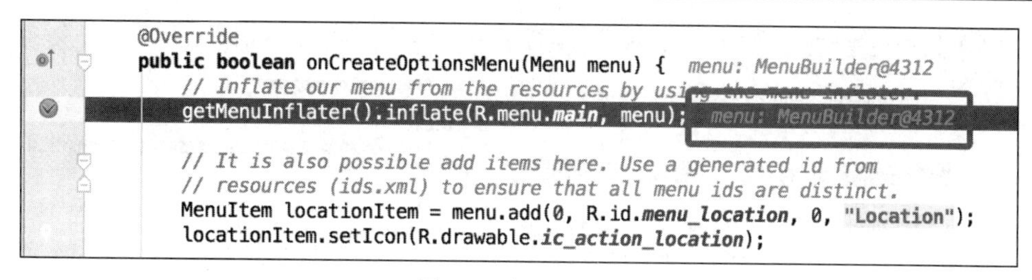

图 1-5 内联变量值

要启用 Debug 调试，请在 Debug 窗口中点击 Settings，然后选中 Show Values Inline 复选框。

1.1.9 性能监视器

Android Studio 提供性能监视器，让用户可以更加轻松地跟踪应用的内存和 CPU 使用情况、查找已解除内存分配的对象、查找内存泄漏以及优化图形性能和分析网络请求。

在设备或模拟器上运行应用时，打开 Android Monitor 工具窗口，然后点击 Monitors 标签。

1.1.10 分配跟踪器

Android Studio 允许在监视内存使用情况的同时跟踪内存分配情况。利用跟踪内存分配功能，可以在执行某些操作时监视对象被分配到哪些位置。了解这些分配后，就可以相应地调整与这些操作相关的方法调用，从而优化应用的性能和内存使用。

1.1.11 数据文件访问

Systrace、logcat 和 Traceview 等 Android SDK 工具可生成性能和调试数据，用于对应用进行详细分析。

要查看已生成的数据文件，请打开 Captures 工具窗口。在已生成的文件列表中，双击相应的文件即可查看数据。右击任何.hprof 文件，即可将其转换为标准.hprof 文件格式。

1.1.12 代码检查

在每次编译程序时，Android Studio 都将自动运行已配置的 Lint 及其他 IDE 检查，帮助轻松识别和纠正代码结构质量问题。

Lint 工具可检查你的 Android 项目源文件是否有潜在的错误，以及在正确性、安全性、性能、可用性、无障碍性和国际化方面是否需要优化改进，如图 1-6 所示。

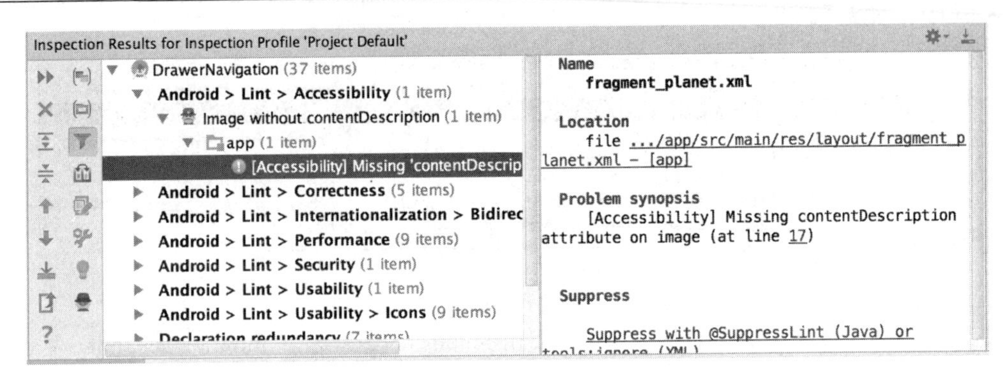

图 1-6 Android Studio 中 Lint 检查的结果

除了 Lint 检查，Android Studio 还可以执行 IntelliJ 代码检查和注解验证，以简化编码工作流程。

1.1.13 日志消息

在使用 Android Studio 构建和运行应用时，点击窗口底部的 Android Monitor 查看 adb 输出和设备日志消息（logcat）。

1.2 下载与安装 Android Studio

在下载 Android Studio 之前，需要先安装 Java 语言的软件开发工具包 JDK，因为 Android 是基于 Java 语言的，并且所有的 App 都是在 Java 虚拟机上运行的。

Oracle 官网下载地址为 http://www.oracle.com/technetwork/java/javase/downloads/jdk8-downloads-2133151.html。

上面这个链接可以下载各个操作系统的 oracle jdk1.8 版本，例如笔者的电脑是 Mac 系统，具体下载地址为 http://download.oracle.com/otn-pub/java/jdk/8u144-b01/090f390dda5b47b9b721c7dfaa008135/jdk-8u144-macosx-x64.dmg。

1.2.1 下载 Android Studio

以前下载 Android Studio 还需要科学上网，或者国内的镜像地址下载，2016 年 12 月 8 日 Google Developers 中国网站（developers.google.cn）正式发布。

Google Developers 中国网站是特别为中国开发者而建立的，汇集了 Google 为全球开发者所提供的开发技术资源，包括 API 文档、开发案例、技术培训的视频。

不需要科学上网的官方下载 Android Studio 地址为 https://developer.android.google.cn/studio/index.html。

打开地址首页，如图 1-7 所示。

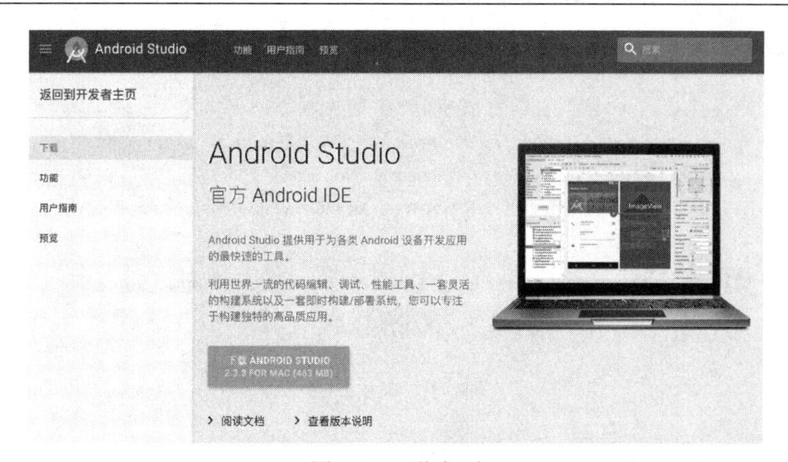

图 1-7　下载主页

它会自动识别用户的操作系统，例如笔者的电脑是 Mac，所以那个下载的按钮上就有 FOR MAC 版本，直接点击“下载 ANDROID STUDIO”按钮即可。

1.2.2　开始安装

安装 Android Studio 的方法如下：

步骤01 双击打开下载的 Android Studio 的安装包，将左侧的 Android Studio 图标拖到右侧，如图 1-8 所示。

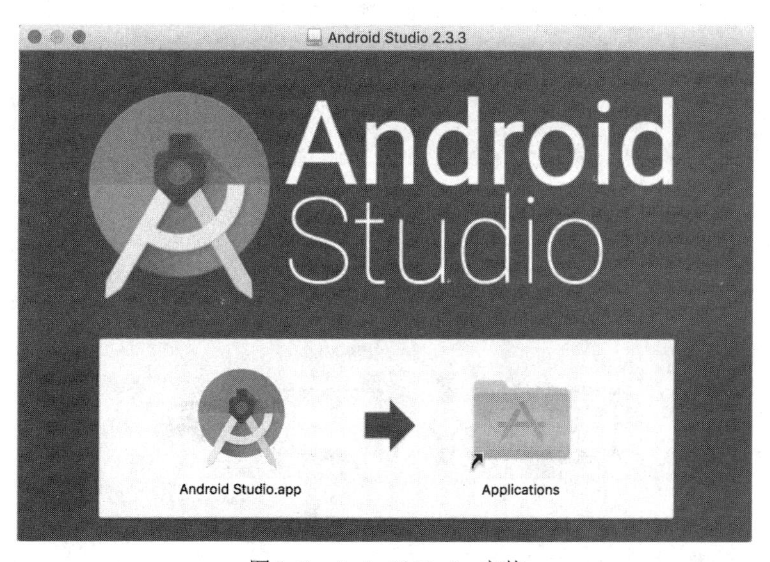

图 1-8　Android Studio 安装

步骤02 在应用列表（Launchpad）中打开 Android Studio，进入如图 1-9 所示的设置界面，其中有两个选项。

- 第一个选项是如果之前使用过 Android Studio，可以导入之前的配置文件。
- 第二个选项是默认的，之前没有安装过 Android Studio 或者我不想导入之前的配置。

一般情况用第二个选项（默认）就行，点击 OK 按钮。

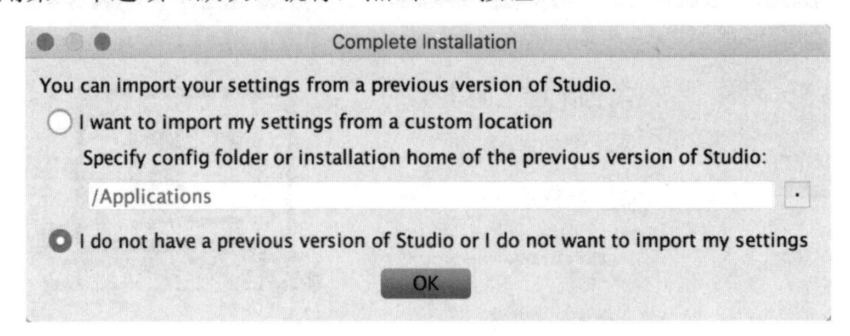

图 1-9　设置界面

步骤03 进入设置代理界面，如图 1-10 所示。

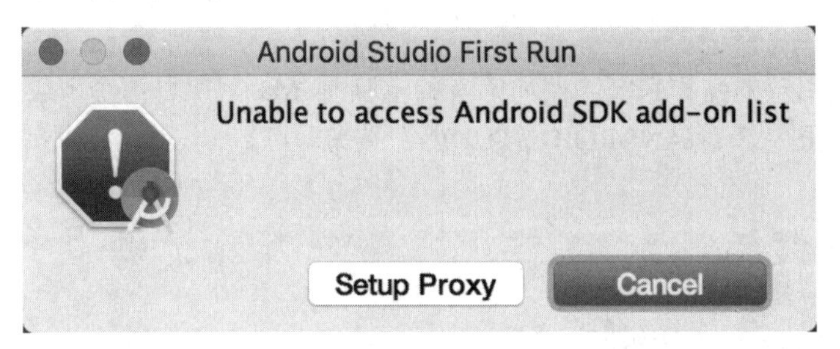

图 1-10　设置代理

步骤04 如果无法访问 Android SDK 列表，这时会弹出一个设置代理的提示对话框。我们不需要设置，直接点击 Cancel 按钮，进入欢迎界面，如图 1-11 所示。

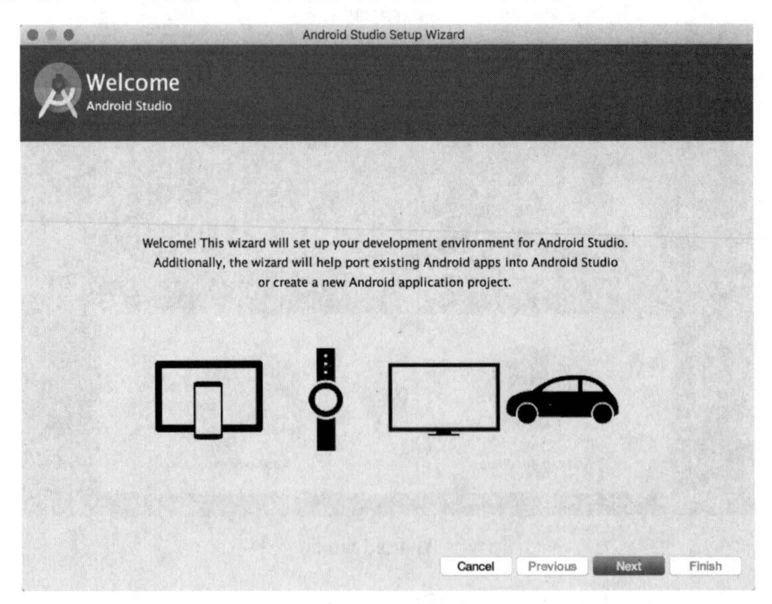

图 1-11　欢迎界面

步骤05 在欢迎界面直接点击 Next 按钮，选择安装类型，如图 1-12 所示，直接选择“Standard（标准）”选项，点击 Next 按钮。

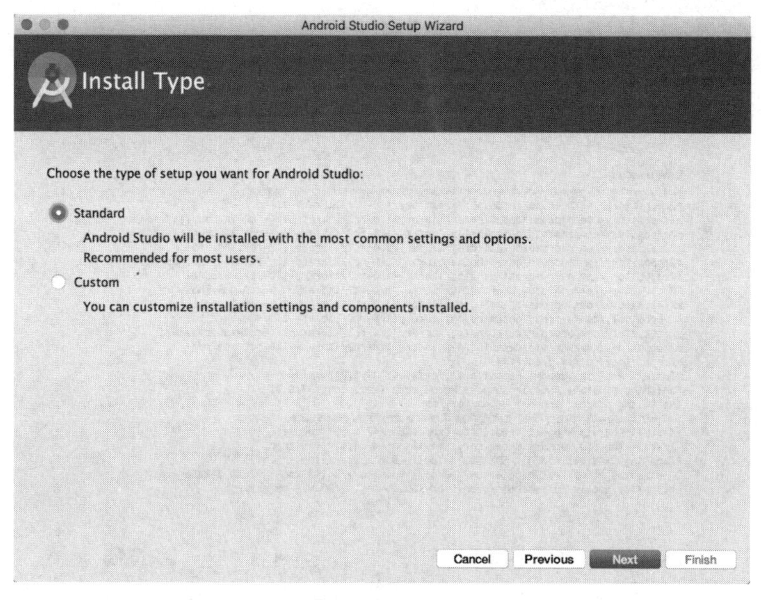

图 1-12　安装类型

步骤06 选择安装类型之后进入 SDK 组件下载列表界面，如图 1-13 所示，直接点击 Finish 按钮。

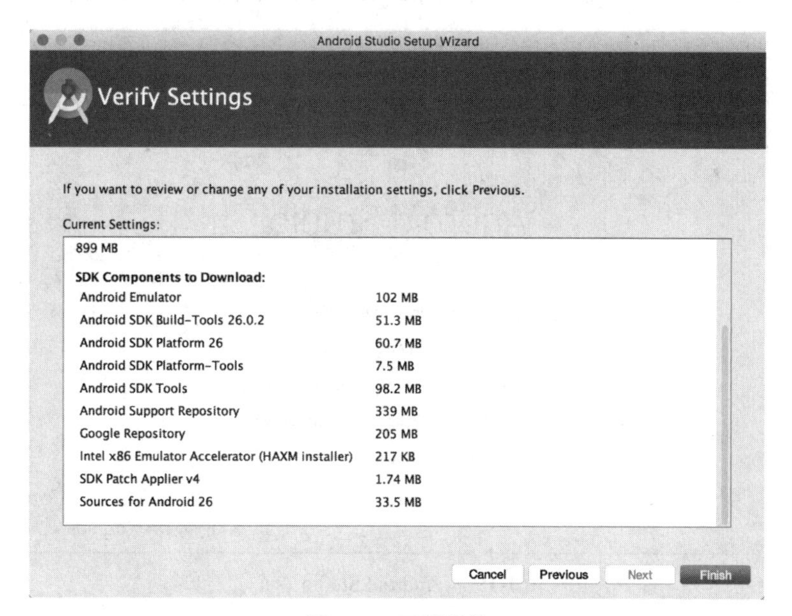

图 1-13　下载组件

步骤07 下载 SDK 组件中，如图 1-14 所示。

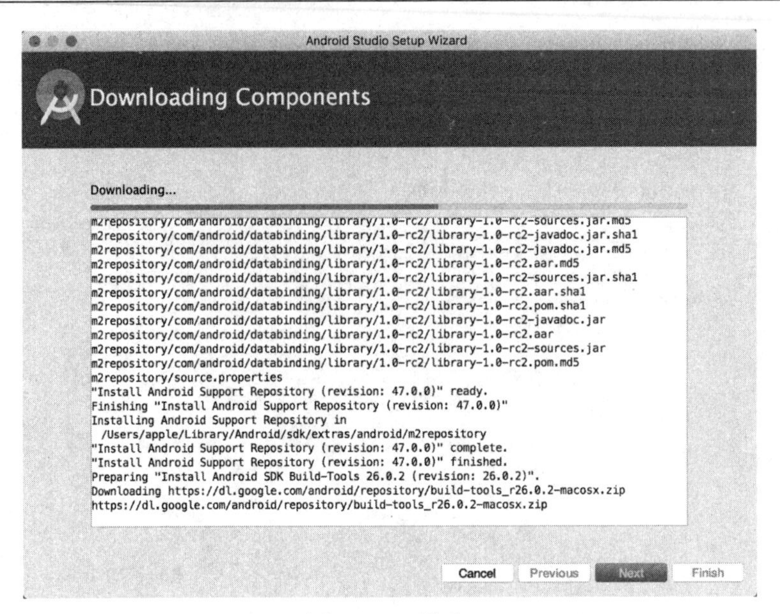

图 1-14 下载中

步骤08 下载这么多组件肯定需要一些时间，等待下载完成直接点击 Finish 按钮。接下来进入 Android Studio 主页，如果 1-15 所示，至此 Android Studio 就算安装配置完成。

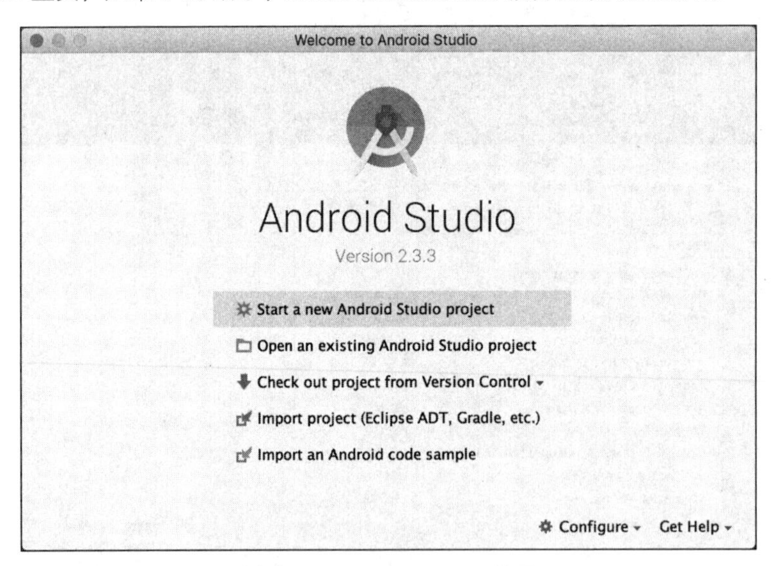

图 1-15 Android Studio 首页

首页中的选项也有不少，但是常用的有以下三个。

- Start a new Android Studio project：新建一个项目。
- Open an existing Android Studio project：打开一个存在的项目。
- Import project（Eclipse ADT,Gradle,etc）：导入项目，包含 Eclipse ADT 项目、Gradle 项目等。

1.3　Android Studio 使用

前面我们介绍了 Android Studio 的下载以及安装，相信你早已按捺不住，是时候开始实战一下了。本节内容包括创建项目、运行以及调试等一系列操作，带你熟悉 Android Studio 的基本使用。

1.3.1 项目结构

Android Studio 的项目包含 App 需要的所有内容，从源代码和资源，到测试代码和构建配置，应有尽有。当创建新项目的时候，Android Studio 会帮助所有的文件创建项目结构，在 IDE 左侧的 Project 窗口中可见。

1. 模块

模块是源文件和构建设置的集合，允许你将项目分成不同的功能单元。一个项目可以有一个或者多个模块，并且一个模块可以对其他模块进行依赖。每个模块可以独立构建、测试和调试。

如果在自己的项目中创建代码库或者希望为不同的设备类型（例如电话和穿戴式设备）创建不同的代码和资源组，但是保留相同项目内的所有文件并共享某些代码，那么增加模块数量将非常有用。

可以点击 File→New→New Module，帮助项目添加新模块。

Android 有两种常用的模块。

（1）Android 应用模块

为应用的源代码、资源文件和应用级设置（例如模块级构建文件和 Android 清单文件）提供容器。在创建新项目时，默认的模块名称是“app”。

在 **Create New Module** 窗口中，Android Studio 提供了以下应用模块：

- Phone & Tablet Module 手机开发
- Android Wear Module 手表开发
- Android TV Module 电视开发
- Glass Module 眼镜开发

每种模块都提供了基础文件和一些代码模板，不同的设备类型对应不同的模板。

（2）库模块

库模块是某个功能的可重用代码，可用作其他项目的依赖或者导入其他项目中。库模块在结构上与应用模块相同，但是在构建时，它将创建一个代码归档文件而不是 APK，因此无法安装到设备上。

在 **Create New Module** 窗口中，Android Studio 提供了以下库模块：

- Android 库：这种类型的库可以包含 Android 项目中支持的所有文件类型，包括源代码、资源和清单文件。构建结果是一个 Android 归档（AAR）文件，可以将其作为 Android 应用模块的依赖项添加。

- Java 库：此类型的库只能包含 Java 源文件。构建结果是一个 Java 归档（JAR）文件，可以将其作为 Andriod 应用模块或其他 Java 项目的依赖项添加。

一些人也将模块称为子项目，完全没有问题，因为 Gradle 也将模块称为项目。

例如，在创建库模块并且希望以依赖项的形式将其添加到你的 Android 应用模块时，必须按照如下形式进行声明：

```
dependencies {
  compile project(':my-library-module')
}
```

2. 项目文件

默认情况下，Android Studio 会在 Android 视图中显示用户的项目文件。此视图无法反映磁盘上的实际文件层次结构，而是按照模块和文件类型组织，简化项目主要源文件之间的导航，同时将不常用的特定文件或目录隐藏。

与磁盘上的结构相比，一些结构变化包括：

- 在顶级 Gradle Script 组中显示项目中与构建相关的所有配置文件。
- 在模块级组（如果不同的产品类型和构建类型使用不同的清单文件）中显示每个模块的所有清单文件。
- 在一个组中显示所有备用资源文件，而不是按照资源限定符在不同的文件夹中显示。例如，所有密度版本的启动器图标将并排显示。
- 项目文件结构如图 1-16 所示。在每个 Android 应用模块内，文件显示在以下组中：
  - manifests 包含 AndroidManifest.xml 文件。
  - java 包含 Java 源代码文件（包括 JUnit 测试代码），这些 java 文件根据包名进行区分。
  - res 包含所有非代码资源，例如 XML 布局、字符串和图片等，这些资源对应不同的文件夹。

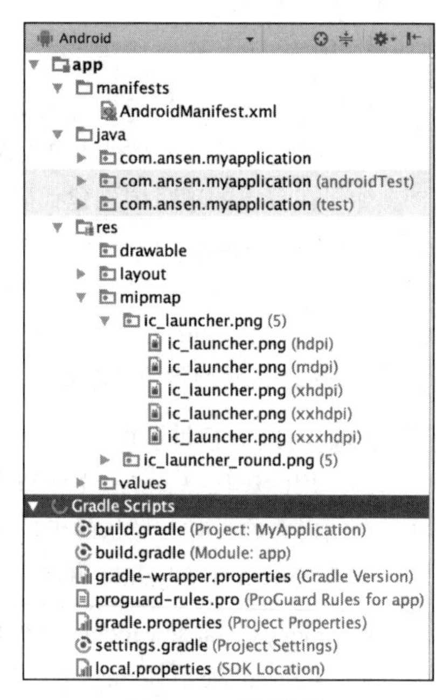

图 1-16 项目文件

3. Android 项目视图

要查看项目的实际文件结构（包括 Android 视图下隐藏的所有文件），请从 Project 窗口顶部的下拉菜单中选择 Project。

选择 Project 视图后，可以看到更多文件和目录，如图 1-17 显示。最重要的一些文件和目录如下：

- **模块名称/**
  - **build/** 包含构建输出。
  - **libs/** 包含私有库。
  - **src/** 包含模块的所有代码和资源文件，分为以下子目录：
    - **androidTest/** 包含在 Android 设备上运行的仪器测试的代码。

◆ **main/** 包含“主”源集文件：所有构建变体共享的Android代码和资源（其他构建变体的文件位于同级目录中，例如调试构建类型的文件位于src/debug/中）。

■ **AndroidManifest.xml** 说明应用及其每个组件的性质。

■ **java/** 包含Java代码源。

■ **jni/** 包含使用Java原生接口（JNI）的原生代码。

■ **gen/** 包含Android Studio生成的Java文件，例如R.java文件以及从AIDL文件创建的接口。

■ **res/** 包含应用资源，例如可绘制对象文件、布局文件和UI字符串。

■ **assets/** 包含原封不动地编译到.apk文件中的文件。可以使用URI像访问文件系统一样访问此目录，以及使用AssetManager以字节流形式读取文件。例如，这个文件夹可以放一种提示音mp3文件。

◆ **test/** 包含在JVM上运行的本地测试的代码。

➢ **build.gradle（模块）** 构建当前模块的配置。

- **build.gradle（项目）** 定义适用于所有模块的构建配置。此文件已集成到项目中，因此应当在所有其他源代码的修订控制中保留这个文件。

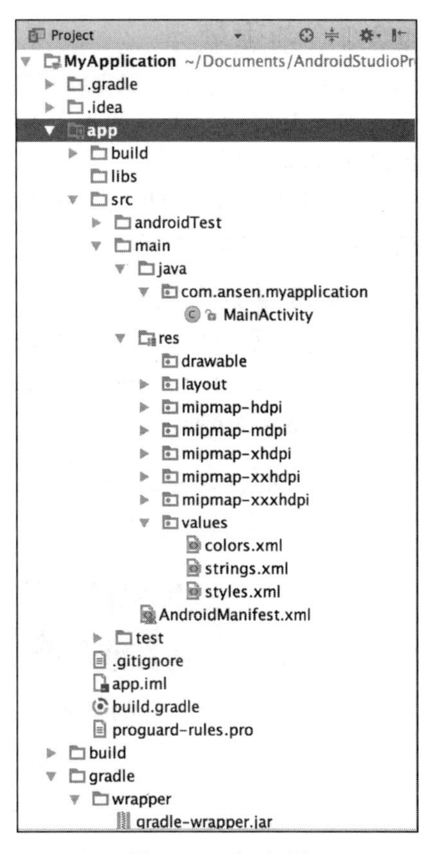

图1-17 项目视图

4. 项目结构设置

要更改Android Studio项目的各种设置，点击File→Project Structure，打开Project Structure对话框。此对话框包含以下部分：

- SDK Location：设置你的项目使用的JDK、Android SDK和Android NDK的位置。
- Project：设置Gradle和Android Plugin for Gradle的版本，以及存储区位置名称。
- Developer Services：包含Google或其他第三方的Android Studio附加组件的设置。
- Modules：允许编辑模块特定的构建配置，包括目标和最低SDK、应用签名和库依赖项。

借助Modules设置部分，可以为项目的每个模块更改配置选项。每个模块的配置页面分成以下标签：

- Properties：指定编译模块所用的SDK和构建工具的版本。
- Signing：指定签名证书。
- Flavors：指定SDK的最低版本、最高版本、版本号、版本名称。我们也可以修改Module的build.gradle文件修改这些配置。

- Build Types：指定编译模式，每个模块都可以设置 release 和 debug 模式，也可以根据需要自定义类型。
- Dependencies：列出此模块的库、文件和模块依赖项。可以在这里添加删除修改依赖库。

1.3.2 创建项目

利用 Android Studio，可以轻松地为各种机型（例如，手机、平板电脑、TV、Wear 和 Google Glass）创建 Android 应用。

新项目向导让用户可以为自己的应用选择机型，并使用启动所需的一切信息填充项目结构。按以下步骤操作来创建新项目。

步骤01 启动并配置项目。

- 如果没有打开项目，在 Android Studio 首页中点击 Start a New Android Studio project 按钮。
- 如果已经打开一个项目，Android Studio 将显示开发环境。要创建新项目时，请点击 File→New→New Project。

这两种方法都能创建项目，点击之后可以在下一个对话框中配置应用的名称、公司主体、软件包名称和项目的位置，如图 1-18 所示。为你的项目输入相应的值后点击 Next 按钮。

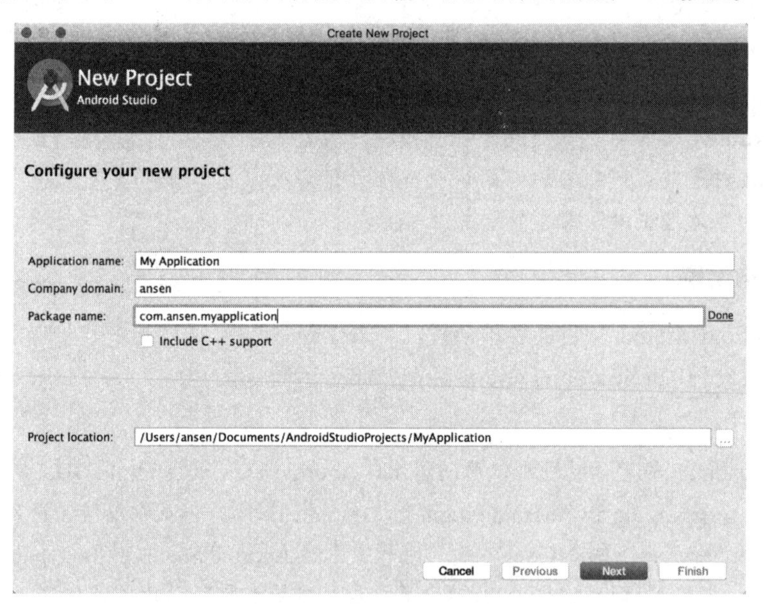

图 1-18 新建项目配置

步骤02 选择机型和 API 级别，如图 1-19 所示。

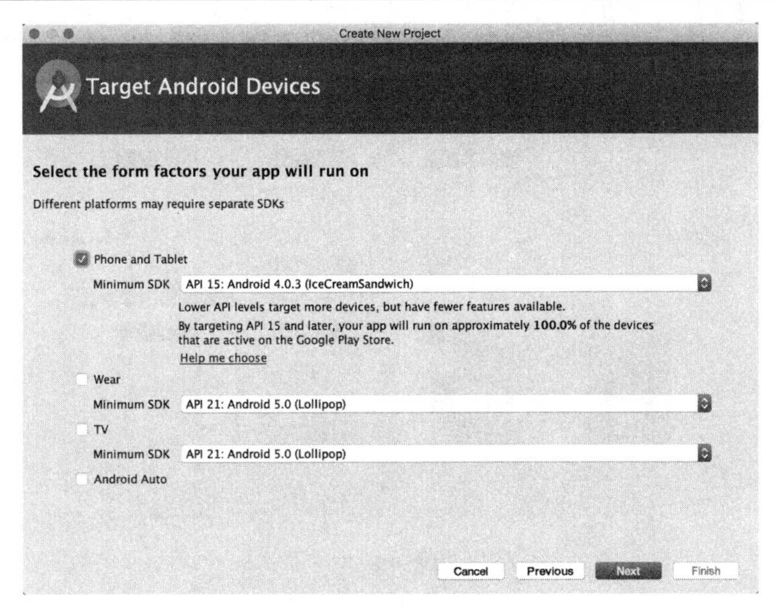

图 1-19　指定目标设备

在这个对话框中选择应用支持的机型，例如手机、平板电脑、TV、Wear 和 Google Glass。

选定的机型将成为项目中的应用模块。对于每种机型，还可以为该应用选择 API 级别。要获取详细信息，可以点击 Help me choose，如图 1-20 所示。

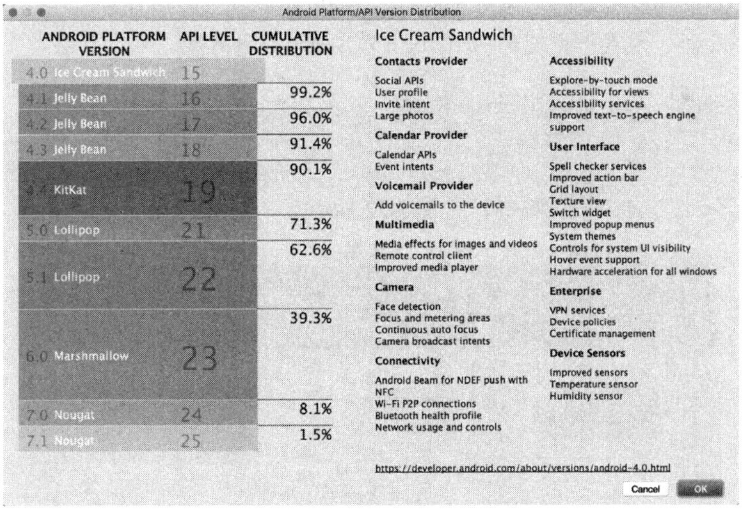

图 1-20　Android Platform Distribution

正常情况下采用默认的就行，直接点击 Next 按钮。

步骤 03 添加 Activity，如图 1-21 所示。可以选择不同的 Activity 类型，就是初始化的 Activity 有什么功能。

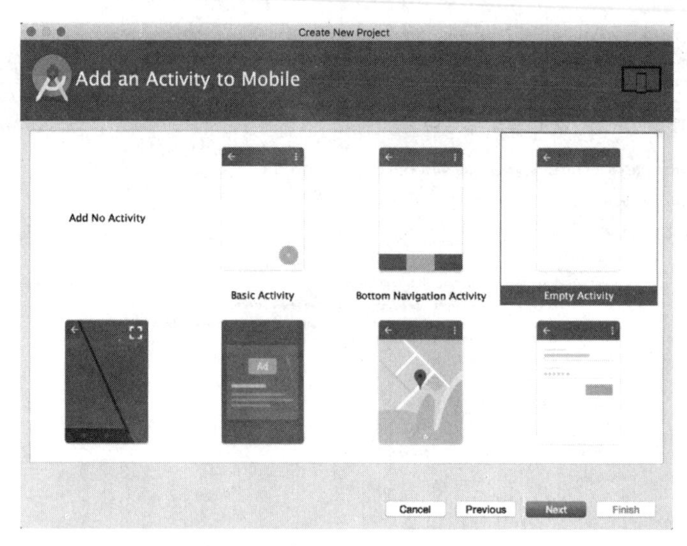

图 1-21 选择 Activity 界面风格

这里有很多选项：

- Add No Activity：就是没有 Activity，这种情况直接点击右下角的 Finish 按钮，项目就创建完成了。
- Basic Activity：基本的 Activity，具有一些基本功能。
- Bottom Navigation Activity：带有导航栏的 Activity。
- Empty Activity：就是一个空的 Activity。官方推荐这种方式。

还有很多初始化的 Activity，就不逐一解释了。其实大部分情况就用官方默认的选项。直接点击 Next 按钮即可。

步骤04 此界面是入口设置界面，如图 1-22 所示。

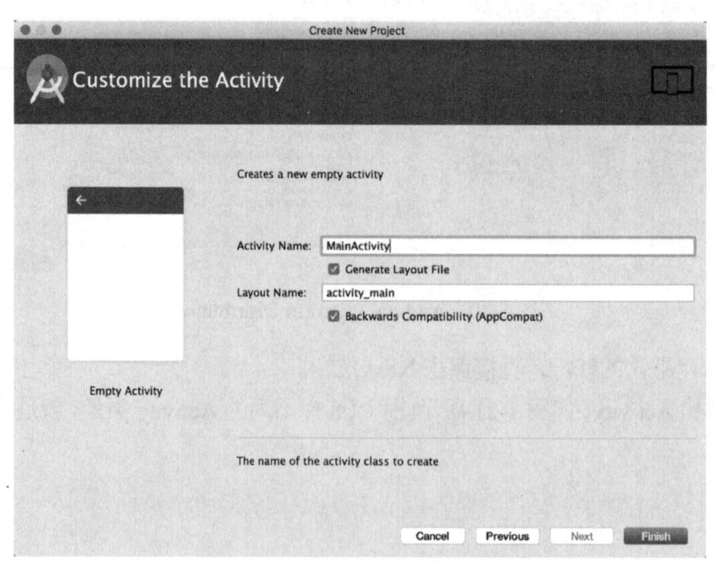

图 1-22 设置入口界面的名称

在该对话框中配置你的 Activity 信息，可以输入活动名称、布局名称。一般情况就用默认的，点击 Finish 按钮开始创建项目，如图 1-23 所示。

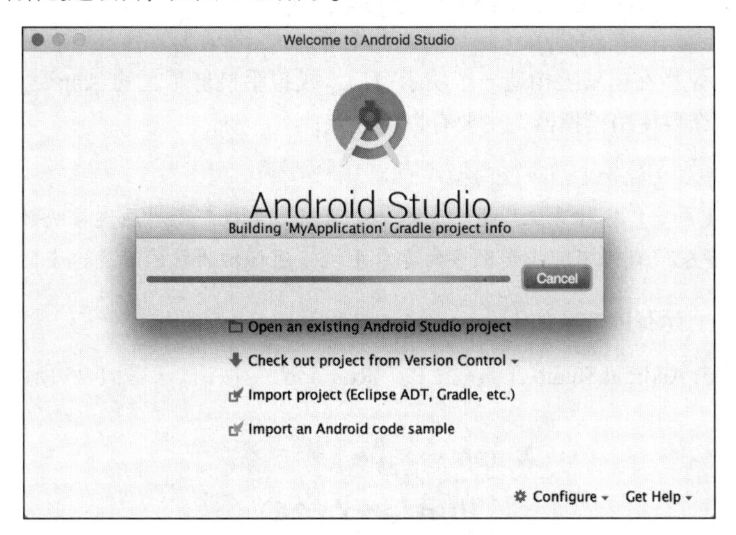

图 1-23　创建中

第一次创建项目的时候会比较慢，这是正常现象。因为第一次需要下载项目对应的 Gradle 版本。下载 Gradle 需要访问外网，下载速度很一般，这时可以先去做别的事情，让它慢慢下载。第二次创建项目时就会很快。

步骤 05 开发应用。Android Studio 会为用户的项目创建默认结构并打开开发环境。如果你的应用支持多种机型，Android Studio 将为每一个机型创建一个包含完整源文件的模块文件夹，如图 1-24 所示。

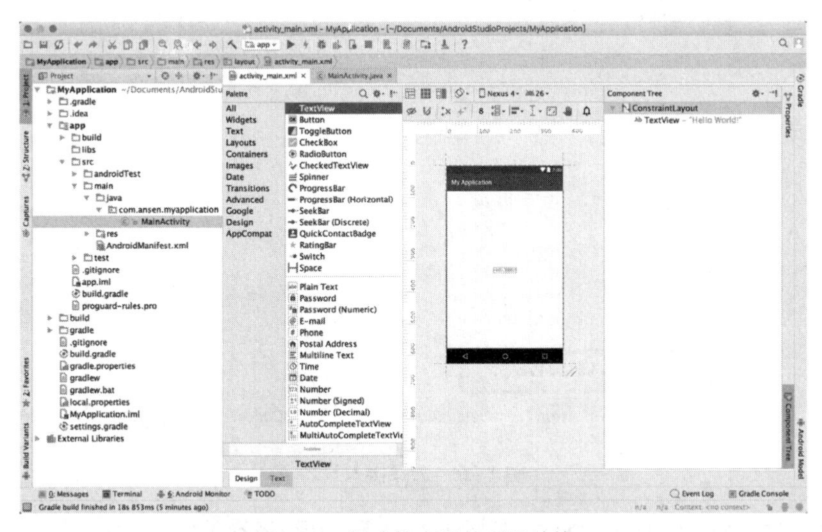

图 1-24　新建的应用的项目结构

至此，你的项目就创建完成了，接下来可以开发自己的应用了。

1.3.3 Android Studio 自带模拟器运行项目

所谓模拟器，是指在电脑上构造一个演示窗口，模拟手机屏幕上的 App 运行效果。首先问自己一个问题，为什么要使用模拟器？主要有以下几点：

- 没有安卓手机也能开发，降低门槛。
- 安卓碎片化严重，各种手机厂商一大堆，并且很多手机厂商对原生系统做了定制。
- 各种屏幕适配。我们不可能买很多的安卓手机，用模拟器就能解决这个问题。

创建模拟器并且运行的方法如下：

步骤01 点击 Android Studio 工具栏上的“Run 'app' ”按钮，如图 1-25 所示。

图 1-25　运行 App

步骤02 Android Studio 会先弹出选择设备界面，如图 1-26 所示。从中可以看到当前没有连接设备，可点击 Create New Virtual Device 按钮创建一个模拟器。

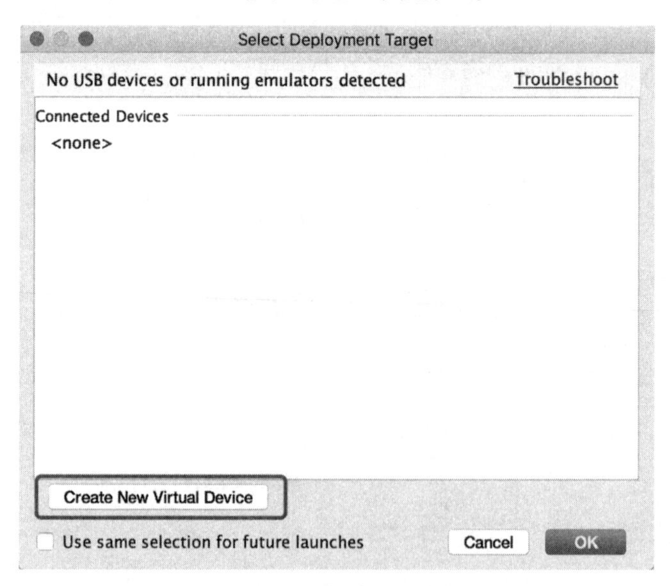

图 1-26　选择接入设备

步骤03 接下来选择模拟的硬件，如图 1-27 所示，可以选择类型、型号，选择默认的 Nexus 5x，点击 Next 按钮。

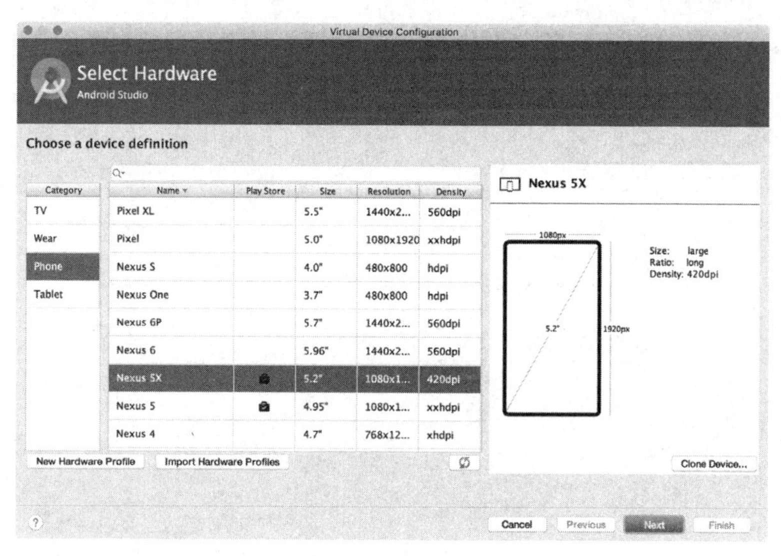

图 1-27　选择型号

步骤04 进入如图 1-28 所示的下载镜像界面，下载 Android 7.0 版本的镜像文件，点击前面的 Download 按钮。下载完成之后点击 Next 按钮。

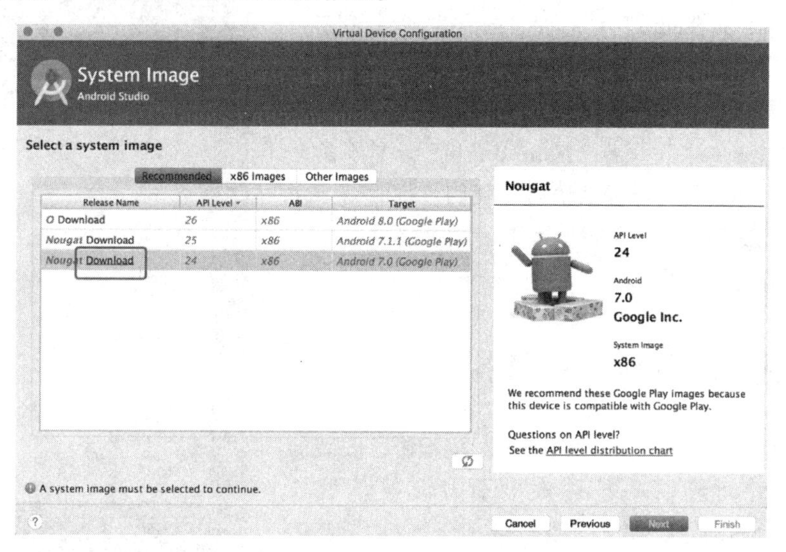

图 1-28　选择镜像

步骤05 设置完镜像之后会进入验证配置界面，可以设置模拟器的名字，选择横屏还是竖屏，如图 1-29 所示，点击 Finish 按钮。

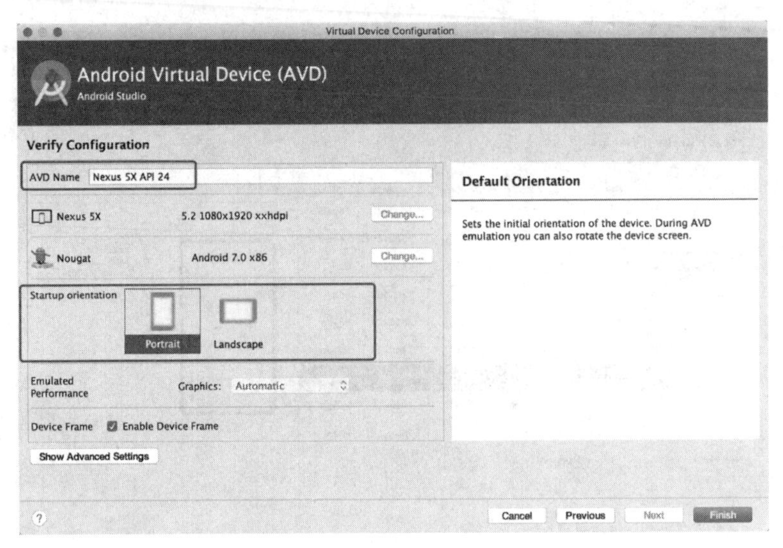

图 1-29 设置模拟器参数

步骤06 再次回到选择设备界面，就会看到创建的模拟器在选择列表中，选择这个模拟器，点击 OK 按钮。选定设备后，Android Studio 自动对项目进行编译、打包成 apk、对 apk 进行临时签名。然后打开我们选择的模拟器，把 apk 文件安装并且运行到模拟器上，效果如图 1-30 所示。

此时看到屏幕的中间写着“Hello World”，是不是有一种久违的熟悉感？无论学什么语言，我们运行的第一个程序总归是 Hello World，希望看本书的读者能够坚持下去，后面的内容更有趣噢！

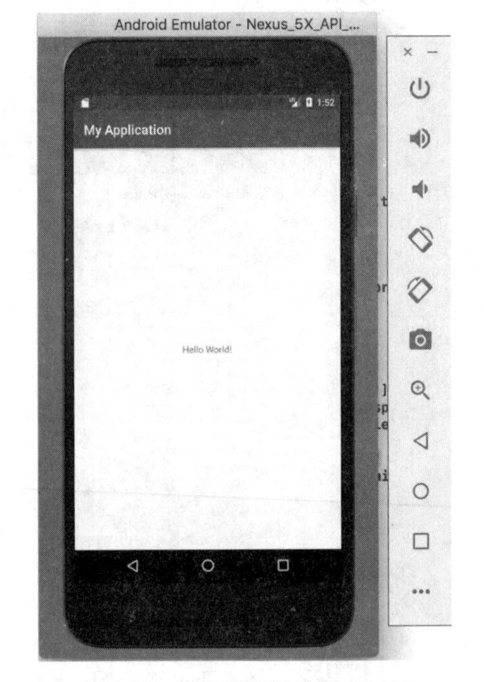

图 1-30 模拟器运行 Hello World

1.3.4 使用 Genymotion 模拟器运行

Android 自带的模拟器运行起来相对比较慢，安装一个 App 花费时间较长，并且效果不太流畅，目前 Genymotion 是比较好用的第三方模拟器。

1. VirtualBox（虚拟机）下载安装

VirtualBox 是一款开源虚拟机软件。VirtualBox 是由德国 Innotek 公司开发、由 Sun Microsystems 公司出品的软件，使用 Qt 编写，在 Sun 被 Oracle 收购后正式更名成 Oracle VM VirtualBox。

Genymotion 依赖于 VirtualBox，两个必须一起使用，所以首先下载 VirtualBox 安装。VirtualBox 的下载地址为 https://www.virtualbox.org/wiki/Downloads。

在下载页面根据自己的操作系统选择不同版本下载，下载完成之后直接安装就行。很简单，一直点击“下一步”即可。

2. Genymotion 下载安装

安装 Genymotion 之前一定记得先安装 VirtualBox，否则 Genymotion 无法运行。官网地址为 http://www.genymotion.net/。

步骤01 下载 Genymotion 模拟器需要注册账号后登录，然后直接打开下载界面，若下载界面找不到，可以直接复制：https://www.genymotion.com/download/。

步骤02 用浏览器下载会比较慢，可以把下载地址粘贴进迅雷中下载。

步骤03 下载完成后先安装 Genymotion，才能进入首页，如图 1-31 所示。

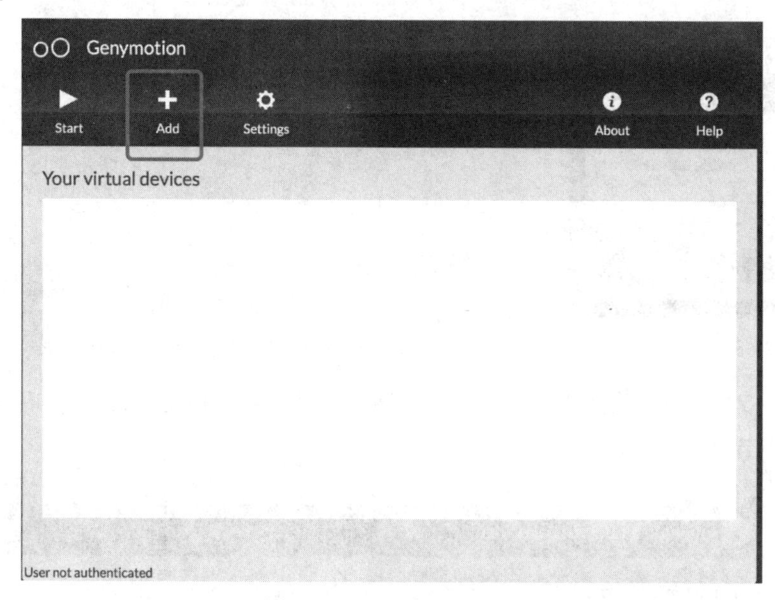

图 1-31 Genymotion 首页

步骤04 由于没有镜像文件，因此点击 Add 按钮进入添加界面，如图 1-32 所示。

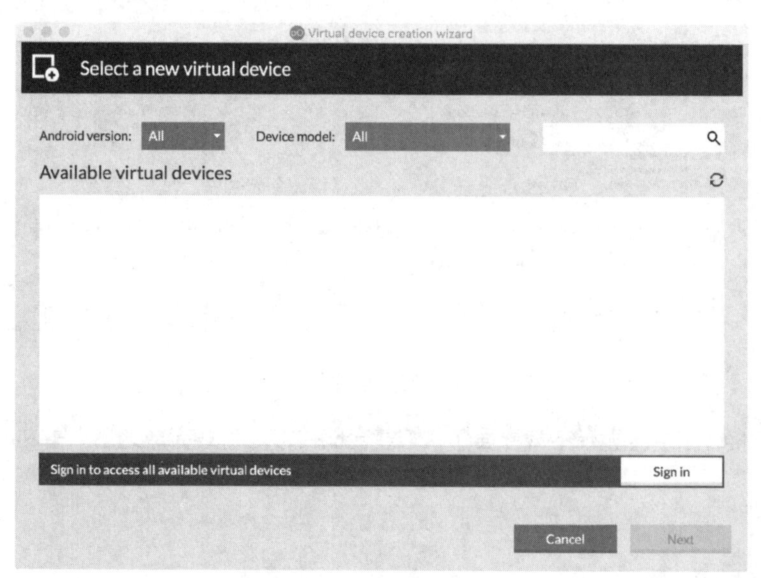

图 1-32 添加模拟器

步骤05 需要登录才能下载镜像文件，点击 Sign in 按钮，进入登录界面，如图 1-33 所示。

步骤06 输入之前在官网注册的账号进行登录。登录成功之后在添加界面会刷新镜像列表，提供各种不同的手机厂商的镜像文件与安卓版本，如图 1-34 所示。

图 1-33　登录

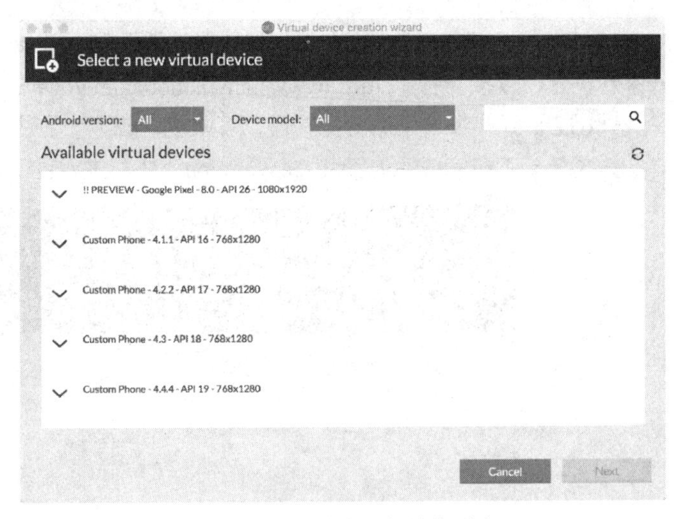

图 1-34　Android 虚拟机镜像列表

步骤07 双击要下载的镜像，进入创建设备界面，可以指定设备的名称，如图 1-35 所示。用默认的也行，点击 Next 按钮。

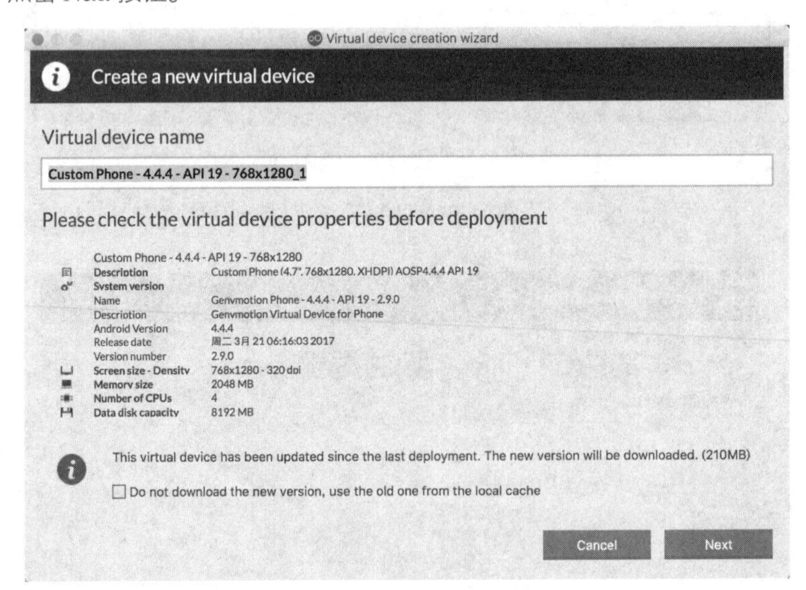

图 1-35　创建设备

步骤08 接下来就会显示下载界面了，下载会比较慢。下载完成之后点击 Finish 按钮，如图 1-36 所示。

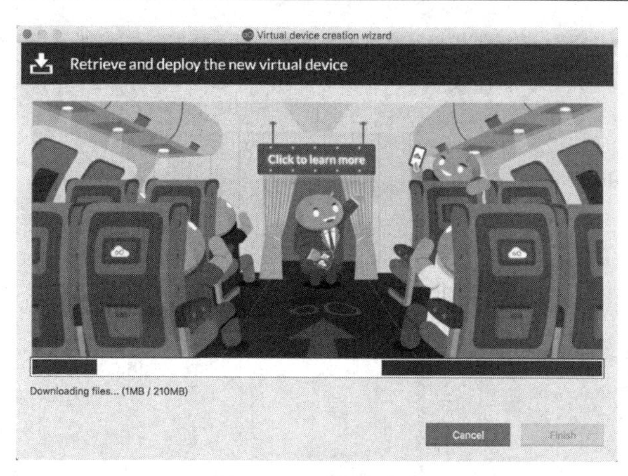

图 1-36　下载镜像中

步骤09 接下来就能在首页看到下载的设备了。选中设备后点击 Start 按钮或者双击这个设备运行模拟器，如图 1-37 所示。

至此，Genymotion 模拟器安装完成了。回到 Android Studio 项目界面，点击工具栏上的“Run 'app'”按钮运行项目。在选择设备中选择刚运行的 Genymotion 模拟器，最后效果如图 1-38 所示。

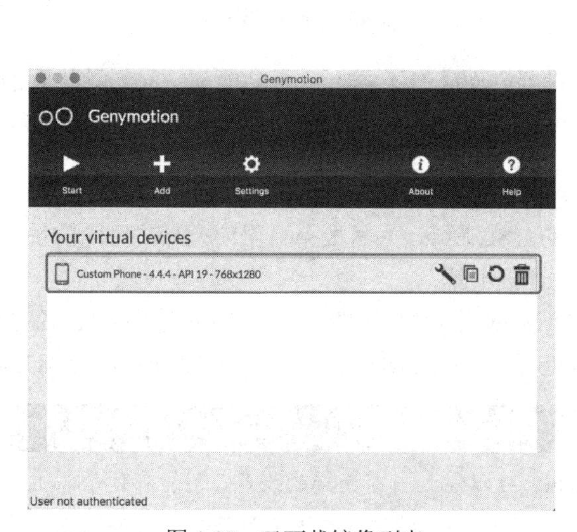

图 1-37　已下载镜像列表

图 1-38　Genymotion 模拟器运行 Hello World

大家用模拟器开发的时候建议用 Genymotion 模拟器，比官方自带的好用很多。

1.3.5　真机运行

除了在模拟器上运行外，还可以直接在手机上运行。

1. 手机开启开发者模式

手机默认是未开启开发者模式的，所以需要在手机的“设置”界面中手动开启。

步骤01 **点击“设置”→“关于手机”→“版本号”**，连续点击版本号 5 次就激活开发者模式，如图 1-39 所示。可能某些国产机界面不一样，但是只要找到版本号连续 5 次点击就对了。

步骤02 进入开发者选项界面，点击“设置”→“全局高级设置”→“开发者选项”界面，开启 USB 调试，如图 1-40 所示。

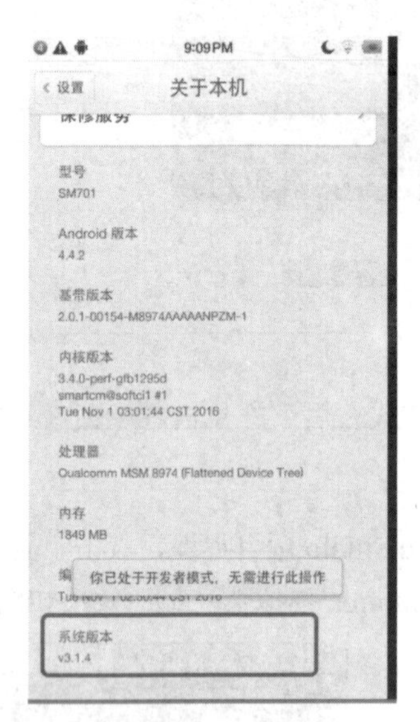

图 1-39　开启开发者模式

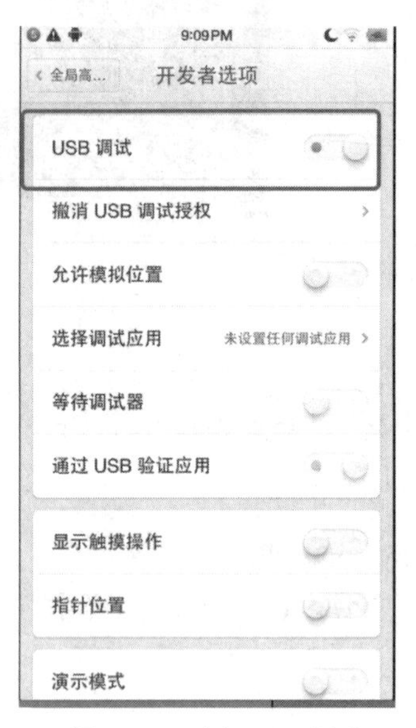

图 1-40　开启 USB 调试

这里的手机是锤子手机，不同的手机厂商设置界面略有不同，但是功能都一样。没找到的读者仔细查找或者求助于网页搜索。

2. 安装 USB 驱动

只有 Windows 系统需要下载这个手机对应的 USB 驱动，大家根据自己的型号去对应的手机厂商官网下载，然后安装。安装之后用手机数据线连接电脑。

如果是 Mac 电脑是不需要下载驱动的，直接用数据线连接电脑即可。有些手机连接之后可能会识别不了（个别情况），可以参考一篇教程：http://blog.csdn.net/lowprofile_coding/article/details/48443249。

如何判断手机是否连接成功

在 Android Studio 底部有一个 Android Monitor，可以看到当前连接的设备，还能看到手机上应用程序打印的 Log，如图 1-41 所示。当然，运行 App 的时候也能看到连接的设备列表。

图 1-41　Android Studio 查看连接设备

3. 运行 App

设备连接成功之后点击 Android Studio 工具栏上的“Run 'app' ”按钮。选择设备（这里是锤子手机），然后就能在手机上看到 Hello World 了。

1.4 调试项目

Android Studio 自带的调试程序让用户能够对运行在模拟器或者真机设备的应用进行调试。

1.4.1 Debug 断点调试

Debug 断点调试是每个开发工具必备的功能，当然 Android Studio 也有，使用 Debug 断点调试可以查看运行中变量的值与表达式的值，可以一行一行代码逐步进行调试。

如果你的程序是逻辑问题（程序本身不报错，但是结果错误），用 Debug 调试进行问题定位非常方便。

1. 设置断点

找到想断点的代码行位置，点击该代码左侧空白处，或者将光标停留在这行代码上，然后按组合键 Ctrl+F8（在 Mac 上，按 Command+F8），如图 1-42 所示。

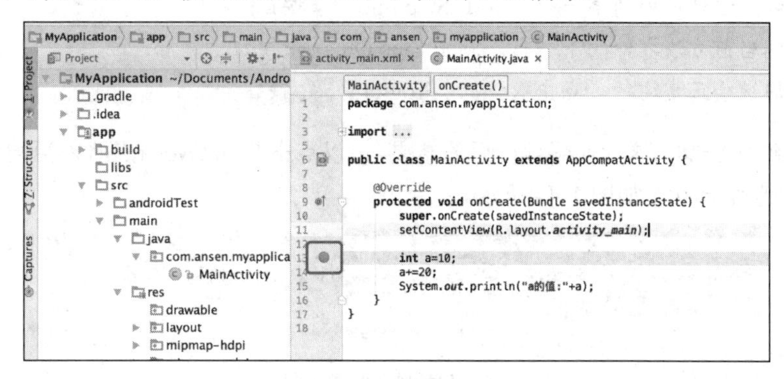

图 1-42 添加断点

例如，想追踪 MainActivity 中变量 a 的值，就可以在声明 a 变量那一行设置一个断点。可以看到左侧空白处有一个酒红色的原点。

2. 调试

给需要调试的代码设置断点之后，点击 Android Studio 工具栏上的 Debug App 按钮运行项目。点击这个按钮之后就会以 Debug 模式运行 App。

当软件运行到“**int a=10;** ”这一行代码时，Android Studio 会暂停应用的运行，并且 Android Studio 会显示 Debugger 标签，如图 1-43 所示。

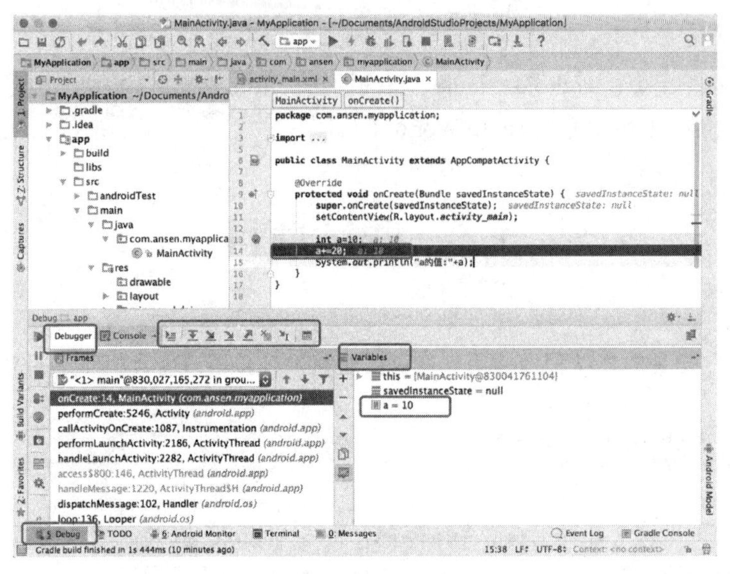

图 1-43 Debug 中

我们可以使用 Debugger 标签中的工具来确定应用的状态：

- 要检查变量的对象树，在 Variables 视图中将其展开。
- 要在当前执行点对某个表达式求值，点击 Evaluate Expression 按钮。
- 要前进到下一行代码（而不进入方法），点击 Step Over 按钮。
- 要前进到方法调用内的第一行，点击 Step Into 按钮。
- 要前进到当前方法之外的下一行，点击 Step Out 按钮。
- 要让应用继续正常运行，点击 Resume Program 按钮。

例如，想看到"**a+=20;**"这行代码运行后的结果，可以点击 Step Over 按钮，然后可以在 Variables 视图中看到 a 的值等于 30，如图 1-44 所示。

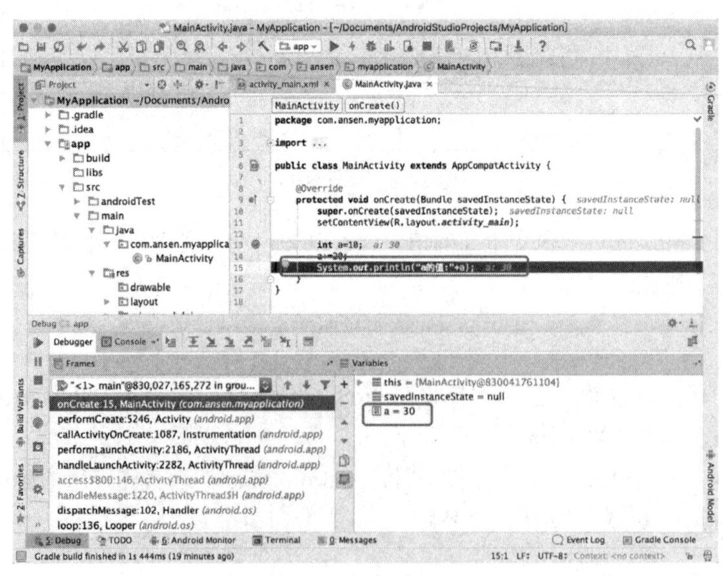

图 1-44 显示代码的运行结果

如果这个断点调试完成，让应用继续正常运行，点击 Resume Program 按钮。

1.4.2　日志调试

1. 写入日志

要在代码中写入日志，使用 Log 类，日志消息可以帮助了解执行流程。android.util.Log 常用的方法有以下 5 个：

- Log.v()：对应 verbose，调试颜色为黑色的，任何消息都会输出。
- Log.d()：对应 debug，仅输出 debug 调试的意思，但它会输出上层的信息。
- Log.i()：对应 info，一般提示性的消息。
- Log.w()：对应 warn，输出为蓝色，可以看作 warning（警告），一般需要注意优化 Android 代码。
- Log.e()：对应 error（异常），输出为红色，红色错误需要认真解决。

这 5 个方法都有两个参数，第一个参数是 tag（为 Log 打上标签），第二个参数是打印内容。

2. 查看日志

首先在 MainActivity 的 onCreate 中加入打印 5 种日志的代码。

```
Log.v("ansen","verbose");
Log.d("ansen","debug");
Log.i("ansen","info");
Log.w("ansen","warn");
Log.e("ansen","error");
```

重新运行程序，然后在 Android Monitor 标签的 logcat 中看到输出的日志，效果如图 1-45 所示。

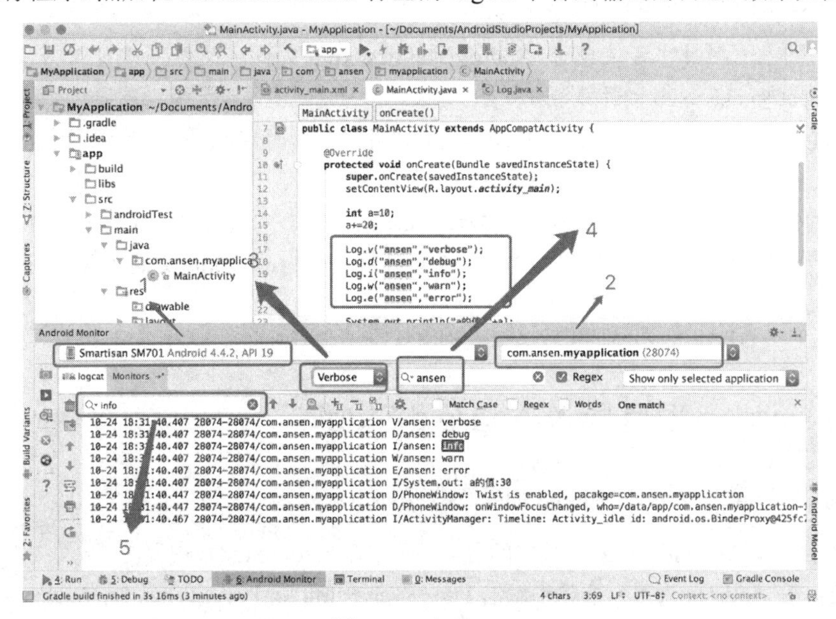

图 1-45　查看 log

① 选择设备。

② 选择运行的程序包名。

③ 选择要显示的日志级别，对应 Log 打印日志的 5 种方法。

④ 根据字符串过滤日志，例如这里是根据 ansen 过滤，日志的 tag 或内容必须要包含 ansen 这个字符串才会显示出来。

⑤ 在过滤之后的日志中进行字符串查找。

我们在 MainActivity 的 oncreate 方法中最后一行用到了 java sdk 里面的 System 类来打印，这样也是可以的，但是不推荐使用。System 类打印日志没有 tag 标签，没有日志级别，当日志过多时过滤不方便。

1.5 Eclipse 项目迁移至 Android Studio

在企业开发中，会经常参考别人写的开源项目，避免重复编写代码。

将项目迁移至 Android Studio 需要适应新的项目结构、构建系统和 IDE 功能。从 Eclipse 迁移至 Android 项目，Android Studio 会提供导入工具，可以将现有代码快速移至 Android Studio 项目，这个项目基于 Gradle 的构建文件。

Android Studio 为使用 Eclipse 创建的现有 Android 项目提供自动导入工具。

1.5.1 Eclipse 项目迁移条件

Eclipse 项目迁移的主要条件如下：

- 确保 Eclipse 项目根目录包含 AndroidManifest.xml 文件。此外，根目录必须包含 Eclipse 的.project 和.classpath 文件或 res/和 src/目录。
- 在需要导入的 project.properties 或.classpath 文件中注释掉对 Eclipse 项目库文件的任何引用。导入之后手动在 build.gradle 文件中添加这些引用。
- 记录工作区目录、路径变量和任何实际路径映射可能会有所帮助，这些内容可用于指定任何未解析的相对路径、路径变量和链接的资源引用。Android Studio 允许在导入过程中手动指定任何未解析的路径。

1.5.2 将 Eclipse 项目导入 Android Studio

将 Eclipse 项目导入 Android Studio 的操作步骤如下：

步骤01 启动 Android Studio，并关闭已经打开的 Android Studio 项目。

步骤02 在 Android Studio 菜单中点击 File→New→Import Project，或在“Welcome”屏幕中点击 Import project (Eclipse ADT, Gradle, etc.)。

步骤03 选择包含 AndroidManifest.xml 文件的 Eclipse 项目文件夹，并点击 OK 按钮，如图 1-46 所示。

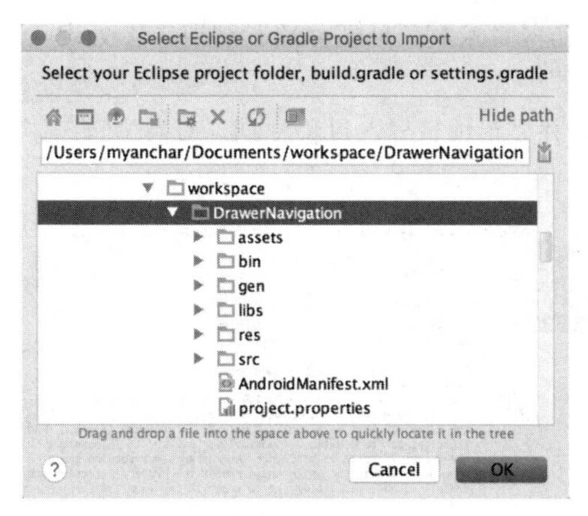

图 1-46　选择项目导入

步骤04 选择目标文件夹，然后点击 Next 按钮，如图 1-47 所示。

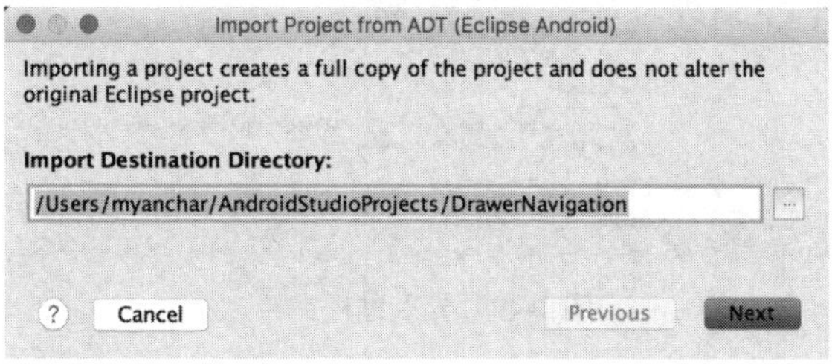

Import Project from ADT (Eclipse Android)

Importing a project creates a full copy of the project and does not alter the original Eclipse project.

Import Destination Directory:

/Users/myanchar/AndroidStudioProjects/DrawerNavigation

Cancel　Previous　Next

图 1-47　选择目标文件夹

步骤05 选择导入选项，然后点击 Finish 按钮。

步骤06 导入过程中会提示将任何库和项目依赖关系迁移到 Android Studio，并将依赖关系声明添加到 build.gradle 文件，如图 1-48 所示。

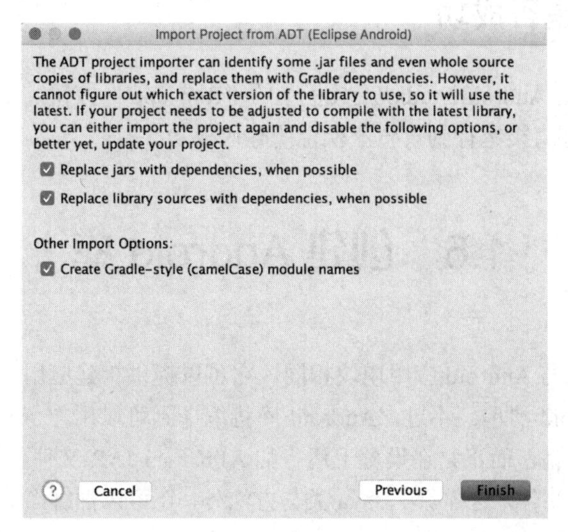

图 1-48　导入设置

导入过程中还将用 Maven 依赖关系替换具有已知 Maven 坐标的任何已知源代码库、二进制库和 JAR 文件，因此无须手动保留这些依赖关系。

导入选项还允许输入工作区目录和任何实际路径映射，以处理任何未解析的相对路径、路径变量和链接的资源引用。

Android Studio 导入应用并显示项目导入过程文档。查看文档，了解项目重组和导入过程的详细信息，如图 1-49 所示。

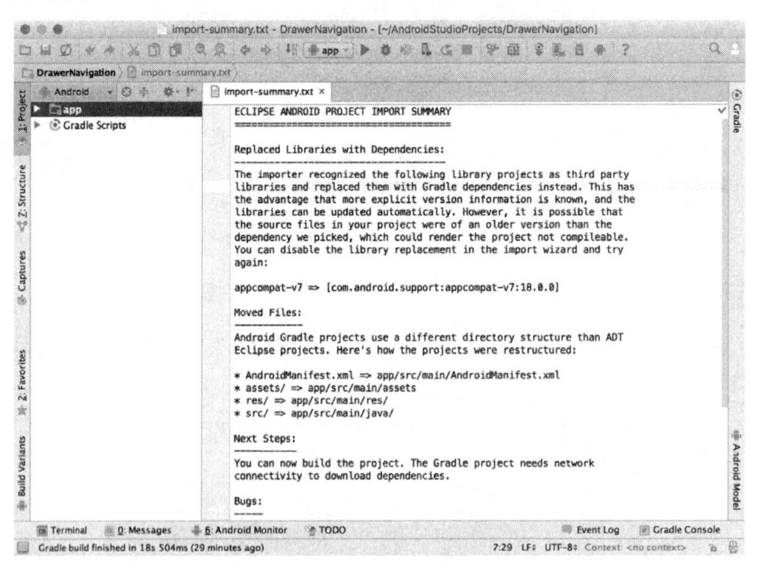

图 1-49　导入过程文档

将项目从 Eclipse 导入 Android Studio 后，Android Studio 中的每个应用模块文件夹都包含该模块的完整源代码集，包括 src/main/和 src/androidTest/目录、资源、构建文件以及 AndroidManifest.xml。在开始应用开发前应该解决项目导入文档说明中显示的所有问题，确保项目重组和导入过程成功完成。

1.5.3　验证导入是否成功

直接运行项目，点击 Android Studio 工具栏上的“Run 'app'”按钮，如果有问题可以根据报错提示进行解决。如果能够直接运行到设备上就说明导入成功。

1.6　创建 Android 库

Android 库在结构上与 Android 应用模块相同。它可以提供构建应用所需的一切内容，包括源代码、资源文件和 Android 清单。不过，Android 库将编译到可以用作 Android 应用模块依赖项的 Android 归档（AAR）文件，而不是在设备上运行的 APK。与 JAR 文件不同，AAR 文件可以包含 Android 资源和一个清单文件，这样除了 Java 类与方法外，还可以捆绑布局和可绘制对象等共享资源。

库模块在以下情况下非常有用：

- 构建使用某些相同组件（例如 Activity、服务或 UI 布局）的多个应用。
- 构建存在多个 APK 变体（例如免费版本和付费版本）的应用并且需要在两种版本中使用相同的核心组件。

就像公司有 10 多个 App，可以把所有 App 都需要用的东西封装到库模块中，例如网络请求、在线加载图片等。

这 10 多个项目都依赖这个库，而不是 10 多个项目都写一遍网络请求的代码，并且用库的方式方便修改。

当访问网络的代码有 bug 的时候，只需要修改这个库文件的代码就好了。

1.6.1 创建库模块

要在项目中创建一个新的库模块，需要进行以下操作：

步骤01 点击 File→New→New Module。

步骤02 在出现的 Create New Module 对话框中，依次点击 Android Library 和 Next。

> **提 示**
>
> 还存在一个用于创建 Java 库的选项，可以构建传统的 JAR 文件。尽管 JAR 文件在大多数项目中都非常实用（尤其在希望与其他平台共享代码时），但这种文件不允许包含 Android 资源或清单文件，而后者对于 Android 项目中的代码重用非常有用。根据需求决定。

步骤03 为你的库命名，并为库中代码选择一个最低的 SDK 版本，然后点击 Finish。在 Gradle 项目同步完成后，库模块将显示在左侧的 Project 面板中。

1.6.2 将库模块导入到项目中

有时我们的项目要用别人项目中依赖的库，手动复制过来太麻烦，Android Studio 支持导入库文件。

步骤01 点击 File→New→Import Module。

步骤02 选择库模块目录的位置，然后点击 Finish。

库模块的代码将会复制到你的项目中，也可以修改库代码。

1.6.3 将应用模块转换为库模块

如果希望把应用模块转换为库模块，可以采用以下步骤：

步骤01 打开现有应用模块的 build.gradle 文件。在顶部看到以下内容：

```
apply plugin: 'com.android.application'
```

步骤02 修改成下面这行代码：

```
apply plugin: 'com.android.library'
```

步骤03 点击 Sync Project with Gradle Files。

将应用模块转换为库模块就是这么简单，只需要修改一行代码。模块的结构是一样的，改了之后构建的是 AAR 文件，而不是可以运行在手机上的 APK 文件。

1.6.4 开发库模块的注意事项

在开发库模块和相关应用时，需要注意以下行为和限制。

将库模块引用添加至你的 Android 应用模块后，可以设置它们的相对优先级。构建时，库会按照一次一个的方式与应用合并，并按照从低到高的优先级顺序进行。

- 资源合并冲突：构建工具会将库模块中的资源与相关应用模块的资源合并。如果在两个模块中均定义了给定资源 ID，将使用应用中的资源。
  如果多个 AAR 库之间发生冲突，将使用依赖项列表首先列出（位于 dependencies 块顶部）库中的资源。
  为了避免常用资源 ID 的资源冲突，请使用在模块（或在所有项目模块）中具有唯一性的前缀或其他一致的命名方案。
- 库模块可以包含 JAR 库：可以开发一个自身包含 JAR 库的库模块。不过，需要手动编辑相关应用模块的构建路径，并添加 JAR 文件的路径。
- 库模块可以依赖外部 JAR 库：可以开发一个依赖于外部库（例如 Maps 外部库）的库模块。在这种情况下，相关应用必须针对包含外部库（例如 Google API 插件）的目标构建。需要注意的是，库模块和相关应用都必须在其清单文件的元素中声明外部库。
- 库模块不得包含原始资源：工具不支持在库模块中使用原始资源文件（保存在 assets/目录中）。应用使用的任何原始资源都必须存储在应用模块自身的 assets/目录中。
- 应用模块的 minSdkVersion 必须大于或等于库定义的版本：库作为相关应用模块的一部分编译，因此，库模块中使用的 API 必须与应用模块支持的平台版本兼容。
- 每个库模块都会创建自己的 R 类：在构建相关应用模块时，库模块将先编译到 AAR 文件中，然后添加到应用模块中。因此，每个库都有其自己的 R 类，并根据库的软件包名称命名。
- 从主模块和库模块生成的 R 类会在所需的所有软件包（包括主模块的软件包和库的软件包）中创建。

1.6.5 AAR 文件详解

AAR 文件的文件扩展名为.aar，Maven 工件类型也应当是.aar。文件本身是一个包含以下强制性条目的 zip 文件：

- /AndroidManifest.xml
- /classes.jar
- /res/
- /R.txt

此外，AAR 文件可能包含以下可选条目中的一个或多个：

- /assets/
- /libs/名称.jar
- /jni/abi 名称/名称.so（其中 abi 名称是 Android 支持的 ABI 之一）
- /proguard.txt - /lint.jar

1.7　项目依赖库

依赖库的方法主要有两种：一种是本地依赖，另一种是在线依赖。

本地依赖库一般是公司内部把一些项目通用的代码封装成库，可以根据业务需求随时修改代码，并且代码都在本地，不会被公开。

在线依赖库一般是个人或者组织对解决某个问题的代码进行开源，例如从服务器请求数据，这是市面上 90%的 App 都需要用到的功能，Android 自带的访问网络 api 太烦琐，于是就需要把网络请求的代码进行封装，这样就有一些公司会把自己 App 中访问网络的代码封装成一个库，提交到远程中央仓库。别人就能通过在线依赖的方式引用这个库，大家都站在巨人的肩膀上，还有一个好处就是，这个库有 bug，只要开源者修复这个问题，然后提交一个新的版本，所有依赖者根本不需要修改代码，只需要修改版本号即可解决 bug。

1.7.1　依赖本地库

依赖本地库（module）就是源代码在你当前电脑上，依赖库有什么问题，可以随时修改。

例如，在自己已打开的项目下新建一个库“my-library-module”，如果想依赖这个库，打开应用模块的 build.gradle 文件，并向 dependencies 块中添加一行如下的新代码：

```
compile project(':my-library-module')
```

点击 Sync Project with Gradle Files。修改后的项目结构如图 1-50 所示。

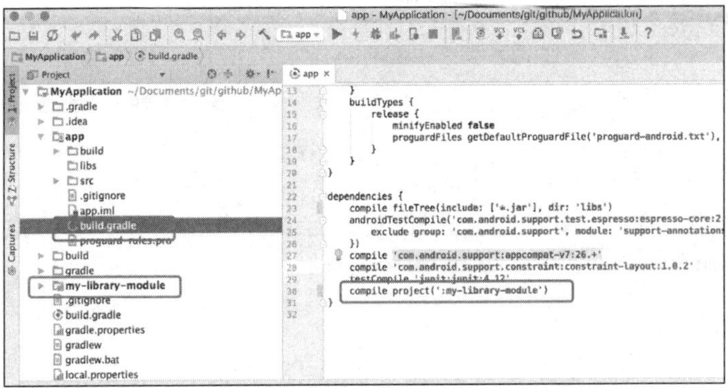

图 1-50　新建 module 并进行依赖

1.7.2　在线依赖库

在线依赖源代码保存在服务器中，当我们第一次依赖时，会从远程仓库中下载 jar 或者 aar 文件，Android Studio 之前默认的在线依赖仓库是 jcenter，从 Android Studio 3.0 之后增加了 Google 自己的仓库。上传到远程仓库上的在线依赖库（module），必须要对代码进行开源。

在线依赖的库可以看到源码，但是不能修改。在后面的章节中我会告诉大家如何让自己的 module 上传到 jcenter 服务器。

在线依赖很简单，跟本地依赖一样，也只需要一行代码。打开应用模块的 build.gradle 文件，并向 dependencies 块中添加一行新代码。例如，新建项目时就有的 v7 包依赖。

```
compile 'com.android.support:appcompat-v7:26.+'
```

在线依赖库的代码能不能不开源

可以不开源，我们可以搭建自己的内网 jcenter 服务器。

1.8　应用清单文件

每个应用的根目录中都必须包含一个清单文件（AndroidManifest.xml）。该文件向 Android 系统提供应用的必要信息，系统必须具有这些信息才能运行应用的任何代码。

此外，清单文件还包含以下功能：

- 为应用设置包名。软件包名称是当前应用的唯一标识符。
- 描述应用的各个组件，包括 Activity、服务、广播接收器和内容提供者。它还为实现每个组件的类命名并发布其功能，例如它们可以处理的 Intent 消息。这些声明向 Android 系统告知有关组件以及可以启动这些组件的条件的信息。
- 确定应用组件的进程。
- 声明应用有哪些权限，还声明其他应用与该应用组件交互所需具备的权限。
- 声明应用需要的最低 Android API 级别。

1.8.1 清单文件结构

下面的代码段显示了清单文件的通用结构以及可以包含的每个元素。

```
<?xml version="1.0" encoding="utf-8"?>

<manifest>
   <uses-permission />
   <permission />
   <permission-tree />
   <permission-group />
   <instrumentation />
   <uses-sdk />
   <uses-configuration />
   <uses-feature />
   <supports-screens />
   <compatible-screens />
   <supports-gl-texture />

   <application>

      <activity>
         <intent-filter>
            <action />
            <category />
            <data />
         </intent-filter>
         <meta-data />
      </activity>

      <activity-alias>
         <intent-filter> . . . </intent-filter>
         <meta-data />
      </activity-alias>

      <service>
         <intent-filter> . . . </intent-filter>
         <meta-data/>
      </service>

      <receiver>
         <intent-filter> . . . </intent-filter>
         <meta-data />
      </receiver>

      <provider>
         <grant-uri-permission />
         <meta-data />
         <path-permission />
      </provider>

      <uses-library />
```

```
    </application>
</manifest>
```

1.8.2 文件约定

适用于清单文件中所有元素和属性的约定和规则。

1. 元素

只有<manifest>和<application>元素是必需的，这两个元素都只能有一个并且必须唯一。其他大部分元素可以出现多次或者根本不出现，但清单文件中必须存在其中某些元素才有效。元素的值通过属性进行设置。

同一级别的元素不区分先后顺序。例如，<activity>、<provider> 和 <service> 元素可以按任何顺序混合在一起。

2. 属性

所有的属性都是可选的，但是有些元素必须要指定属性才能有效。

除了根 <manifest> 元素的一些属性外，所有属性名称均以 android:前缀开头。例如，android:alwaysRetainTaskState。

3. 声明类名

有很多元素可以指定 java 类，例如<application> 、Activity(<activity>)、服务(<service>)、广播接收器 (<receiver>) 以及内容提供者 (<provider>)。

```
<manifest package="com.ansen.myapplication" ... >

    <application
        ...>
        <activity android:name=".MainActivity">
            ...
        </activity>
    </application>
</manifest>
```

如果我们没有给 manifest 指定 package 属性，写法如下:

```
<manifest ... >
    <application
        ...>
        <activity android:name="com.ansen.myapplication.MainActivity">
            ...
        </activity>
    </application>
</manifest>
```

4. 多个值

如果指定多个值，就一直重复增加这个元素，而不是给一个元素赋多个值。例如，intent 过滤器可以有多个：

```
<intent-filter . . . >
```

```
    <action android:name="android.intent.action.EDIT" />
    <action android:name="android.intent.action.INSERT" />
    <action android:name="android.intent.action.DELETE" />
    . . .
</intent-filter>
```

5. 权限

权限是一种限制，用于限制对部分代码或设备数据的访问。增加限制是为了保护可能被误用以致破坏或损害用户体验的关键代码。

如果应用需要访问受权限保护的功能，就必须在清单中使用<uses-permission>元素声明应用需要该权限。

将应用安装到设备上之后，安装程序会通过检查签署应用证书的颁发机构并（在某些情况下）询问用户，确定是否授予请求的权限。如果授予权限，则应用能够使用受权限保护的功能。否则，访问这些权限保护的功能会失败，并且不会向用户发送任何通知。

1.9　常用快捷键

Android Studio 为了方便操作，提供了很多快捷键。我们也可以为某个功能自定义快捷键。表 1-3 列出了常用的快捷键。

表 1-3　常用的快捷键

说明	Windows/Linux	Mac
全部保存	Ctrl + S	Command + S
打开项目结构对话框	Ctrl + Alt + Shift + S	Command + ;（英文分号）
查找	Ctrl + F	Command + F
搜索全部内容（包括代码和菜单）	按两次 Shift	按两次 Shift
替换	Ctrl + R	Command + R
生成代码（getter、setter、构造函数、hashCode/equals、toString、新文件、新类）	Alt + Insert	Command + N
复制当前行或选择	Ctrl + D	Command + D
Command + D	Ctrl + 空格键	Control + 空格键
重命名	Shift + F6	Shift + F6
缩进/取消缩进行	Tab/Shift + Tab	Tab/Shift + Tab
删除插入符处的行	Ctrl + Y	Command + 退格键

1.9.1　自定义快捷键

首先打开快捷键设置界面，只需在 Android Studio 中点击 File→Settings→Keymap（在 Mac 上，点击 File→Properties→Keymap），如图 1-51 所示。

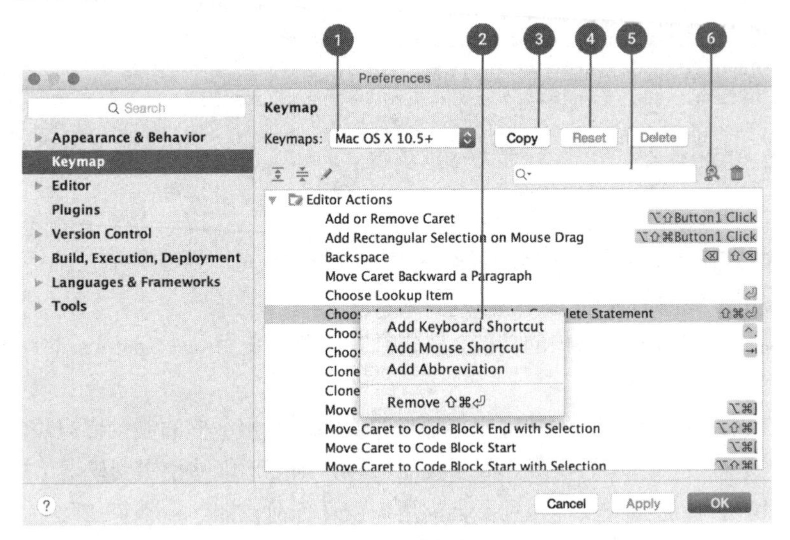

图 1-51 设置快捷键

① **快捷键类型：**从此列表中选择某个操作系统或者某个 IDE 工具对应的快捷键。

② **快捷菜单：**右击可以对其进行修改。可以为这个操作添加更多键盘快捷键，添加鼠标快捷键以将某个操作与鼠标点击关联，或者移除当前快捷键。

③ **Copy 按钮：**从快捷键类型下拉菜单中选择一个要复制的类型，然后点击 Copy 按钮以创建新的快捷键类型。可以修改类型名称和快捷键。

④ **Reset 按钮：**从下拉菜单中选择一个类型，然后点击 Reset 恢复成默认配置。

⑤ **搜索框：**输入文字按操作名称进行搜索。

⑥ **按快捷键搜索：**点击 Find Actions by Shortcut，输入快捷键进行搜索。

1.10 应用签名

Android 要求所有 APK 必须先使用证书进行数字签名，然后才能安装。

1.10.1 证书和密钥库

公钥证书（也称为数字证书或身份证书）包含公钥/私钥对的公钥，以及可以标识密钥所有者的一些其他元数据（例如名称和位置）。证书的所有者持有对应的私钥。

在签名 APK 时，签名工具会将公钥证书附加到 APK。公钥证书充当“指纹”，用于将 APK 唯一关联到你以及对应的私钥。这有助于 Android 确保 APK 的任何更新都是原版更新并来自原始作者。用于创建此证书的密钥称为应用签名密钥。

密钥库是一种包含一个或多个私钥的二进制文件。

每个应用在其整个生命周期内必须使用相同的证书，以便用户能够以应用更新的形式安装新版本。

1.10.2 调试项目时签名

当点击 Android Studio 工具栏上的“Run 'app'”按钮时，Android Studio 将自动使用通过 Android SDK 工具生成的测试证书签名你的 APK。

当在 Android Studio 中首次运行或调试项目时，IDE 将自动在 $HOME/.android/debug.keystore 中创建调试密钥库和证书，并设置密钥库和密钥密码。

由于测试证书通过构建工具创建并且在设计上不安全，大多数应用商店（包括 Google Play 商店）都不接受使用调试证书签名要发布的 APK。

Android Studio 会自动将你的测试签名信息存储在签名配置中，因此不必在每次测试时都输入此信息。签名配置是一种包含签名 APK 所需全部必要信息的对象，这些信息包括密钥库位置、密钥库密码、密钥名称和密钥密码。

1.10.3 正式签名

1. 生成密钥和密钥库

使用 Android Studio 生成应用签名或上传密钥，具体操作步骤如下：

步骤01 在菜单栏中点击 Build→Generate Signed APK。

步骤02 从下拉列表中选择一个模块，然后点击“Next”。

步骤03 点击“Create new”以创建一个新密钥和密钥库。

步骤04 在 New Key Store 对话框中，需要填写以下信息，用来生成密钥文件，如图 1-52 所示。

图 1-52 新建密钥库文件

- Key store path：选择创建密钥库的位置。
- Password：为密钥库创建并确认一个安全的密码。

- Alias：为密钥输入一个标识名。
- Password：为密钥创建并确认一个安全的密码。此密码应当与密钥库选择的密码不同。
- Validity（years）：以年为单位设置密钥的有效时长。密钥的有效期应至少为 25 年，可以在应用的整个生命期内使用相同的密钥签名应用更新。
- Certificate：为证书输入一些关于自己的信息。例如城市、国家、姓名等基本信息。这些信息不会显示在应用中，但会作为 APK 的一部分包含在证书中。

填写完表单后，点击 OK 按钮。这时证书已经生成了（在选择的证书保存路径下有个 xxx.jks 文件）。如果想继续签名 App，就点击 Next 按钮；如果只是生成一个签名文件，点击 Cancel 按钮。

2. 手动签名 APK

Android Studio 可以手动签名 APK，也可以在 Gradle 配置文件中构建签名信息，运行 App 时对 APK 自动签名。

要在 App 中手动签名 APK，操作步骤如下：

步骤01 点击 Build→Generate Signed APK，打开 Generate Signed APK 对话框，如图 1-53 所示。选择刚生成的 jks 文件、密钥库密码、密钥标示、密钥密码，然后点击 Next 按钮。

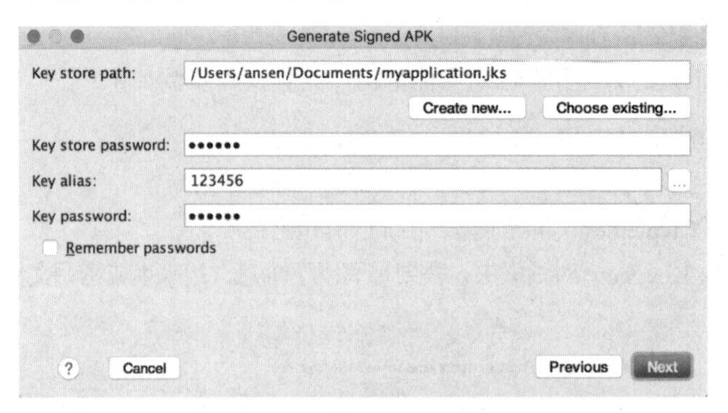

图 1-53　签名 APK

步骤02 在下一个对话框中，选择 APK 签名之后保存路径，同时可以选择渠道，由于我们没有配置渠道列表，所以 Flavors 一栏是空的；接下来选择签名版本，V1 和 V2 都选中，然后点击 Finish 按钮，如图 1-54 所示。

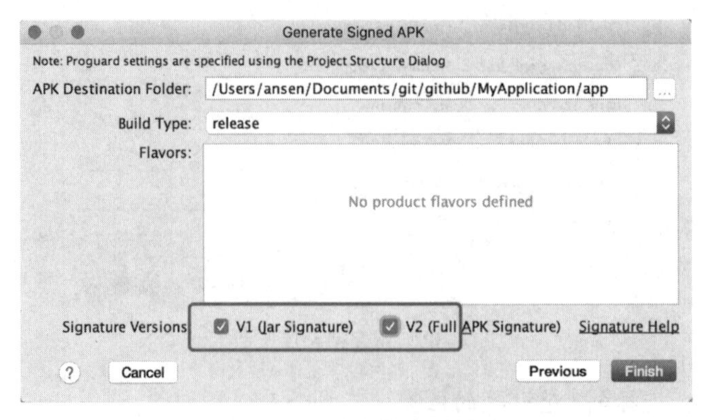

图 1-54　生成签名配置

签名完成之后，就会在选择的签名文件保存路径下生成一个签名之后的 APK 文件，可以用这个 APK 发布到各大应用市场。保管好 jks 文件密码，当每次需要更新版本到应用市场时，需要用这个 jks 文件进行签名。

3. Gradle 自动签名

手动签名很不方便，每次都要创建 APK，Google 早就想到了这个问题，只需要在 build.gradle 文件中加几行代码就可以了，以后每次运行 App 的时候就会用生成的 jks 文件签名，而不是临时手动签名。

使用自动签名需要修改 app/build.gradle 文件。在 Android 标签下增加 signingConfigs 标签（debug 与 release 版本签名信息设置），如图 1-55 所示。

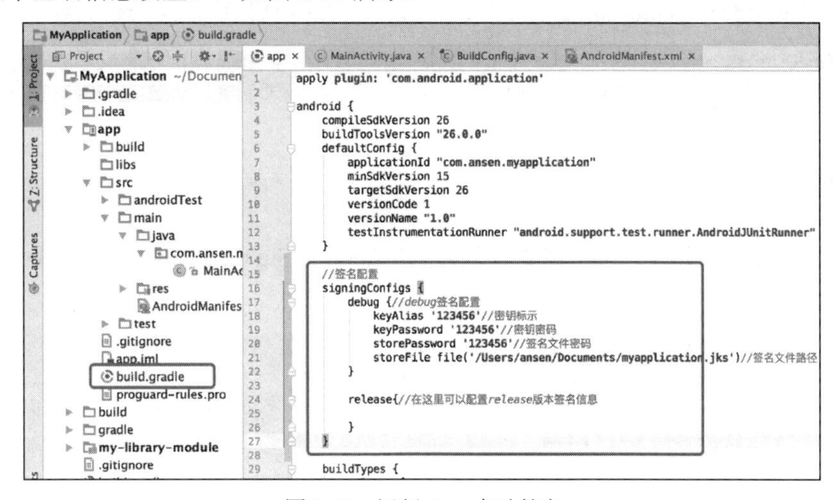

图 1-55　运行 App 自动签名

1.11　多渠道打包

国内提供了许多应用市场，例如 360、百度、应用宝、豌豆荚以及各手机厂商的市场等。

当需要去统计 App 的下载量、激活量的时候，不能对单个市场的流量进行统计。推广部门也不知道推广效果如何。例如，今天 App 在应用宝进行了首发，需要统计今天应用宝有多少激活设备，有多少注册用户，这样才知道推广有没有效果。

为了解决这个问题就出现了多渠道打包，一份源码给不同的市场编译出不同的 APK 文件，每个 APK 文件中都包含了当前市场的渠道码（自己指定一个字符串）。

1.11.1　代码实现

修改 app/build.gradle 文件，在 Android 标签下增加 productFlavors 标签，内容如下：

```
//不同渠道
productFlavors {
```

```
    wandoujia {//豌豆荚市场包
        //自定义变量  参数1:变量类型  参数2:变量名称  参数3:变量值
        buildConfigField "String", "FR", "\"wandoujia\""
    }
    baidu {//百度市场包
        buildConfigField "String", "FR", "\"baidu\""
    }
    oppo {//oppo市场包
        buildConfigField "String", "FR", "\"oppo\""
    }
}
```

我们就自定义了一个变量 FR，不同的渠道赋值不同的字符串。

这个自定义变量会在 BuildConfig 类中自动生成，在 Java 代码中取这个值只要一行代码即可。

```
String fr=BuildConfig.FR;//取到当前的渠道码，然后上传到服务器，就能根据不同的渠道进行统计
Log.i("ansen","当前渠道码:"+fr);
```

1.11.2 测试

打渠道包必须要手动进行，选择需要的渠道，在 Android Studio 菜单栏中点击 Build→Generate Signed APK→Next，如图 1-56 所示。

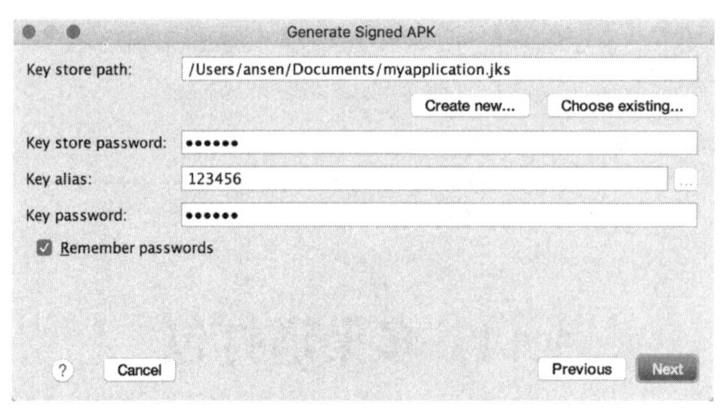

图 1-56 签名 APK

输入签名信息，点击 Next 按钮。可以选择 APK 文件输出路径、编译类型、渠道包。渠道包可以多选，这里全选了，然后点击 Finish 按钮，如图 1-57 所示。

图 1-57 签名时选择渠道码

Android Studio 编译会需要一点时间，打包完成后在选择的 APK 保存路径下会生成三个 APK 文件，对应不同的渠道，如图 1-58 所示，可以依次安装这三个软件包。看打印 Logo，会发现安装不同的包打印的 fr 值是不一样的。

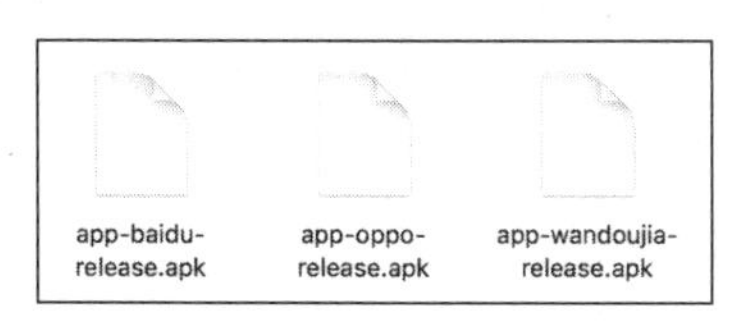

图 1-58　不同渠道的 APK 文件

多渠道还能干什么

其实多渠道打包还能干很多事情，比如给不同的渠道配置不同的 applicationId、生成不同应用名称或图标，还可以指定不同渠道包的名字。但是大部分人只需要打渠道包，如果想实现上面列出的功能，可以参考 Google 官方文档。

1.12　ADB 详解

ADB 的全称为 Android Debug Bridge，是一个标准的 CS 结构工具，用于连接模拟器或真机进行调试。身为 Android 开发者，熟练使用 ADB 命令将会大大提升开发效率。

在电脑上会运行一个 adb 进程，用于扫描 5555~5585 之间的奇数端口来搜索模拟器或真机。一旦发现 adb 守护进程，就通过此端口进行连接。需要说明的是，每一个模拟器或真机使用一对端口，奇数端口用于 adb 连接，偶数端口用于控制台连接。

如果模拟器与 adb 在 5555 端口连接，则控制台的连接端口将是 5554。

1.12.1　Mac 下 adb 加入环境变量（Windows 电脑自行搜索）

首先打开 terminal 终端命令窗口，使用命令[cd ~]到 home 目录下：

```
cd ~
```

接着使用 touch 命令，这个命令有两个功能：

- 如果文件存在，把已存在文件的时间标签更新为系统当前的时间。
- 如果文件不存在，就创建新的空文件。

```
touch .bash_profile
```

然后输入如下命令打开文件：

```
open -e .bash_profile
```

在打开的文件最后增加如下两行代码再保存。

```
export ANDROID_SDK=/Users/ansen/Library/Android/sdk
export PATH=${PATH}:${ANDROID_SDK}/platform-tools
```

ANDROID_SDK 指向的路径需要替换成自己的 sdk 路径。随便打开一个项目，在项目结构页面选中左侧的 SDK Location 就能看到本地的 sdk 路径：

```
sdk_location
sdk_location
```

最后用[source .bash_profile]命令使用修改后的：

```
source .bash_profile
```

验证 adb 环境变量是否配置成功。在终端输入“adb version”，如果显示类似这样的内容就成功了：

```
Android Debug Bridge version 1.0.39
Revision 3db08f2c6889-android
Installed as /Users/ansen/Library/Android/sdk/platform-tools/adb
```

1.12.2 adb 常用命令

- adb version：查看 adb 版本。
- adb install：安装 App。
- adb uninstall：卸载 App。
- adb push：从电脑复制东西到手机设备上。
- adb pull：从设备复制东西到电脑上。
- adb logcat：设备的日志。
- adb bugreport：查看 bug 报告。
- adb shell：进入设备的 shell 命令。
- adb devices：列出所有连接的设备，实际列出的就是设备的 serialnumber，可以通过-s 指定列出的 serialNumber 找到对应的设备。
- adb start-server：启动 adb server。
- adb kill-server：停止 adb server。
- adb get-state：列出设备状态，即 offline | bootloader | device。
- adb root：获取管理员权限。
- adb shell dumpsys activity activities：获取当前运行的 Activity。

1.13 Android Studio 3.0 新特性

如果你的电脑上已经安装了 Android Studio，想要获取最新版本，点击 Help→ Check for update（如果是 Mac 系统，点击 Android Studio→Check for updates）。

如果检测到有新版本，会弹出一个对话框，提示当前的版本与可以更新的版本。点击升级并且重启 Android Studio 这个功能。接下来会自动下载，安装完成后自动重启。

如果还没有安装过，可从官网页面下载：https://developer.android.google.cn/studio/index.html。

目前的 Android Studio 3.0 是一个重要版本，包含许多新功能以及旧功能改进。

MAC 用户在更新 Android Studio 时，可能会遇到一个更新错误对话框，指出“在安装过程中发生冲突”。不需要管它，直接点击“取消”继续安装即可。

1.13.1 Android Gradle 插件 3.0.0

Gradle 3.0 包含新功能并且改进了旧功能，可为包含大量 module 的项目提高构建性能。使用 Gradle 3.0 版本开发大型项目，主要具有以下优点：

- 对代码或资源进行简单修改，编译时间更快。
- 支持 Android 8.0。
- 支持基于语言资源构建单独的 APK。
- 支持 Java 8。
- 改进了 ndk-build 和 cmake 的构建速度。
- 改进 Gradle 同步速度。

1.13.2 手动更新 Gradle 版本

（1）修改 gradle-wrapper.properties 文件中 distributionUrl 的值：

```
distributionUrl=https\://services.gradle.org/distributions/gradle-4.1-all.
zip
```

（2）修改项目根目录 build.gradle 文件，把 Gradle 插件版本改成 3.0.0：

```
buildscript {
    repositories {
        jcenter()
    }
    dependencies {
        classpath 'com.android.tools.build:gradle:3.0.0'

        // NOTE: Do not place your application dependencies here; they belong
        // in the individual module build.gradle files
    }
}
```

1.13.3 Kotlin 支持

正如 Google I/O 2017 宣布的那样，Kotlin 编程语言在 Android 上正式得到支持。因此，在这个版本中，Android Studio 包含了 Android 开发的 Kotlin 语言支持。

通过将 Java 文件转换为 Kotlin（点击代码→Convert Java File to Kotlin File）或者使用 New Project 创建一个新的 Kotlin 的项目，可以将 Kotlin 合并到项目中，如图 1-59 所示。

```
override fun onCreate(savedInstanceState: Bundle?) {
    super.onCreate(savedInstanceState)
    setContentView(R.layout.activity_main)
    val toolbar = findViewById(R.id.toolbar) as Toolbar
    setSupportActionBar(toolbar)
Enter action or option name:                     s FloatingActionButton
Convert Java File to
Convert Java File to Kotlin File ⌥⇧⌘K   Code   icked", Snackbar.LENGTH_LONG)
'package.html' may be converted to 'package-i...   Preferences   ll).show()
Press ^↑ or ^↓ to navigate through the history

    val drawer = findViewById(R.id.drawer_layout) as DrawerLayout
    val toggle = ActionBarDrawerToggle(
```

图 1-59　Java 转 Kotlin

1.13.4　Java 8 支持

现在可以使用 Java 8 的某些语法，并且可以使用 Java 8 构建的库。

如果想要项目支持 Java 8，点击 File→Project Structure。在 Project Structure 对话框中将 Source Compatibility 与 Target Compatibility 都选择 1.8，如图 1-60 所示。

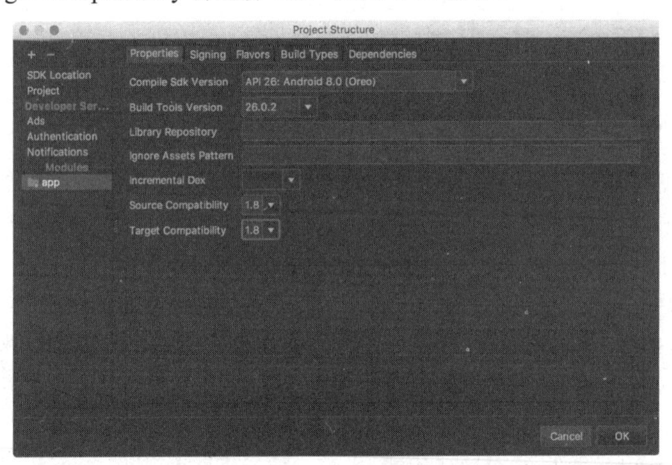

图 1-60　项目支持 Java 8

1.13.5　Android Profiler

新的 Android Profiler 替代了 Android Monitor，提供一套新的工具，实时测试应用程序的 CPU、内存、网络使用情况，如图 1-61 所示。还可以取代抓包工具，能够查看网络传输的具体细节。

要打开这个工具，点击 View→Tool Windows→Android Profiler（如果 toolbar 上有，直接点击 Android Profiler）。

提　示
当 Android Profiler 工具显示时，Logcat 会隐藏，在 Toolbar 上可以看到。

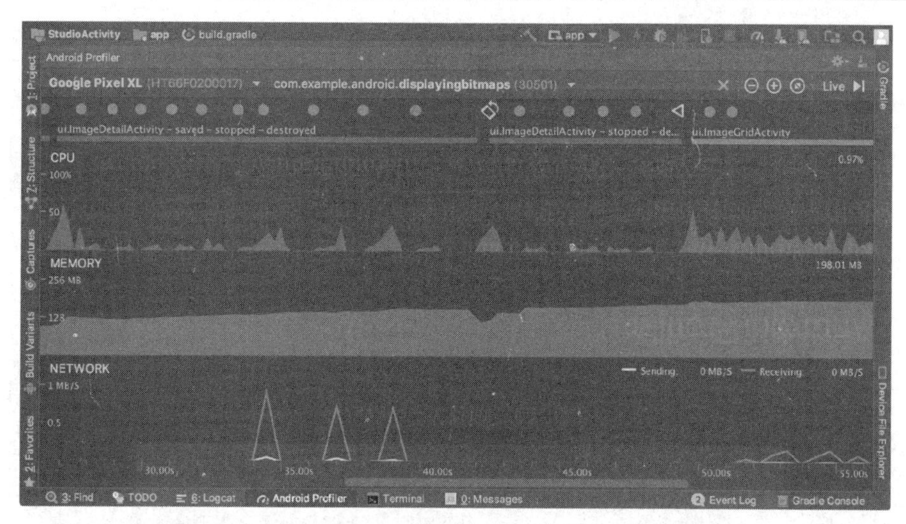

图 1-61　Android Profiler 查看 CPU、内存、网络使用情况

从上到下一共分三块：CPU、内存、网络。如果想分享具体的某一个，点击就会显示具体细节。

1.13.6　CPU Profiler

CPU Profiler 主要用于分析应用程序的 CPU 线程使用情况，如图 1-62 所示。

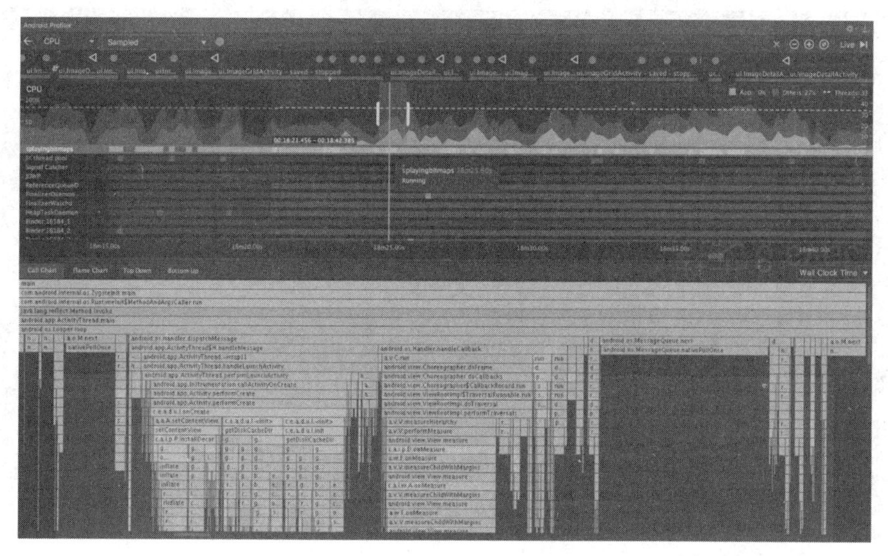

图 1-62　CPU 使用分析

1.13.7　Memory Profiler

Memory Profiler 显示了应用程序内存使用情况，并且用图形界面表示，可以捕捉堆的存储、垃圾内存回收以及内存分配跟踪，如图 1-63 所示。

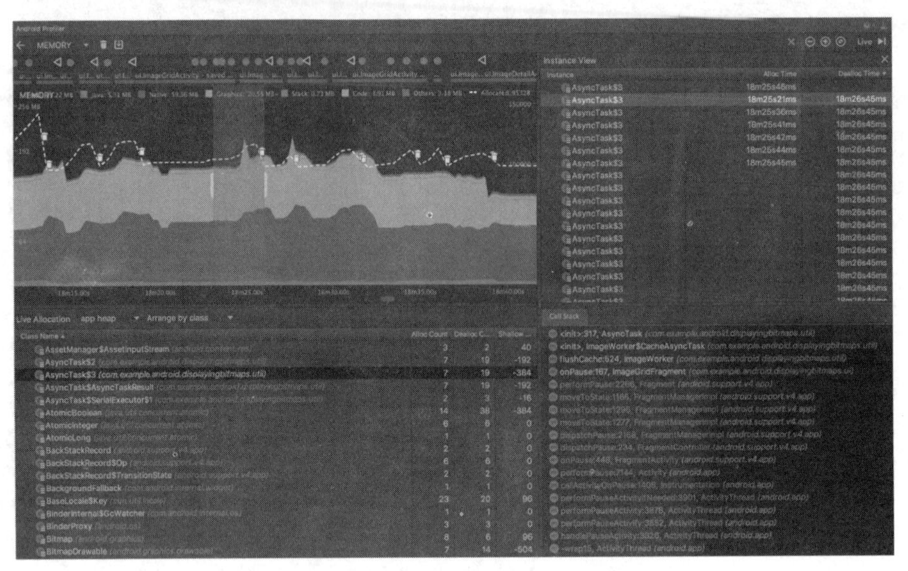

图 1-63　内存分析

1.13.8　Network Profiler

Network Profiler 显示请求链接地址、时间、状态码以及请求回来的数据，如图 1-64 所示。完全可以用这个替代抓包工具。

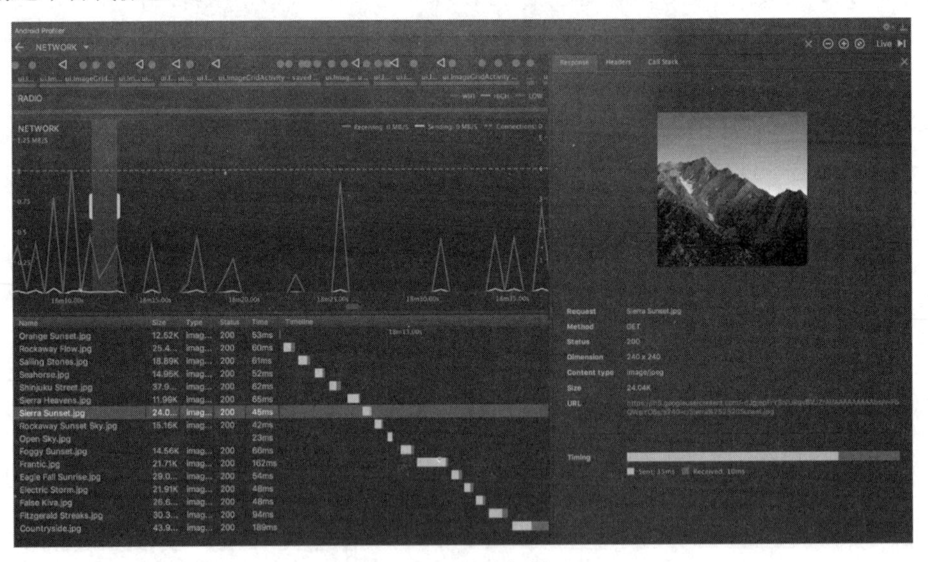

图 1-64　网络访问情况

1.13.9　APK profiling

如果想看 apk 文件的资源文件，不需要用 apktool 工具了，Android Studio 3.0 支持直接打开 apk 文件，只要双击 apk 文件即可，如图 1-65 所示。

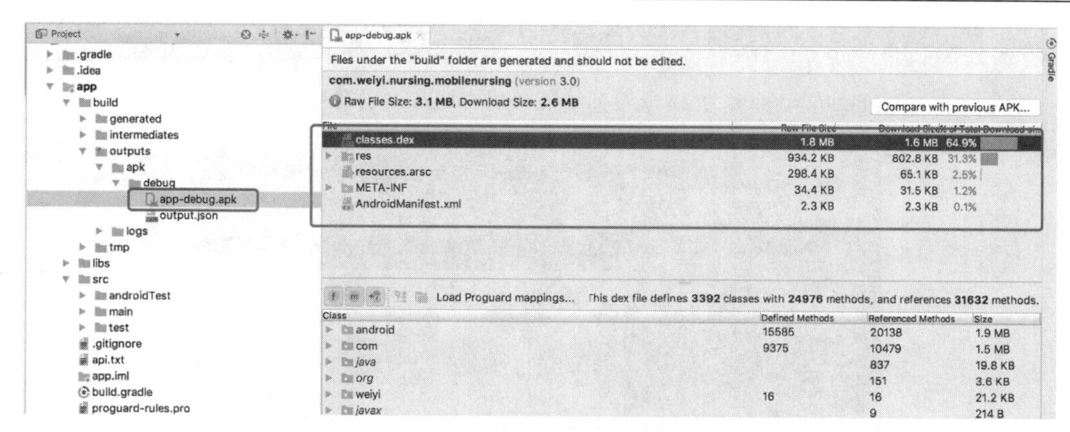

图 1-65　apk 文件分析

可以看到 apk 文件中 res 文件夹下的资源，还能够看到各个文件占比大小。

1.13.10　Device File Explorer

新的设备文件管理器允许设备与计算机之间进行文件传输。如果要打开手机上的文件，双击文件即可。选择方便，不像之前还要用 adb 命令。

如果要打开设备文件管理，点击 View→Tool Windows→Device File Explorer，如图 1-66 所示。

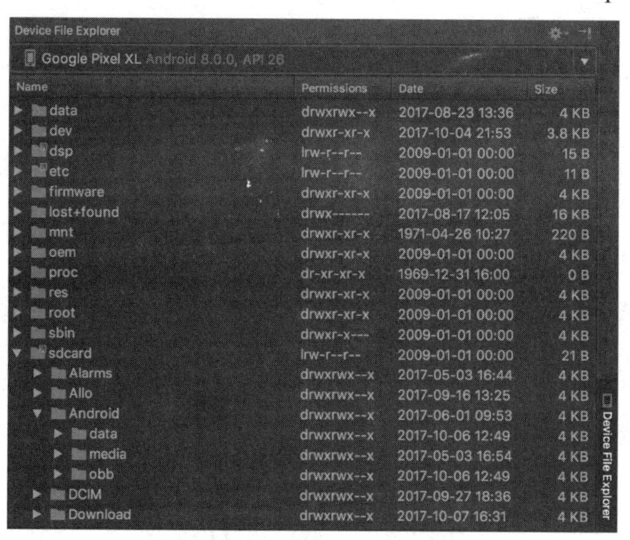

图 1-66　设备文件管理

1.13.11　Adaptive Icons wizard

Image Asset Studio 现在支持矢量绘图，可以为 Android 8.0 创建自适应启动 Icon，同时可以为 8.0 以下的手机创建传统图标。

右击项目中的 res 文件夹，选择 **New→Image Asset**。在 Asset Studio 对话框中，选择 Launcher Icons（Adaptive and Legacy）作为图标类型，如图 1-67 所示。

注　意

必须设置 compileSdkVersion 为 26 或更高才能使用自适应启动器图标。

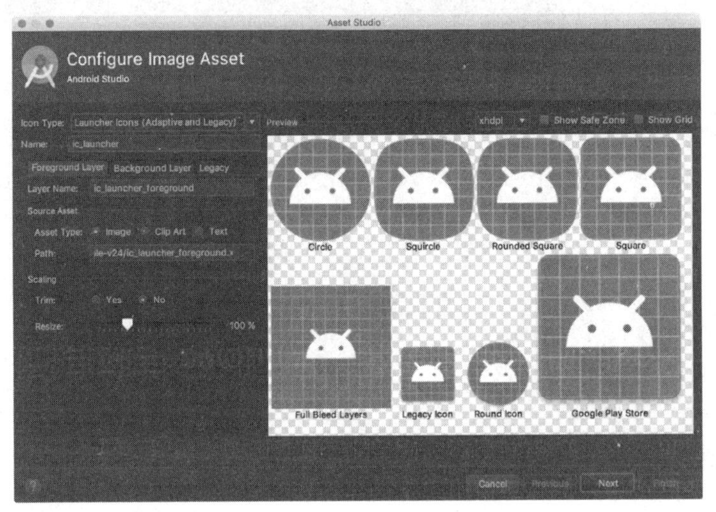

图 1-67　矢量绘图

1.13.12　Google 的 Maven 存储库

Android Studio 现在默认使用 Google 的 Maven 存储库，而不是依赖 Android SDK Manager 来获取 Android 支持库、Google Play 服务、Firebase 和其他依赖项的更新。这样更新更加方便，特别是在使用持续集成（CI）系统时。

现在新项目默认会添加 Google Maven 存储库。如果需要更新之前的项目，打开项目下的 build.gradle 文件。在 allprojects 标签中增加 google()。

```
allprojects {
    repositories {
        google()//增加这行代码
        jcenter()
    }
}
```

1.14　本章小结

本章主要学习 Android Studio 的基本使用方法，包括如何创建项目、运行项目、调试项目。接下来慢慢熟悉项目结构，用 Android Studio 运行第一个应用程序 Hello World，当这个 Hello World 显示在 Android 手机上时，有没有觉得很有趣呢？当然，跟市场上的应用程序相比，这个应用太简单了。不过，很多人学习 Java 时，也是从 Hello World 开始一步一步过来的。学完本章也算加入 Android 开发的大军啦，大家继续努力加油，后面的内容更加精彩！

第2章

Android 控件

本章主要介绍 Android 中常用的控件及其使用方法，Android SDK 本身给我们提供大量的 UI 控件，合理熟练地使用这些控件才能做出优美的界面。有时候 Android 自带的控件不一定能满足业务需求，所以本章还会介绍自定义控件。

2.1 View 介绍

在 Android 开发中，Android 的 UI 界面都是由 View 及其派生类组合而成的。View 类几乎包含了所有的屏幕类型，每一个 View 都有一个用于绘图的画布。画布可以进行任意扩展，只需要重写 onDraw 方法，就能绘制界面显示。界面既可以是复杂的 3D 效果，也可以只是简单的文本显示。表 2-1 描述了 View 常用的 XML 属性及相关方法。

表 2-1 View 常用的 XML 属性及相关方法

XML 属性	相关方法	说明
android:id	setId(int)	给当前 View 设置一个在当前 xxx.xml 中的唯一编号，可以通过调用 View.findViewById()或 Activity.findViewById()根据编号查找到对应的 View。不同的 layout.xml 之间定义相同的 id 不会冲突
android:clickable	setClickable(boolean clickable)	是否响应点击事件 ● clickable=true：允许 ● clickable=false：禁止
android:longClickable	setLongClickable(boolean clickable)	是否响应长点击事件 ● clickable=true：允许 ● clickable=false：禁止

（续表）

XML 属性	相关方法	说明
android:backgroud	setBackgroundColor	设置背景颜色
android:visibility	setVisibility	是否可见 ● Visible：默认值，可见 ● Invisible：不可见 ● Gone：不可见，并且在屏幕中不占位置
android:scrollbars		设置滚动条显示 ● None：隐藏，不可见 ● Horizontal：水平 ● Vertical：垂直
android:onClick	在 xml 中给 onclik 设置什么值，在对应的 Activity 中写对应的方法。例如：android:onClick="test"，在 Activity 中写对应的方法 public void test(View view){}	设置点击事件
android:padding	setPadding(int,int,int,int)	设置上下左右内边距
android:paddingTop		设置顶部内边距
android:paddingBottom		设置底部内边距
android:paddingLeft		设置左边内边距
android:paddingRight		设置右边内边距
android:layout_margin		设置外边距
android:layout_marginTop		设置顶部外边距
android:layout_marginBottom		设置底部外边距
android:layout_marginLeft		设置左边外边距
android:layout_marginRight		设置右边外边距
android:minHeight	setMinimumHeight(int)	设置视图最小高度
android:minWidth	setMinimumWidth(int)	设置视图最小宽度

View是所有UI组件的基类。一般来说，开发Android应用程序的UI界面都不会直接使用View，而是使用派生类。

View 派生出的直接子类有 ImageView、TextView、ViewGroup、ProgressBar 等。

View 派生出的间接子类有 AbsListView、Button、Edittext、CheckBox 等。

使用系统给我们提供的 View 派生类能实现开发中的大部分需求，但是有时候需要针对业务需求去定制 View，例如画饼状图、折线图、贝塞尔曲线等。

2.1.1 自定义 View

我们要实现的效果就是一个画圆的过程，在增加圆形弧度的同时改变圆的颜色（颜色随机生成）。如果弧度大于 360°，就重新开始画，一直循环。每次增加的弧度根据 xml 文件设置，不需要修改 Java 代码，运行效果如图 2-1 所示。

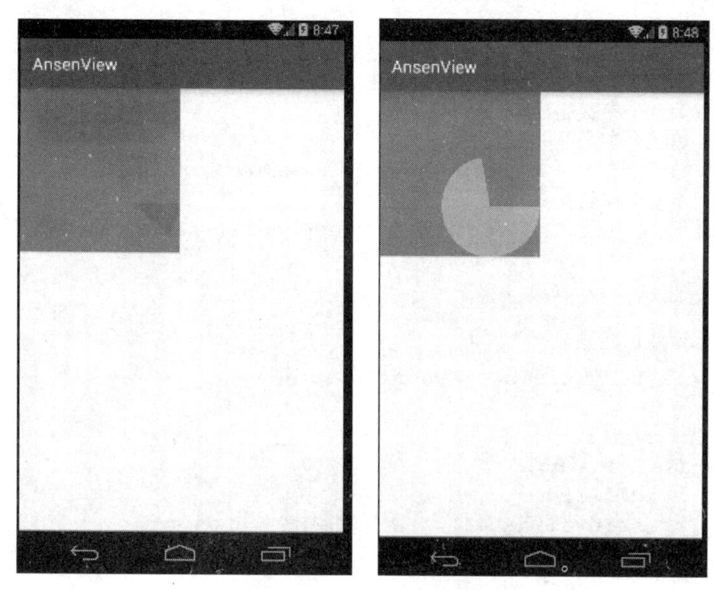

图 2-1　自定义 View 运行效果

新建一个项目，在项目下建一个 MyView 类，继承自 View。重写构造方法和 onDraw 方法。

```
public class MyView extends View{
    private MyThread myThread;

    private Paint paint;//画笔

    private RectF rectF=new RectF(150,150,380,380);
    private int sweepAngle=0;//弧的当前度数
    private int sweepAngleAdd=20;//弧每次增加度数
    private Random random=new Random();
    private boolean running=true;//控制循环

    public MyView(Context context) {
        this(context,null);
    }

    public MyView(Context context,AttributeSet attrs) {
        super(context, attrs);
        init(context,attrs);
    }

    //初始化
    private void  init(Context context,AttributeSet attrs){
        paint=new Paint();
        paint.setTextSize(60);
    }

    @Override
    protected void onDraw(Canvas canvas){
```

```
        Log.i("MyView","onDraw");
        if(null==myThread){
            myThread=new MyThread();
            myThread.start();
        }else{
            //第一个参数是 RectF  左上的 x y 坐标  右下的 x y 坐标
            //第二个参数是 弧形的开始角度
            //第三个参数是 弧形的结束角度
            //第四个参数是 true:画扇形  false:画弧线
            //第五个参数是 画笔
            canvas.drawArc(rectF, 0, sweepAngle, true, paint);
        }
    }

    //开启一个子线程绘制 ui
    private class MyThread extends Thread{
        @Override
        public void run() {
            while(running){
                logic();
                postInvalidate();//重新绘制,会调用 onDraw
                try {
                    Thread.sleep(200);
                } catch (InterruptedException e) {
                    e.printStackTrace();
                }
            }
        }
    }

    protected void logic() {
        sweepAngle+=sweepAngleAdd;//每次增加弧度

        //随机设置画笔的颜色
        int r=random.nextInt(255);
        int g=random.nextInt(255);
        int b=random.nextInt(255);
        paint.setARGB(255, r, g, b);

        if(sweepAngle>=360){//如果弧度大于 360°就从头开始
            sweepAngle=0;
        }
    }

    @Override
    protected void onDetachedFromWindow() {
        running=false;//销毁 View 的时候设置成 false,退出无限循环
        super.onDetachedFromWindow();
    }
}
```

我们在构造方法中初始化画笔，View 第一次绘制在界面上时会调用 onDraw 方法。我们首先判断当前的线程是否为空，第一次运行 myThread 肯定是空的，于是开启线程。如果不为空就绘制圆。

启动 MyThread 线程时，会调用 run 方法。在 run 方法中用一个变量控制循环，默认是 true，只要我们没有改变这个变量的值为 false，就是一个死循环。在循环中调用 logic 方法，同时调用 View 自带的 postInvalidate 方法重新绘制界面。最后延迟 200 毫秒。

logic 方法中增加圆形的弧度，设置画笔的颜色。最后判断一下弧度是不是大于 360°，如果大于 360° 就从 0 重新开始。

onDetachedFromWindow 方法是 View 生命周期里面的一个方法，在 View 销毁的时候调用。改变 running 变量的值为 false，这样线程就会退出死循环。

1. View 测量（重写 onMeasure 方法）

讲到 View 的 measure 测量，会涉及 View 的一个静态内部类 MeasureSpec，MeasureSpec 类封装了父 View 传递给子 View 的布局（layout）要求，每个 MeasureSpec 实例代表宽度或者高度（只能是其一）要求。MeasureSpec 的字面意思是测量规格或者测量属性，在 measure 方法中有两个参数 widthMeasureSpec 和 heightMeasureSpec。如果使用 widthMeasureSpec，我们就可以通过 MeasureSpec 计算出宽度模式 Mode 和宽度的实际值。

测量的模式分以下三种：

- EXACTLY：精确值模式。当 View 的 layout_width 或者 layout_height 属性设置为具体的数值（例如，android:layout_width = "100dp"）或者指定为 “match_parent” 时（系统会自动分配为父布局的大小）使用的模式。
- AT_MOST：最大值模式。当 View 的 layout_width 或者 layout_height 属性设置为 “wrap_content”时使用的模式。
- UNSPECIFIED：可以将视图按照自己的意愿设置成任意大小，没有任何限制。这种情况比较少见，很少用到。

2. 为什么要重写 onMeasure 方法

我们修改一下 activity_main.xml：

```
    <?xml version="1.0" encoding="utf-8"?>
    <RelativeLayout
xmlns:android="http://schemas.android.com/apk/res/android"
        android:layout_width="match_parent"
        android:layout_height="match_parent">

        <com.ansen.view.MyView

android:background="@android:color/holo_green_dark"
            android:layout_width="wrap_content"

android:layout_height="wrap_content"/>
    </RelativeLayout>
```

最外层用 RelativeLayout，里面用了我们自定义的 View，运行效果如图 2-3 所示。

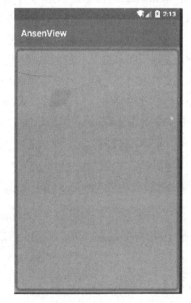

图 2-3 未重写 onMeasure 方法运行效果图

从布局文件中可以看到明明设置了 android:layout_ width="wrap_content"，但是背景颜色的宽高跟屏幕一样大。所以我们得重写 onMeasure 方法：

```
@Override
protected void onMeasure(int widthMeasureSpec, int heightMeasureSpec){
    //获得父容器器为它设置的测量模式和大小
    int widthMode = MeasureSpec.getMode(widthMeasureSpec);
    int widthSize = MeasureSpec.getSize(widthMeasureSpec);

    int heightMode = MeasureSpec.getMode(heightMeasureSpec);
    int heightSize = MeasureSpec.getSize(heightMeasureSpec);

    int width;
    int height ;
    if (widthMode == MeasureSpec.EXACTLY){//指定宽度或者 match_parent
        width = widthSize;
    } else{
        width = (int) (getPaddingLeft() + getPaddingRight() + rectF.width()*2);
    }

    if (heightMode == MeasureSpec.EXACTLY){//指定高度或者 match_parent
        height = heightSize;
    } else{
        height = (int) (getPaddingTop()+getPaddingBottom()+rectF.height()*2);
    }
    setMeasuredDimension(width, height);
}
```

首先获取宽高模式以及父容器器为这个 View 分配的宽高，然后判断宽高的类型，因为在布局文件中设置的是 wrap_content，所以都会执行 else 语句。宽度是圆形的 2 倍，高度也是圆形的 2 倍，View 的宽高只有是圆形的两倍才能看到圆形以外的绿色背景。最后调用 setMeasuredDimension 传入测量之后的宽高。重新运行代码，效果如图 2-4 所示。从效果图中可以看到显示正常。

为什么我们的圆形画在背景颜色区域的右下角？因为 RectF 的 left 跟 top 都是从 190 开始的。如果想画在左上角应该怎么做呢？其实很简单，只需要修改 RectF 的初始值即可。

```
private RectF rectF=new RectF(0,0,190,190);
```

这样改了之后就从（0,0）坐标画圆，并且圆形的宽高还是 190。View 的宽高是圆形的两倍，还是 380。

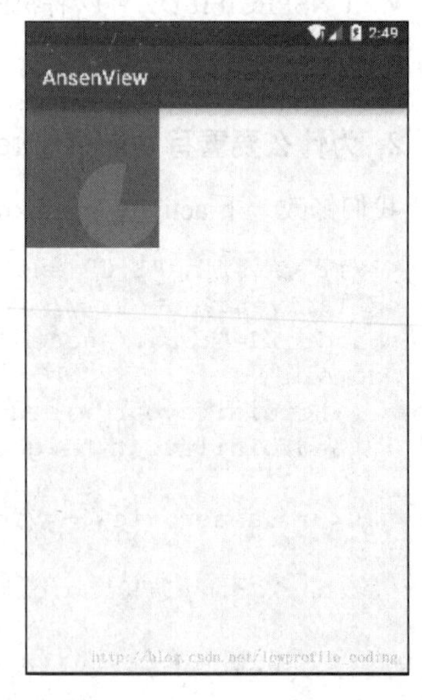

图 2-4　重写 onMeasure 方法后运行效果图

2.1.2　自定义属性

首先在 res/values 下新建一个 attrs.xml 文件。内容如下：

```
<?xml version="1.0" encoding="utf-8"?>
<resources>
    <declare-styleable name="customStyleView">
        <attr name="sweepAngleAdd" format="integer"/>
    </declare-styleable>
</resources>
```

其中，name 为属性集的名字，主要用途是标识属性集；attr 标签可以有多个；format 属性对应的类型也有很多，例如 string、integer、dimension、reference、color、enum，这里使用 integer。

在布局 xml 文件中引用我们刚刚自定义的属性。activity_main.xml 在修改之后的内容如下：

```
<?xml version="1.0" encoding="utf-8"?>
<RelativeLayout xmlns:android="http://schemas.android.com/apk/res/android"
    xmlns:custom="http://schemas.android.com/apk/res-auto"
    android:layout_width="match_parent"
    android:layout_height="match_parent">

    <com.ansen.view.MyView
        android:background="@android:color/holo_green_dark"
        android:layout_width="wrap_content"
        android:layout_height="wrap_content"
        custom:sweepAngleAdd="10"/>
</RelativeLayout>
```

这里只增加了两行代码，即最外层的 RelativeLayout 增加了自定义 View 的命名空间。之后，在自定义 View 中就可以使用 custom:sweepAngleAdd 属性了。

接下来在 MyView 类的 init 方法中获取 sweepAngleAdd 的值。

```
    //获取自定义属性的值
    TypedArray typedArray=context.obtainStyledAttributes(attrs,
R.styleable.customStyleView);
    sweepAngleAdd=typedArray.getInt(R.styleable.customStyleView_sweepAngleAdd,
0);
    typedArray.recycle();
```

这样做的好处就是如果我们想改变圆形每次增加的弧度大小，只需要修改 xml 文件即可，不需要修改自定义 View 的 Java 代码，达到封装的效果。

2.2　ViewGroup 介绍

做过 Android 应用开发的朋友都知道，Android 的 UI 界面都是由 View 和 ViewGroup 及其派生类组合而成的。其中，View 是所有 UI 组件的基类，而 ViewGroup 是容纳这些组件的容器，其本身也是从 View 派生出来的。

AndroidUI 界面的一般结构可参见图 2-5。

可见，作为容器的 ViewGroup 可以包含作为叶子节点的 View，也可以包含作为更低层次的子 ViewGroup，而子 ViewGroup 又可以包含下一层的叶子节点的 View 和 ViewGroup。事实上，这种灵活的 View 层次结构可以形成非常复杂的 UI 布局，开发者可据此设计、开发非常精致的 UI 界面。

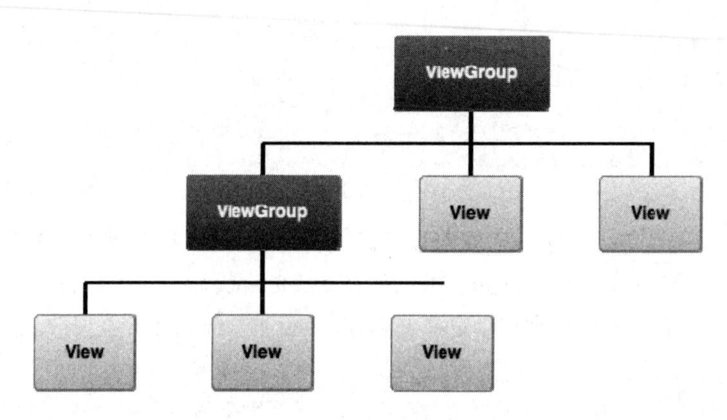

图 2-5 ViewGroup 如何在布局中形成分支并容纳其他 View 的图解

一般来说，开发 Android 应用程序的 UI 界面都不会直接使用 ViewGroup，而是使用它的派生类。

ViewGroup 派生出的直接子类有 CoordinatorLayout、DrawerLayout、RecyclerView，FrameLayou、LinearLayout、RelativeLayout、SwipeRefreshLayout、ViewPager 等。

ViewGroup 派生出的间接子类有 GridView、ListView，WebView 等。

2.2.1 自定义 ViewGroup

在前面的内容中，我们学习过如何自定义 View。自定义 ViewGroup 跟自定义 View 有点类似。一般情况下自定义 ViewGroup 会重写以下几个方法：

- onMeasure 测量子 View 的宽高，根据子 View 的宽高与自己的测量模式来决定自己的宽高。
- onLayout 决定子 View 的摆放位置。
- generateLayoutParams 根据子 View 的布局参数决定子 View 在当前容器的摆放位置，这个方法根据需求决定要不要重写。

1. 首先实现 LinearLayout 布局的水平排列效果

实现这种效果只需要重写 onLayout 和 onMeasure 方法，这里新建一个类 MyViewGroup，继承自 ViewGroup。

```
public class MyViewGroup extends ViewGroup{
    public MyViewGroup(Context context) {
        super(context);
    }

    public MyViewGroup(Context context, AttributeSet attrs) {
        super(context,attrs);
    }

    @Override
    protected void onMeasure(int widthMeasureSpec, int heightMeasureSpec) {
        //获取 ViewGroup 宽高
        int sizeWidth = MeasureSpec.getSize(widthMeasureSpec);
        int sizeHeight = MeasureSpec.getSize(heightMeasureSpec);
```

```
        measureChildren(widthMeasureSpec,heightMeasureSpec);
                                                    //测量所有子 View 的宽高

        Log.i("ansen","测量宽度:"+sizeWidth+" 测量高度:"+sizeHeight);
        setMeasuredDimension(sizeWidth,sizeHeight);
    }

    @Override
    protected void onLayout(boolean changed, int l, int t, int r, int b) {
        int count=getChildCount();//获取子 View 数量
        int left=0;
        for(int i=0;i<count;i++){
            View child=getChildAt(i);

            int childWidth=child.getMeasuredWidth();//获取子 View 宽度
            int childHeight=child.getMeasuredHeight();//获取子 View 高度

            child.layout(left,0,left+childWidth,childHeight);//设置摆放位置
            left+=childWidth;//
        }
    }
}
```

执行流程是在 onMeasure 方法调用两遍之后再调用 onLayout 方法。首先我们看 onMeasure 方法，跟之前自定义 View 时有点类似，没有使用过的就只有 measureChildren 方法。measureChildren 方法的作用是测量所有子 View 的宽高。

接下来看 onLayout。先获取所有的子 View，用 for 循环进行迭代，调用 View 的 getMeasuredWidth 与 getMeasuredHeight 方法获取子 View 的测量宽度和高度。注意，这里不能调用 getWidth 和 getHeight，这两个方法必须在 onLayout 执行完才能调用，不然返回都是 0。最后两行代码连起来看，在 child.layout 这行代码中，layout 方法有 4 个参数，分别是 Left、Top、Right、Bottom，分别表示当前 View 在 Viewgoup 中显示的位置。普及一下，在 Android 手机上屏幕的左上角坐标是[0,0]，测量之后给 left 加上当前 View 的宽度，这样就能水平进行排列。

然后看布局文件 activity_main.xml，将布局换成了我们自己写的 MyViewGroup，里面放了 3 个 TextView。

```
    <com.ansen.views.MyViewGroup
xmlns:android="http://schemas.android.com/apk/res/ android"
        android:layout_width="match_parent"
        android:layout_height="match_parent"
        android:background="@android:color/darker_gray">

        <TextView
            android:layout_width="60dp"
            android:layout_height="60dp"
            android:background="@android:color/holo_red_light"
            android:padding="10dp"
            android:text="1"/>

        <TextView
            android:layout_width="60dp"
```

```
        android:layout_height="80dp"
        android:background="@android:color/holo_blue_light"
        android:padding="10dp"
        android:text="2" />

    <TextView
        android:layout_width="40dp"
        android:layout_height="60dp"
        android:background="@android:color/holo_orange_light"
        android:padding="10dp"
        android:text="3"
        android:textColor="@android:color/white"/>
</com.ansen.views.MyViewGroup>
```

运行代码，效果如图 2-6 所示。

图 2-6 自定义 ViewGroup 实现 LinearLayout 布局水平排列效果

提 示

如果所有子 View 的宽度之和超过屏幕宽度，就会显示非正常效果。这个案例主要以学习为主。

2. 测量 ViewGroup 宽高

如果我们把 activity_main.xml 文件中 MyViewGroup 控件的 android:layout_width 和 android:layout_height 属性的值改成 wrap_content，就会发现 ViewGroup 的宽高还是填充整个屏幕，所以需要重新修改 onMeasure 方法。

```
@Override
protected void onMeasure(int widthMeasureSpec, int heightMeasureSpec) {
    //获得此 ViewGroup 计算模式
    int widthMode = MeasureSpec.getMode(widthMeasureSpec);
    int heightMode = MeasureSpec.getMode(heightMeasureSpec);
```

```
        //获取 ViewGroup 宽高
        int sizeWidth = MeasureSpec.getSize(widthMeasureSpec);
        int sizeHeight = MeasureSpec.getSize(heightMeasureSpec);

        measureChildren(widthMeasureSpec,heightMeasureSpec);//测量所有子 View 的宽高

        if(getChildCount()<=0){//如果没有子 View 当前 ViewGroup 的宽高直接设置为 0
            setMeasuredDimension(0,0);
        }else if(heightMode==MeasureSpec.AT_MOST&&widthMode==
                MeasureSpec.AT_MOST){                       //宽和高是 wrap_content
            int measuredWidth=0;//测量宽度
            int maxMeasuredHeigh=0;//测量高度子 View 最大的高度
            for(int i=0;i<getChildCount();i++){
                View child=getChildAt(i);

                measuredWidth+=child.getMeasuredWidth();

                if(child.getMeasuredHeight()>maxMeasuredHeigh){
                    //如果当前的 View 大于之后 View 的高度
                    maxMeasuredHeigh=child.getMeasuredHeight();
                }
            }
            setMeasuredDimension(measuredWidth,maxMeasuredHeigh);
        }else{
            setMeasuredDimension(sizeWidth,sizeHeight);
        }
    }
```

首先获取 ViewGroup 计算模式，如果宽和高都是 wrap_content，迭代子 View，所有子 View 的宽度加起来等于 ViewGroup 的宽度，高度取子 View 中最高的值。重新运行代码，效果如图 2-7 所示，因为给 ViewGroup 设置了一个灰色的背景，所以明显地看到 ViewGroup 的宽高正是我们想要的。

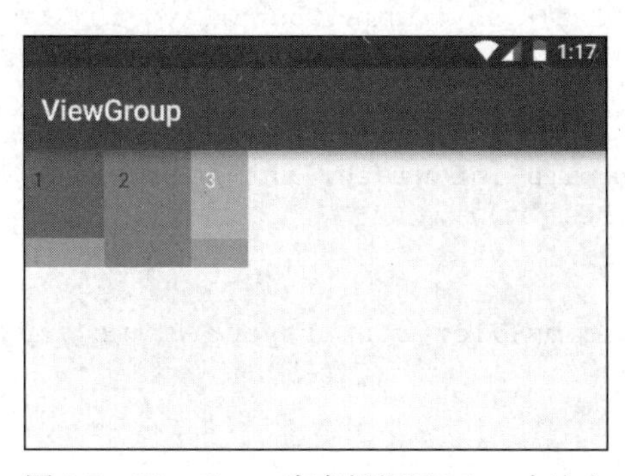

图 2-7 ViewGroup 宽高根据子 View 来决定

3. 自定义 LayoutParams

我们在前面用 LinearLayout 布局时，子 View 使用了 android:layout_weight 属性，通过 layout_weight 属性来决定当前 View 在 LinearLayout 中的占比，这也是因为 LinearLayout 源码中重写了 generateLayoutParams 方法。

还有一种情况就是所有的控件都有 android:layout_margin 属性，可以通过这个属性来设置控件之间的间距，这是因为 ViewGroup 中封装了 MarginLayoutParams 静态类。

首先在 res/values 文件夹下新建 attrs.xml 文件，自定义属性，有两个值 right 和 bottom，right 让当前 View 显示在 ViewGroup 右边，bottom 让当前 View 显示在 ViewGroup 左边，内容如下：

```
<?xml version="1.0" encoding="utf-8"?>
<resources>
    <declare-styleable name="CustomLayoutLP">
        <attr name="layout_position">
            <enum name="right" value="1" />
            <enum name="bottom" value="2" />
        </attr>
    </declare-styleable>
</resources>
```

在 MyViewGroup 中新建静态内部类，因为我们自定义的 ViewGroup 肯定也要设置子 View 之间的间距，所以直接继承 MarginLayoutParams，然后重写 generateLayoutParams 返回我们自己重写的 MyLayoutParams。在构造方法中把 layout_position 的值取出来赋值给 position 变量。

```
@Override
public ViewGroup.LayoutParams generateLayoutParams(AttributeSet attrs){
    return new MyLayoutParams(getContext(), attrs);
}

public static class MyLayoutParams extends MarginLayoutParams {
    public static int POSITION_RIGHT = 1;//右边
    public static int POSITION_BOTTOM = 2;//底部

    public int position = -1;//

    public MyLayoutParams(Context c, AttributeSet attrs) {
        super(c, attrs);

        TypedArray a=c.obtainStyledAttributes(attrs, R.styleable.CustomLayoutLP);
        position=a.getInt(R.styleable.CustomLayoutLP_layout_position, position);
        a.recycle();
    }

    public MyLayoutParams(int width, int height) {
        super(width, height);
    }

    public MyLayoutParams(ViewGroup.LayoutParams source) {
        super(source);
    }
}
```

LayoutParams 重写完了，但是我们的 MyViewGroup 如果想要支持 android:layout_margin 属性以及自定义属性，还得继续修改 onMeasure 和 onLayout 方法。

onMeasure 方法只需要修改一行代码，累加宽度的时候顺便加上左右边距：

```
MyLayoutParams lp= (MyLayoutParams) child.getLayoutParams();
measuredWidth+=child.getMeasuredWidth()+lp.leftMargin+lp.rightMargin;//加上
左右边距
```

onLayout 方法修改代码比较多，修改后的代码如下：

```
    @Override
    protected void onLayout(boolean changed, int l, int t, int r, int b) {
        Log.i("ansen","onLayout");
        int count=getChildCount();//获取子View数量
        int left=0;

        for(int i=0;i<count;i++){
            View child=getChildAt(i);

            MyLayoutParams lp= (MyLayoutParams) child.getLayoutParams();

            int childWidth=child.getMeasuredWidth();//获取子View宽度
            int childHeight=child.getMeasuredHeight();//获取子View高度

            if(lp.position==MyLayoutParams.POSITION_RIGHT){
                                 //当前子View显示在ViewGroup右边
                child.layout(getWidth()-childWidth,0,getWidth(),childHeight);
                                 //设置摆放位置
            }else if(lp.position==MyLayoutParams.POSITION_BOTTOM){
                                 ////当前子View显示在ViewGroup底部
                child.layout(left+lp.leftMargin,getHeight()-childHeight,left+
                             childWidth+lp.leftMargin,getHeight());
            }else{//没有设置位置的View
                child.layout(left+lp.leftMargin,0,left+childWidth+lp.leftMargin,
                             child.getMeasuredHeight());
            }

            Log.i("ansen","left:"+left+" top:"+0+" right:"+(left+childWidth)+"
bottom:"+childHeight);
            left+=childWidth+lp.leftMargin+lp.rightMargin;
        }
    }
```

（1）我们先看position=-1的View，也就是else里面的代码。这个有点绕，要仔细地看几遍，最好自己打印log，或者把View的宽高在纸上画出来。Top和Bottom跟之前一样，下面我们来看Left和Right。

- 当i=0的时候，Left=左边距，Right=宽度+左边距。
- 当i=1的时候，Left=上一个View到ViewGroup的宽度距离+上一个View的右边距+当前View的左边距，Right=上一个View到ViewGroup的宽度距离+上一个View的右边距+当前View的宽度+当前View的左边距。

（2）接下来我们看position有值的View，只要理解了position=-1的情况，position=1和position=2里面的代码就很好理解了。

- position=1时，当前View显示在ViewGroup的右边，需要改变onLayout方法中的Left和Right参数，既然是显示在ViewGroup的右边，Left的值就是ViewGroup的宽度-当前View的宽度，Right的值就是ViewGroup的宽度。
- position=2时，当前View显示在ViewGroup的底部，需要改变onLayout方法中的Top和Bottom，原理同上。

最后重新修改 activity_main.xml 文件，给控件设置外间距，让第一个 TextView 显示在底部、最后一个 TextView 显示在右边。运行代码，效果如图 2-8 所示。

```
    <com.ansen.views.MyViewGroup
xmlns:android="http://schemas.android.com/apk/res/android"
       android:layout_width="wrap_content"
       android:layout_height="wrap_content"
       xmlns:app="http://schemas.android.com/apk/res-auto"
       android:background="@android:color/darker_gray">

       <TextView
          android:layout_width="60dp"
          android:layout_height="60dp"
          android:background="@android:color/holo_red_light"
          android:padding="10dp"
          android:layout_marginLeft="10dp"
          android:layout_marginRight="10dp"
          app:layout_position="bottom"
          android:text="1"/>

       <TextView
          android:layout_width="60dp"
          android:layout_height="80dp"
          android:layout_marginLeft="10dp"
          android:layout_marginRight="20dp"
          android:background="@android:color/holo_blue_light"
          android:padding="10dp"
          android:text="2" />

       <TextView
          android:layout_width="40dp"
          android:layout_height="60dp"
          android:background="@android:color/holo_orange_light"
          android:padding="10dp"
          app:layout_position="right"
          android:text="3"
          android:textColor="@android:color/white"/>
    </com.ansen.views.MyViewGroup>
```

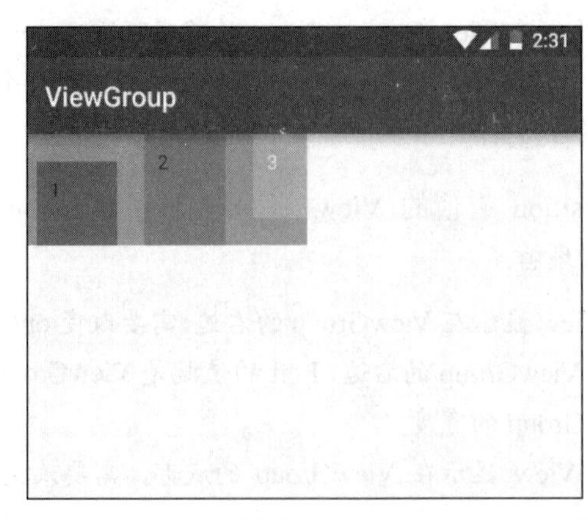

图 2-8　ViewGroup 重写 generateLayoutParams 方法

第三个 TextView 设置了 right 但是却看不出效果，需要修改一下 xml 文件，把 MyViewGroup 控件的 android:layout_width 和 android:layout_height 改成 match_parent，再次运行，效果如图 2-9 所示。

图 2-9　ViewGroup 重写 generateLayoutParams 方法

2.3　几种常用的布局

Android 中系统 SDK 有 5 种布局，所有的布局都继承自 ViewGroup，分别是 LinearLayout（线性布局）、FrameLayout（框架布局）、AbsoluteLayout（绝对布局）、RelativeLayout（相对布局）、TableLayout（表格布局）。但是从这几年的开发经验来看，AbsoluteLayout 与 TableLayout 几乎没有用到，所以就给大家讲解其他三个常用布局。

2.3.1　LinearLayout（线性布局）

LinearLayout 是线性布局控件，是 ViewGroup 的子类，会按照 android: orientation 属性的值对子 View 进行排序，可以将子 View 设置为垂直或水平方向布局。LinearLayout 的每个子视图都会按照它们各自在 XML 中的出现顺序显示在屏幕上。其他两个属性 android:layout_width 和 android:layout_height 则是所有视图的必备属性，用于指定它们的尺寸。

1. 设置排列方式

```
android:orientation="vertical" ; 垂直排列
android:orientation="horizontal"; 水平排列
```

例如，在 LinearLayout 布局中放三个 TextView 并且垂直显示，布局如下：

```
<?xml version="1.0" encoding="utf-8"?>
<LinearLayout xmlns:android="http://schemas.android.com/apk/res/android"
    android:layout_width="match_parent"
    android:layout_height="match_parent"
    android:orientation="vertical">

    <TextView
        android:layout_width="wrap_content"
        android:layout_height="wrap_content"
        android:text="文本 1"/>

    <TextView
        android:layout_width="wrap_content"
        android:layout_height="wrap_content"
        android:layout_marginTop="10dp"
        android:layout_marginBottom="10dp"
        android:text="文本 2"/>

    <TextView
        android:layout_width="wrap_content"
        android:layout_height="wrap_content"
        android:text="文本 3"/>
</LinearLayout>
```

垂直显示效果如图 2-10 所示，让大家有一个直观感受，其实就是从上往下一行一行显示。

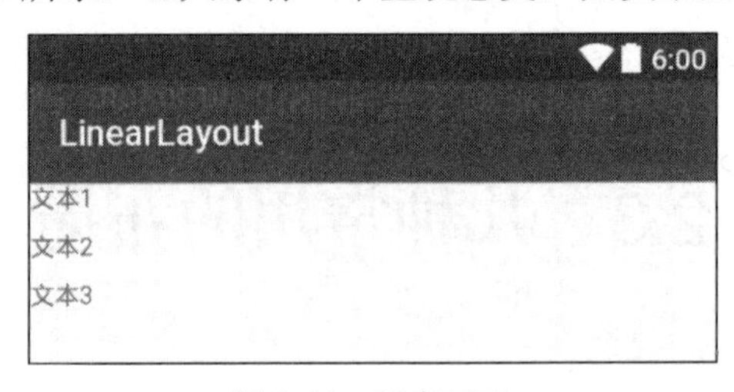

图 2-10　垂直显示

如果将排列方式改成 horizontal（水平），查看水平显示效果，可以看到依次从左到右显示，如图 2-11 所示。

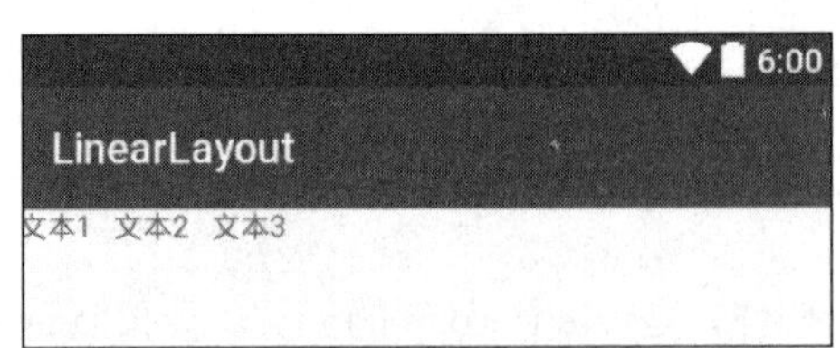

图 2-11　水平显示

2.android:layout_gravity（对齐方式）

layout_gravity 是 LinearLayout 子元素的特有属性。对于 layout_gravity，该属性用于设置控件相对于容器的对齐方式，可选项有 top、bottom、left、right、center_vertical、center_horizontal、center、fill 等。

接着上面的布局进行修改，给 LinearLayout 中文本 2 设置一个 layout_gravity 属性：

```
android:layout_gravity="center_horizontal"
```

其效果如图 2-12 所示。

图 2-12　文本 2 设置 layout_gravity 属性

从效果图中可以看到，显示文本 2 的 TextView 水平居中了。

3. weight（权重）

LinearLayout 布局中 layout_weight 属性用来分配子 View 在 LinearLayout 中占用的空间（显示大小），只有 LinearLayout 包裹的 View 才有这个属性。将上面水平显示的布局文件修改一下，其效果如图 2-13 所示。

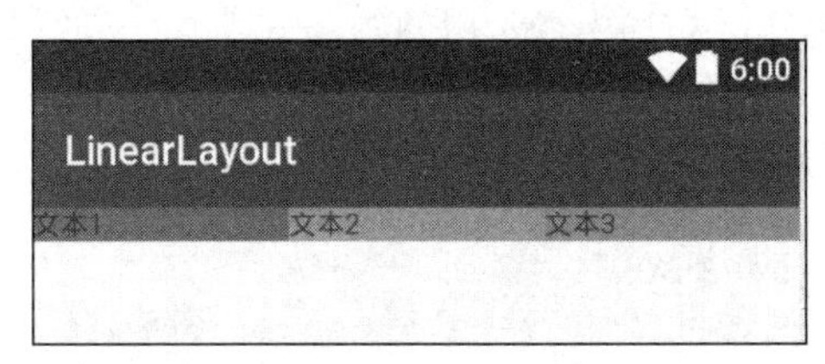

图 2-13　三个 TextView 的宽度一致

从上面的效果图可以看出三个 TextView 的宽度是一致的。我们来看看源码有什么变化。

```
<?xml version="1.0" encoding="utf-8"?>
<LinearLayout xmlns:android="http://schemas.android.com/apk/res/android"
    android:layout_width="match_parent"
    android:layout_height="match_parent"
    android:orientation="horizontal">

    <TextView
        android:layout_width="wrap_content"
        android:layout_height="wrap_content"
        android:background="@android:color/holo_red_light"
        android:layout_weight="1"
        android:text="文本 1"/>

    <TextView
        android:layout_width="wrap_content"
        android:layout_height="wrap_content"
        android:background="@android:color/holo_green_light"
        android:layout_weight="1"
        android:text="文本 2"/>

    <TextView
        android:layout_width="wrap_content"
        android:layout_height="wrap_content"
```

```
        android:background="@android:color/holo_orange_dark"
        android:layout_weight="1"
        android:text="文本 3"/>
</LinearLayout>
```

LinearLayout 的宽度匹配父类，也就是屏幕的宽度，然后来看三个 TextView 都设置了 android:layout_weight="1"，其实就是整个屏幕分成三份，显示三个文本。因为怕大家不好区分，所以顺便给 TextView 加上了背景颜色。

在学习过程中大家也可以自己修改 android:layout_weight 的值，看看会出现什么效果。例如，把 1:1:1 改成 2:3:4，当 android:layout_weight 的值改成 2:3:4 时，通俗点解释就是把屏幕的宽度分成 9 份，第一个 TextView 的宽度是 2/9，第二个 TextView 的宽度是 3/9，第三个 TextView 的宽度是 4/9。

2.3.2 RelativeLayout（相对布局）

RelativeLayout 是相对布局，控件的位置是按照相对位置来计算的，一个控件需要依赖另外一个控件或者依赖父控件。这是实际布局中最常用的布局方式之一。它灵活性大很多，当然属性也多，操作难度也大。

RelativeLayout 常用的一些属性如下：

第一类：属性值为 true 或 false

```
android:layout_centerHorizontal  相对于父元素水平居中
android:layout_centerVertical  相对于父元素垂直居中
android:layout_centerInparent  相对于父元素完全居中(水平垂直都居中)
android:layout_alignParentBottom  贴紧父元素的下边缘
android:layout_alignParentLeft  贴紧父元素的左边缘
android:layout_alignParentRight  贴紧父元素的右边缘
android:layout_alignParentTop  贴紧父元素的上边缘
```

第二类：属性值必须为 id 的引用名"@+id/name"

```
android:layout_below  在某元素的下方
android:layout_above  在某元素的上方
android:layout_toLeftOf  在某元素的左边
android:layout_toRightOf  在某元素的右边
android:layout_alignTop  本元素的上边缘和某元素的上边缘对齐
android:layout_alignLeft  本元素的左边缘和某元素的左边缘对齐
android:layout_alignBottom  本元素的下边缘和某元素的下边缘对齐
android:layout_alignRight  本元素的右边缘和某元素的右边缘对齐
```

第三类：属性值为具体的值，如 30dp、40dp

```
android:layout_marginBottom  离某元素底边缘的距离
android:layout_marginLeft  离某元素左边缘的距离
android:layout_marginRight  离某元素右边缘的距离
android:layout_marginTop  离某元素上边缘的距离
```

由于涉及的属性比较多，就不一一详细讲解了，用 RelativeLayout 布局写了一个登录界面，大家可以学习在布局中如何运用这些属性。

```
<?xml version="1.0" encoding="utf-8"?>
```

```
<RelativeLayout xmlns:android="http://schemas.android.com/apk/res/android"
    android:layout_width="match_parent"
    android:layout_height="match_parent">

    <TextView
        android:id="@+id/tv_nickname"
        android:layout_marginTop="10dp"
        android:layout_width="wrap_content"
        android:layout_height="wrap_content"
        android:textSize="20sp"
        android:text="用户名:" />

    <EditText
        android:id="@+id/et_nickname"
        android:layout_width="match_parent"
        android:layout_height="wrap_content"
        android:layout_toRightOf="@+id/tv_nickname"
        android:textSize="20sp"
        android:hint="请输入用户名" />

    <TextView
        android:id="@+id/tv_password"
        android:layout_marginTop="20dp"
        android:layout_width="wrap_content"
        android:layout_height="wrap_content"
        android:layout_below="@+id/tv_nickname"
        android:layout_alignBottom="@+id/et_password"
        android:textSize="20sp"
        android:text="密    码:" />

    <EditText
        android:id="@+id/et_password"
        android:layout_marginTop="10dp"
        android:layout_width="match_parent"
        android:layout_height="wrap_content"
        android:layout_below="@+id/tv_nickname"
        android:layout_toRightOf="@+id/tv_password"
        android:textSize="20sp"
        android:hint="请输入密码" />

    <Button
        android:layout_below="@+id/et_password"
        style="@style/Widget.AppCompat.Button.Colored"
        android:layout_width="wrap_content"
        android:layout_height="wrap_content"
        android:layout_alignParentRight="true"
        android:text="登录"/>
</RelativeLayout>
```

从布局中可以看到，通过 android:layout_toRightOf="@+id/tv_nickname"属性把输入框放到文本的后面，或者通过 android:layout_below="@+id/tv_nickname"属性把密码输入框放到用户名的下面。还可以通过 android:layout_marginTop 设置上面的外边距，并为 Button 设置 android:layout_alignParentRight="true"属性，就是放在父布局的右边。最后效果如图 2-14 所示。

图 2-14　RelativeLayout 效果

2.3.3　FrameLayout（框架布局）

FrameLayout 是 Android 中比较简单的布局之一，该布局直接在屏幕上开辟出了一块空白区域。向其中添加控件时，所有的组件都会置于这块区域的左上角。如果所有的组件都一样大的话，同一时刻只能看到最上面的那个组件。当然，可以为组件添加 layout_gravity 属性，从而指定对齐方式。

写一个简单的 Demo，效果如图 2-15 所示。

布局文件最外层是 FrameLayout 布局，第一个控件是 TextView，宽高都是 220dp，第二个和第三个的宽高依次减少，可以看到第三个会覆盖第二个，第二个会覆盖第一个，第四个 TextView 因为设置了 android:layout_gravity="center_horizontal|bottom"属性，所以相对父布局水平居中并且位于底部。

图 2-15　FrameLayout 布局效果

```
    <?xml version="1.0" encoding="utf-8"?>
    <FrameLayout
xmlns:android="http://schemas.android.com/apk/res
/android"
       android:layout_width="match_parent"
       android:layout_height="match_parent">

       <TextView
          android:layout_width="220dp"
          android:layout_height="220dp"
          android:background="@android:color/holo_red_light" />

       <TextView
          android:layout_width="180dp"
          android:layout_height="180dp"
          android:background="@android:color/holo_blue_light" />

       <TextView
```

```
        android:layout_width="120dp"
        android:layout_height="120dp"
        android:background="@android:color/holo_orange_light"/>

    <TextView
        android:layout_width="50dp"
        android:layout_height="50dp"
        android:layout_gravity="center_horizontal|bottom"
        android:background="@android:color/link_text_holo_dark"/>
</FrameLayout>
```

2.3.4　三大布局嵌套以及动态添加 View

1. 多层嵌套

当布局比较复杂时，就需要多层嵌套来解决问题了，但是也不要嵌套过多，要灵活运用学会的这几种布局。当熟悉之后看到一个 Android 界面，就知道哪里应该用什么布局。多层嵌套所有的布局都支持的，LinearLayout 嵌套 FrameLayout，FrameLayout 也可以嵌套 LinearLayout，这里就用 LinearLayout 实现登录界面。

先看效果，如图 2-16 所示。

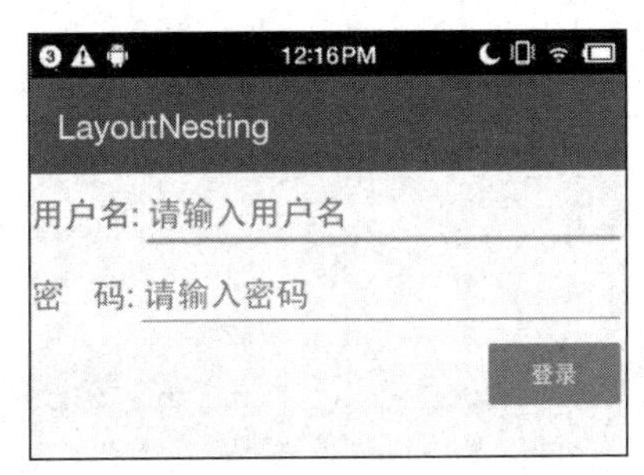

图 2-16　嵌套效果

从宏观的角度来看，第一行、第二行、第三行可以用 LinearLayout 的垂直排列实现，接下来细分到每一行，第一行与第二行用到了两个控件，所以可以嵌套一层，然后设置水平排列方式。第三行是一个按钮，可以设置 android:layout_gravity="right"属性。

```
<?xml version="1.0" encoding="utf-8"?>
<LinearLayout xmlns:android="http://schemas.android.com/apk/res/android"
    android:layout_width="match_parent"
    android:layout_height="match_parent"
    android:orientation="vertical">

    <LinearLayout
        android:layout_width="match_parent"
        android:layout_height="wrap_content"
        android:orientation="horizontal">

        <TextView
            android:layout_width="wrap_content"
            android:layout_height="wrap_content"
            android:text="用户名:"
```

```
            android:textSize="20sp" />

        <EditText
            android:id="@+id/et_nickname"
            android:layout_width="match_parent"
            android:layout_height="wrap_content"
            android:hint="请输入用户名"
            android:textSize="20sp" />
    </LinearLayout>

    <LinearLayout
        android:layout_width="match_parent"
        android:layout_height="wrap_content"
        android:orientation="horizontal">

        <TextView
            android:layout_width="wrap_content"
            android:layout_height="wrap_content"
            android:text="密    码:"
            android:textSize="20sp" />


        <EditText
            android:id="@+id/et_password"
            android:layout_width="match_parent"
            android:layout_height="wrap_content"
            android:hint="请输入密码"
            android:textSize="20sp" />
    </LinearLayout>

    <Button
        style="@style/Widget.AppCompat.Button.Colored"
        android:layout_width="wrap_content"
        android:layout_height="wrap_content"
        android:layout_gravity="right"
        android:text="登录"/>
</LinearLayout>
```

2. 动态添加 View

有时布局文件的 View 数量是不固定的，需要根据逻辑来判断要添加多少 View。这种情况下，只能在代码中来添加 View 了。我们就在登录的代码上增加。

给 LinearLayout 设置一个 id，这样才能在代码中查找这个控件。

```
android:id="@+id/ll_root_view"
```

在布局中增加一个按钮：

```
<Button
    android:id="@+id/btn_add_view"
    style="@style/Widget.AppCompat.Button.Colored"
    android:layout_width="wrap_content"
    android:layout_height="wrap_content"
    android:layout_gravity="right"
    android:text="给 LinearLayout 动态添加 View"/>
```

Activity（活动页面）中的代码比较简单，查找控件，给按钮设置点击事件，在点击回调中动态添加 View。

```
    public class MainActivity extends AppCompatActivity implements
View.OnClickListener{
        private LinearLayout llRootView;
        @Override
        protected void onCreate(Bundle savedInstanceState) {
            super.onCreate(savedInstanceState);
            setContentView(R.layout.activity_main);

            //通过 id 查找 LinearLayout
            llRootView= (LinearLayout) findViewById(R.id.ll_root_view);
            //给按钮设置点击事件
            findViewById(R.id.btn_add_view).setOnClickListener(this);
        }

        @Override
        public void onClick(View v) {
            TextView textView=new TextView(this);
            textView.setText("动态添加 View");
            llRootView.addView(textView);//通过 addView 方法动态添加控件
        }
    }
```

运行修改后的代码，其效果如图 2-17 所示。

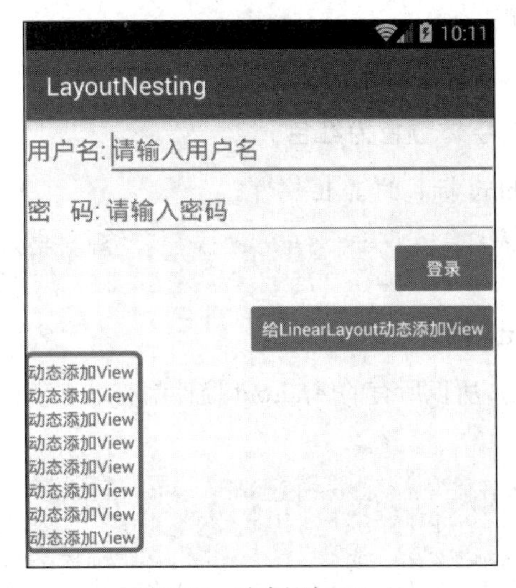

图 2-17 动态添加 View

2.4 初级控件的使用

本节将学习 Android 基础控件的使用，了解常用属性并且熟练运用。

2.4.1 TextView（文本视图）

TextView 显示一行或者多行文本，也能显示 html。在 Android 开发中，TextView 是最常用的组件之一，基本上每天都会使用。

1. 设置背景颜色

```
<TextView
      android:layout_width="match_parent"
      android:layout_height="wrap_content"
      android:background="#FF00FF"
      android:layout_marginTop="10dp"
      android:text="设置背景颜色" />
```

2. 在程序中动态赋值

这里既可以直接是字符串，也可以是字符串资源 id。

```
TextView tv0=(TextView) findViewById(R.id.tv0);
tv0.setText("如何在程序里面动态赋值");
```

3. 实现多字符串的动态处理

（1）在 strings.xml 文件中写上字符串

```
<string name="testing">这是一个数：%1$d,这是两位数：%2$d，这是三位数：%3$s</string>
```

（2）在 Java 代码中设置值

```
tv1.setText(getString(R.string.testing, new Integer[]{11,21,31}));
```

4. TextView 显示 html，字体颜色为红色

需要注意的是，不支持 html 标签的 style 属性。

```
String html="<font color ='red'>TextView 显示 html 字体颜色为红色</font><br/>";
tv3.setText(Html.fromHtml(html));
```

5. 给 TextView 设置点击事件

这个事件是父类 View 的，所以所有的 Android 控件都有这个事件，这里为了方便就采用了内部类的方式。

```
tv4.setOnClickListener(new OnClickListener() {
    @Override
    public void onClick(View v) {
        Toast.makeText(MainActivity.this, "点击了 TextView4",
Toast.LENGTH_LONG).show();
    }
});
```

6. 给 TextView 文字加粗并设置阴影效果

字体阴影需要 4 个相关参数：

- android:shadowColor：阴影的颜色。
- android:shadowDx：水平方向上的偏移量。

- android:shadowDy：垂直方向上的偏移量。
- android:shadowRadius：阴影的半径大小。

```
<TextView
        android:id="@+id/tv5"
        android:layout_width="wrap_content"
        android:layout_height="wrap_content"
        android:layout_marginTop="10dp"
        android:textStyle="bold"

        android:shadowColor="#ff000000"
        android:shadowDx="10"
        android:shadowDy="10"
        android:shadowRadius="1"
        android:text="文字阴影,文字加粗" />
```

7. TextView显示文字加图片

设置图片相关的属性主要有以下几个：

- drawableBottom：在文本框内文本的底端绘制指定图像。
- drawableLeft：在文本框内文本的左边绘制指定图像。
- drawableRight：在文本框内文本的右边绘制指定图像。
- drawableTop：在文本框内文本的顶端绘制指定图像。
- drawablePadding：设置文本框内文本与图像之间的间距。

以下代码在文字左边显示一张图片，并且设置文字跟图片之间的间距为10dp。

```
<TextView
        android:id="@+id/tv6"
        android:layout_width="wrap_content"
        android:layout_height="wrap_content"
        android:layout_marginTop="10dp"
        android:drawableLeft="@drawable/ic_launcher"
        android:drawablePadding="10dp"
        android:gravity="center_vertical"
        android:text="文字+图片" />
```

8. TextView的样式类Span的使用

首先新建一个SpannableString对象，构造方法中传入要显示的内容，调用SpannableString的setSpan方法实现字符串各种风格的显示。setSpan方法有四个参数。参数1表示格式，可以是前景色、背景色等，我们这里用的是背景色。参数2设置格式的开始index。参数3结束index。参数4是一个常量，有以下四个值：

- Spannable. SPAN_INCLUSIVE_EXCLUSIVE：前面包括，后面不包括，即在文本前插入新的文本会应用该样式，而在文本后插入新文本不会应用该样式。
- Spannable. SPAN_INCLUSIVE_INCLUSIVE：前面包括，后面包括，即在文本前插入新的文本会应用该样式，而在文本后插入新文本也会应用该样式。
- Spannable. SPAN_EXCLUSIVE_EXCLUSIVE：前面不包括，后面不包括。
- Spannable. SPAN_EXCLUSIVE_INCLUSIVE：前面不包括，后面包括。

最后调用 TextView 的 setText 把 SpannableString 对象设置进去。

```
    SpannableString spannableString = new SpannableString("TextView 的样式类 Span
的使用详解") ;
            BackgroundColorSpan backgroundColorSpan = new
BackgroundColorSpan(Color.RED);
            //0 到 10 的字符设置红色背景
            spannableString.setSpan(backgroundColorSpan, 0, 10, Spannable.
             SPAN_EXCLUSIVE_EXCLUSIVE) ;
            tv7.setText(spannableString);
```

9. TextView 设置点击事件 Spannable

除了给 TextView 设置背景颜色之外，还可以给 TextView 中某一段文字设置点击效果，调用 SpannableString.setSpan 方法时第一次参数传入 ClickableSpan 格式。使用 ClickableSpan 时，在点击链接时凡是有要执行的动作，必须要给 TextView 设置 MovementMethod 对象。

```
    SpannableString spannableClickString = new SpannableString("TextView 设置点击
事件 Span") ;
            ClickableSpan clickableSpan = new ClickableSpan() {
                @Override
                public void onClick(View widget) {
                        Toast.makeText(MainActivity.this,"TextView 设置点击事件 Span",
Toast.LENGTH_LONG).show();
                }
            };
            spannableClickString.setSpan(clickableSpan,11,15,
Spannable.SPAN_EXCLUSIVE_INCLUSIVE) ;
            tv8.setMovementMethod(LinkMovementMethod.getInstance());
            tv8.setText(spannableClickString);
```

10. TextView 设置点击背景

步骤01 新建一个 selector_textview.xml 文件，放到 drawable 目录下。

```
    <?xml version="1.0" encoding="utf-8"?>
    <selector xmlns:android="http://schemas.android.com/apk/res/android">

        <item android:drawable="@color/textview_click_background"
android:state_focused="true"/>
        <item android:drawable="@color/textview_click_background"
android:state_pressed="true"/>
        <item android:drawable="@color/textview_default"/>

    </selector>
```

步骤02 在 TextView 的 xml 布局中设置背景。

```
    android:background="@drawable/selector_textview"
```

步骤03 设置点击事件。

```
    //必须要给 TextView 加上点击事件，点击之后才能改变背景颜色
    findViewById(R.id.tv9).setOnClickListener(new OnClickListener() {
        @Override
        public void onClick(View v) {
```

```
            Toast.makeText(MainActivity.this,"点击了 TextView9",
Toast.LENGTH_LONG).show();
        }
    });
```

11. 跑马灯效果

当一行文本的内容太多，导致无法全部显示，也不想分行演示时，可以让文本从左到右滚动显示，类似于跑马灯。

```
    <!-- 跑马灯效果 -->
    <TextView
        android:id="@+id/tv12"
        android:layout_width="match_parent"
        android:layout_height="wrap_content"
        android:layout_margin="10dp"
        android:ellipsize="marquee"
        android:marqueeRepeatLimit="marquee_forever"
        android:scrollHorizontally="true"
        android:focusable="true"
        android:focusableInTouchMode="true"
        android:singleLine="true"
        android:text="跑马灯效果 学好 android 开发就关注公众号  android 开发 666 经常
推送原创文章"/>
```

最后效果如图 2-18 所示。

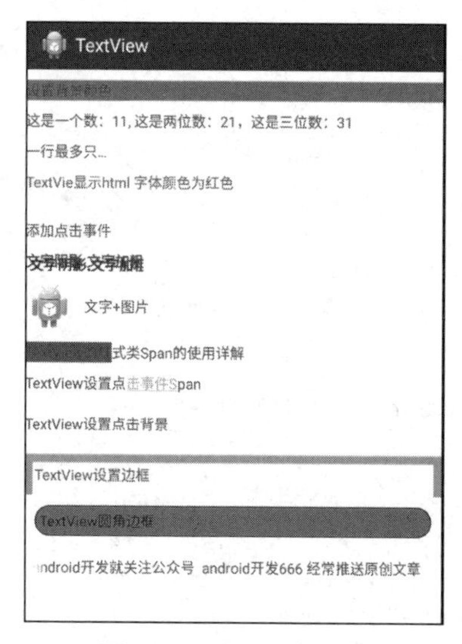

图 2-18　设置文本视图

2.4.2 Button（按钮）

Button 继承自 TextView。在 Android 开发中，Button 是常用的控件，用起来很简单，既可以写在 xml 布局文件中，也可以在 Java 代码中手动创建后加入到布局管理器中，其效果都是一样的。不过，最好是在 xml 文档中定义，因为一旦要改变界面的话，直接修改 xml 就行了，不用修改 Java 程序，并且在 xml 中定义层次分明，一目了然。Button 支持的 XML 属性及相关方法如表 2-2 所示。

表 2-2　Button 支持的 XML 属性及相关方法

XML 属性	相关方法	说明
android:clickable	setClickable(boolean clickable)	设置是否允许点击 ● clickable=true：允许点击 ● clickable=false：禁止点击
android:background	setBackgroundResource(int resid)	通过资源文件设置背景色 resid：资源 xml 文件 ID 按钮默认背景为 android.R.drawable.btn_default
android:text	setText(CharSequence text)	设置文字
android:textColor	setTextColor(int color)	设置文字颜色
android:onClick	setOnClickListener(OnClickListener l)	设置点击事件

下面通过实例来给大家介绍 Button 的常用效果。

首先看一下布局文件 activity_main.xml。

```
<?xml version="1.0" encoding="utf-8"?>
<LinearLayout xmlns:android="http://schemas.android.com/apk/res/android"
    android:layout_width="match_parent"
    android:layout_height="match_parent"
    android:layout_marginLeft="10dp"
    android:orientation="vertical">

    <Button
        android:id="@+id/btn_click_one"
        android:layout_width="wrap_content"
        android:layout_height="wrap_content"
        android:text="Button 点击事件写法 1" />

    <Button
        android:id="@+id/btn_click_two"
        android:layout_width="wrap_content"
        android:layout_height="wrap_content"
        android:onClick="click"
        android:text="Button 点击事件写法 2" />

    <Button
        android:layout_width="wrap_content"
        android:layout_height="wrap_content"
        android:layout_marginTop="10dp"
        android:background="@mipmap/icon_button_bg"
        android:padding="10dp"
        android:text="Button 设置背景图片" />

    <Button
        android:layout_width="wrap_content"
        android:layout_height="wrap_content"
        android:layout_marginTop="10dp"
        android:background="@android:color/holo_red_dark"
        android:padding="10dp"
        android:text="Button 设置背景颜色" />
```

```
    <Button
        android:layout_width="wrap_content"
        android:layout_height="wrap_content"
        android:layout_marginTop="10dp"
        android:background="@drawable/shape_button_test"
        android:padding="10dp"
        android:text="Button 设置 shape" />

    <TextView
        style="@style/Widget.AppCompat.Button.Colored"
        android:layout_width="match_parent"
        android:layout_height="50dp"
        android:layout_marginLeft="20dp"
        android:layout_marginRight="20dp"
        android:layout_marginTop="10dp"
        android:text="V7 包按钮样式"
        android:textColor="#ffffffff"
        android:textSize="20sp" />

</LinearLayout>
```

布局文件对应的效果如图 2-19 所示。

图 2-19　布局了 6 个按钮

上面布局文件中定义了 6 个 Button，指定的规则如下。

（1）按钮 1：给 Button 指定了 android:id="@+id/btn_click_one"，在 MainActivity.xml 根据 id 进行查找并且设置点击事件。

```
//给第一个按钮设置点击事件
findViewById(R.id.btn_click_one).setOnClickListener(onClickListener);
```

点击之后进行 Toast 提示。

```
private View.OnClickListener onClickListener=new View.OnClickListener() {
        @Override
        public void onClick(View v){
            Toast.makeText(MainActivity.this,"Button 点击事件 1",
             Toast.LENGTH_LONG).show();
        }
```

```
    };
```

（2）按钮 2：给 xml 中的 button 增加了 android:onClick="click"属性，然后在该布局文件对应的 Acitivity 中实现该方法。需要注意的是，这个方法必须符合三个条件：

① 方法的修饰符是 public。

② 返回值是 void 类型。

③ 只有一个参数 View，这个 View 就是被点击的控件。

```
    public void click(View v){
        switch (v.getId()){
            case R.id.btn_click_two:
                Toast.makeText(MainActivity.this,"Button 点击事件 2",
                  Toast.LENGTH_LONG).show();
                break;
        }
    }
```

（3）按钮 3：设置一张背景图片用 android:background 属性。

```
android:background="@mipmap/icon_button_bg"
```

（4）按钮 4：设置背景颜色用 android:background 属性。

```
android:background="@android:color/holo_red_dark"
```

（5）按钮 5：设置背景 shape，android:background="@drawable/shape_button_test"，可以自定义 Button 的外观，从效果图中可以看到 Button 背景透明、有边框、有弧度。shape_button_test.xml 文件如下：

```
<?xml version="1.0" encoding="utf-8"?>
<shape xmlns:android="http://schemas.android.com/apk/res/android" >
    <!-- 默认背景色 -->
    <solid android:color="@android:color/transparent"/>
    <!-- 边框 -->
    <stroke
        android:width="1dp"
        android:color="@android:color/black" />
    <!-- 设置弧度 -->
    <corners
        android:radius="20dp"/>
</shape>
```

（6）按钮 6：设置按钮的样式。

```
style="@style/Widget.AppCompat.Button.Colored"
```

这是 V7 包中自带的 Style 样式。按钮的颜色是 ButtonTest/app/src/main/res/values/colors.xml 下 name="colorAccent"的颜色。

Button 使用注意事项：

（1）Button 的 setOnClickListener 优先级比 xml 中 android:onClick 高，如果同时设置点击事件，只有 setOnClickListener 才有效。

（2）能用 TextView 就尽量不要用 Button，TextView 灵活性更高。

2.4.3 EditText（文本编辑框）

EditText 在开发中是经常用到的控件，也是一个比较重要的组件，可以说它是用户与 Android 应用进行数据传输的窗户，比如实现一个登录界面，需要用户输入账号和密码，然后获取用户输入的内容，提交给服务器进行判断。EditText 支持的 XML 属性及相关方法如表 2-3 所示。

表 2-3 EditText 支持的 XML 属性及相关方法

XML 属性	相关方法	说明
android:text	setText(CharSequence text)	设置文本内容
android:textColor	setTextColor(int color)	字体颜色
android:hint	setHint(int resid)	内容为空时显示的文本
android:textColorHint	void setHintTextColor(int color)	为空时显示的文本颜色
android:inputType	setInputType(int type)	限制输入类型 ● number：整数类型 ● numberDecimal：小数点类型 ● date：日期类型 ● text：文本类型（默认值） ● phone：拨号键盘 ● textPassword：密码 ● textVisiblePassword：可见密码 ● textUri：网址
android:maxLength		限制显示的文本长度，超出部分不显示
android:minLines	setMaxLines(int maxlines)	设置文本的最小行数
android:gravity	setGravity(int gravity)	设置文本位置，如设置成“center”，文本将居中显示
android:drawableLeft	setCompoundDrawables(Drawable left,Drawable top,Drawable right, Drawable bottom)	在 text 的左边输出一个 drawable，如图片
android:drawablePadding		设置 text 与 drawable（图片）的间隔，与 drawableLeft、drawableRight、drawableTop、drawableBottom 一起使用，可设置为负数，单独使用没有效果
android:digits		设置允许输入哪些字符，如“1234567890”
android:ellipsize		设置当文字过长时该控件该如何显示 ● start：省略号显示在开头 ● end：省略号显示在结尾 ● middle：省略号显示在中间 ● marquee：以跑马灯的方式显示（动画横向移动）
android:lines	setLines(int lines)	设置文本的行数，设置两行就显示两行，即使第二行没有数据

（续表）

XML 属性	相关方法	说明
android:lineSpacingExtra		设置行间距
android:singleLine	setSingleLine()	true：单行显示 false：可以多行
android:textStyle		设置字形，可以设置一个或多个，用"\

1. EditText 实现登录界面

首先查看布局文件 activity_main.xml。

```
<?xml version="1.0" encoding="utf-8"?>
<LinearLayout xmlns:android="http://schemas.android.com/apk/res/android"
    android:layout_width="match_parent"
    android:layout_height="match_parent"
    android:orientation="vertical">

    <EditText
        android:id="@+id/et_phone"
        android:layout_width="match_parent"
        android:layout_height="wrap_content"
        android:layout_marginLeft="20dp"
        android:layout_marginRight="20dp"
        android:background="@null"
        android:inputType="number"
        android:maxLength="11"
        android:hint="请输入手机号"
        android:drawablePadding="10dp"
        android:padding="10dp"
        android:drawableLeft="@mipmap/icon_phone"
        android:drawableBottom="@drawable/shape_et_bottom_line"
        android:layout_marginTop="20dp"/>

    <EditText
        android:id="@+id/et_password"
        android:layout_width="match_parent"
        android:layout_height="wrap_content"
        android:layout_marginLeft="20dp"
        android:layout_marginRight="20dp"
        android:layout_marginTop="10dp"
        android:background="@null"
        android:inputType="textPassword"
        android:maxLength="16"
        android:padding="10dp"
        android:drawablePadding="10dp"
        android:hint="请输入密码"
        android:drawableBottom="@drawable/shape_et_bottom_line"
        android:drawableLeft="@mipmap/icon_password"/>

    <TextView
        android:id="@+id/tv_login"
        style="@style/Widget.AppCompat.Button.Colored"
        android:layout_width="match_parent"
```

```
        android:layout_height="50dp"
        android:layout_marginLeft="10dp"
        android:layout_marginRight="10dp"
        android:layout_marginTop="30dp"
        android:text="登 录"
        android:textColor="#ffffffff"
        android:textSize="18sp" />
</LinearLayout>
```

运行效果如图 2-20 所示。

图 2-20 EditText 实现登录界面

这两个输入框用的大部分属性都在上面的表格中，这里介绍一下没有提到的属性。

android:background="@null"表示输入框无背景。android:drawableBottom="@drawable/shape_et_bottom_line"表示底部引入一个 shape 布局文件，这个布局文件就是输入框的下划线。shape_et_bottom_ line.xml 内容如下：

```
<shape xmlns:android="http://schemas.android.com/apk/res/android"
    android:shape="rectangle" >

    <solid android:color="#1E7EE3" />
    <size android:height="1dp" android:width="500dp"/>

</shape>
```

2. EditText 监听输入内容

当使用 Google 或者百度进行网页搜索时，只要我们在输入框中输入内容，就会对关键词进行联想匹配。实现的原理很简单，就是监听输入框内容的变化，把内容上传给服务器进行关键词查询，然后客户端把结果展示出来。

下面的代码实现了监听 EditText 内容的变化，首先通过 id 找到 EditText 控件，调用 EditText 控件的 addTextChangedListener 添加文本改变监听，这里我们用内部类方式实现 TextWatcher 接口，重写 TextWatcher 接口的三个方法（beforeTextChanged、onTextChanged、afterTextChanged）。可以在 onTextChanged 方法中把用户输入的结果上传给服务器。

```
EditText etOne= (EditText) findViewById(R.id.et_phone);
etOne.addTextChangedListener(new TextWatcher() {
        @Override
```

```
            public void beforeTextChanged(CharSequence s, int start, int count,
int after) {
                Log.i("Ansen","内容改变之前调用:"+s);
            }

            @Override
            public void onTextChanged(CharSequence s, int start, int before, int
count) {
                Log.i("Ansen","内容改变，可以去告诉服务器:"+s);
            }

            @Override
            public void afterTextChanged(Editable s) {
                Log.i("Ansen","内容改变之后调用:"+s);
            }
    });
```

2.4.4 ImageView（图像视图）

ImageView 用于显示图片的 View，是开发中频繁使用的一个控件，毕竟现在 4G 网络普及了，手机加载一张图片很快，所以很多 App 都使用了大量的图片。

只需在 xml 文件加入 ImageView，就能够显示图片：

```
<?xml version="1.0" encoding="utf-8"?>
<RelativeLayout xmlns:android="http://schemas.android.com/apk/res/android"
    android:layout_width="match_parent"
    android:layout_height="match_parent">

    <ImageView
        android:id="@+id/iv_one"
        android:layout_width="wrap_content"
        android:layout_height="wrap_content"
        android:src="@mipmap/ic_launcher"/>

</RelativeLayout>
```

这里给 ImageView 设置了 src 属性，引用的图片在 mipmap 文件夹中，是用 Android Studio 创建一个项目时自带的图片。在真实的企业开发中，一般会替换掉这个图片的内容，因为这张图片的名字一般用作程序的启动图标。程序运行效果如图 2-21 所示。

图 2-21　添加图片

接下来，给 ImageView 设置点击事件，点击图片之后通过代码动态改变显示的图片。

```
public class MainActivity extends AppCompatActivity {
```

```
    private ImageView imageView;

    @Override
    protected void onCreate(Bundle savedInstanceState) {
        super.onCreate(savedInstanceState);
        setContentView(R.layout.activity_main);

        imageView=(ImageView) findViewById(R.id.iv_one);//通过 id 查找到图片控件
        imageView.setOnClickListener(onClickListener);//设置点击监听事件
    }

    private View.OnClickListener onClickListener=new View.OnClickListener() {
        @Override
        public void onClick(View v) {
            imageView.setImageResource(R.mipmap.coffee);//改变显示的图片
        }
    };
}
```

在 onCreate 方法中，首先通过 id 查找 xml 中的图片控件，然后设置点击事件，点击之后修改显示的图片，其运行结果如图 2-22 所示。

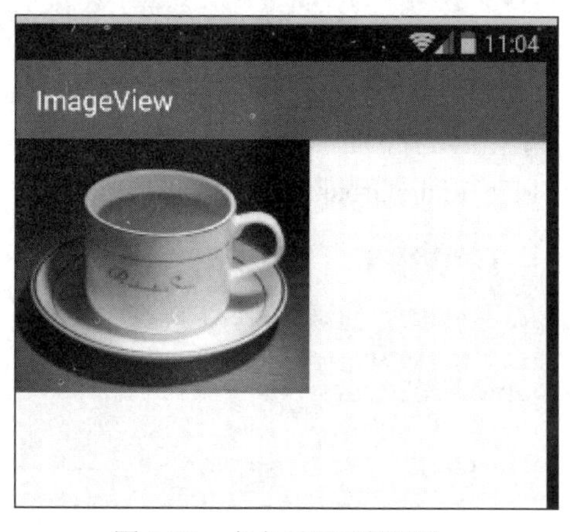

图 2-22 点击后显示新图片

可以看到点击之后图片改变了，这张咖啡的图片是事先复制到 res/mipmap-hdpi 文件夹中的。

2.4.5 RadioButton（单选按钮）

RadioButton 在 Android 开发中也是比较常见的控件，从多个选项中选择一项时会用到。实现单选按钮需要将 RadioButton 和 RadioGroup 配合使用。

RadioGroup 是单选组合框（相当于容器），用于将 RadioButton 框起来，在没有 RadioGroup 的情况下，RadioButton 可以全部选中；当多个 RadioButton 被 RadioGroup 包含的情况下，RadioButton 只可以选择一个。

开发中比较常见的就是 App 注册界面需要选择性别，一般都会有两个选项：男 / 女。

首先查看布局文件，就是外层一个 RadioGroup，其中包含两个 RadioButton。

```
<?xml version="1.0" encoding="utf-8"?>
<RelativeLayout xmlns:android="http://schemas.android.com/apk/res/android"
    android:layout_width="match_parent"
    android:layout_height="match_parent">

    <RadioGroup
        android:id="@+id/radiogroup"
        android:layout_width="wrap_content"
        android:layout_height="wrap_content"
        android:orientation="horizontal">

        <RadioButton
            android:id="@+id/rb_male"
            android:layout_width="wrap_content"
            android:layout_height="wrap_content"
            android:text="男"/>

        <RadioButton
            android:id="@+id/rb_girl"
            android:layout_width="wrap_content"
            android:layout_height="wrap_content"
            android:text="女"/>
    </RadioGroup>
</RelativeLayout>
```

布局文件只能展示 UI 效果，但是我们的程序肯定要增加交互性，需要把用户选中的结果记录下来，所以还得在活动页面中监听 RadioGroup 的选中事件。

```
public class MainActivity extends AppCompatActivity {
    @Override
    protected void onCreate(Bundle savedInstanceState) {
        super.onCreate(savedInstanceState);
        setContentView(R.layout.activity_main);

        RadioGroup radioGroup= (RadioGroup) findViewById(R.id.radiogroup);
        radioGroup.setOnCheckedChangeListener(onCheckedChangeListener);
    }

    private OnCheckedChangeListener onCheckedChangeListener=new
OnCheckedChangeListener() {
        @Override
        public void onCheckedChanged(RadioGroup group, int checkedId) {
            if(checkedId==R.id.rb_male){
                Toast.makeText(MainActivity.this,"您的性别是男",
                Toast.LENGTH_SHORT).show();
            }else if(checkedId==R.id.rb_girl){
                Toast.makeText(MainActivity.this,"您的性别是女",
                Toast.LENGTH_SHORT).show();
            }
        }
    };
}
```

在 onCreate 方法中通过 id 获取到 RadioGroup 对象，然后将其设置为选中改变监听器，在监听器的回调方法中通过 checkedId 来判断选中了哪个选项。接下来通过 Toast 进行显示。

给大家补充一个新的知识点：Toast 是 Android 中用来显示信息的一种机制，与 Dialog 不一样的是，Toast 没有焦点，而且 Toast 显示的时间有限，会根据用户设置的显示时间后自动消失，主要用于向用户显示提示信息。

运行程序，点击性别“男”，其效果如图 2-23 所示。

图 2-23　单选按钮显示效果

2.4.6　Checkbox（复选框）

CheckBox 与 RadioButton 一样，只有两种状态，即选中与未选中，两者的区别就是 CheckBox 可以实现多选。例如，做一个在线考试系统，选择题有单选题和多选题，单选题可以用 RadioButton 实现，多选题可以用 CheckBox 实现。

上一节已经学习了 RadioButton，接下来学习 CheckBox，首先查看 xml 布局文件:

```
<?xml version="1.0" encoding="utf-8"?>
<LinearLayout xmlns:android="http://schemas.android.com/apk/res/android"
   android:layout_width="match_parent"
   android:layout_height="match_parent"
   android:orientation="vertical">

   <TextView
      android:layout_width="wrap_content"
      android:layout_height="wrap_content"
      android:text="你会几种编程语言?" />

   <TextView
      android:id="@+id/tv_result"
      android:layout_margin="10dp"
      android:layout_width="wrap_content"
      android:layout_height="wrap_content"
      android:textSize="18sp"
      android:textColor="@android:color/holo_blue_light"/>

   <CheckBox
      android:id="@+id/cb_java"
      android:layout_width="wrap_content"
      android:layout_height="wrap_content"
      android:text="Java"/>

   <CheckBox
      android:id="@+id/cb_php"
      android:layout_width="wrap_content"
      android:layout_height="wrap_content"
      android:text="Php"/>
```

```
    <CheckBox
        android:id="@+id/cb_c"
        android:layout_width="wrap_content"
        android:layout_height="wrap_content"
        android:text="C"/>
</LinearLayout>
```

前面用到了两个 TextView：第一个用来显示提问的文字，第二个用来展示选中之后的结果。后面是三个复选框，代表三个选项。

按照前面的例子，大家应该明白了 xml 只有 UI 效果，所有的逻辑需要在活动页面中处理。

```
public class MainActivity extends AppCompatActivity {
    private TextView tvResult;//用来显示结果

    private String javaResult="",phpResult="",cResult="";

    @Override
    protected void onCreate(Bundle savedInstanceState) {
        super.onCreate(savedInstanceState);
        setContentView(R.layout.activity_main);

        tvResult= (TextView) findViewById(R.id.tv_result);

        //查找控件并且设置选中改变监听
        CheckBox cbJava= (CheckBox) findViewById(R.id.cb_java);
        CheckBox cbPhp= (CheckBox) findViewById(R.id.cb_php);
        CheckBox cbC= (CheckBox) findViewById(R.id.cb_c);
        cbJava.setOnCheckedChangeListener(onCheckedChangeListener);
        cbPhp.setOnCheckedChangeListener(onCheckedChangeListener);
        cbC.setOnCheckedChangeListener(onCheckedChangeListener);
    }

    private CompoundButton.OnCheckedChangeListener
onCheckedChangeListener=new CompoundButton.OnCheckedChangeListener() {
        @Override
        public void onCheckedChanged(CompoundButton buttonView, boolean
isChecked){
            if(buttonView.getId()==R.id.cb_java){//通过 id 区分不同的复选框
                //如果选中了 Java 就把"Java"赋值给 javaResult，否则""赋值给 javaResul
                //这里用到了三元表达式，如果不会请先去学习 java 基础
                javaResult=isChecked?"Java":"";
            }else if(buttonView.getId()==R.id.cb_php){
                phpResult=isChecked?"Php":"";
            }else if(buttonView.getId()==R.id.cb_c){
                cResult=isChecked?"C":"";
            }
            //展示选中结果
            tvResult.setText(javaResult+" "+phpResult+" "+cResult);
        }
    };
}
```

首先在 onCreate 中查找控件，并设置选中改变监听，在监听函数中根据 id 区分不同的复选框，

然后根据是否选中来赋值，最后把结果给 TextView 显示在屏幕上。运行之后，选中了 Java 和 C 复选框，效果如图 2-24 所示。

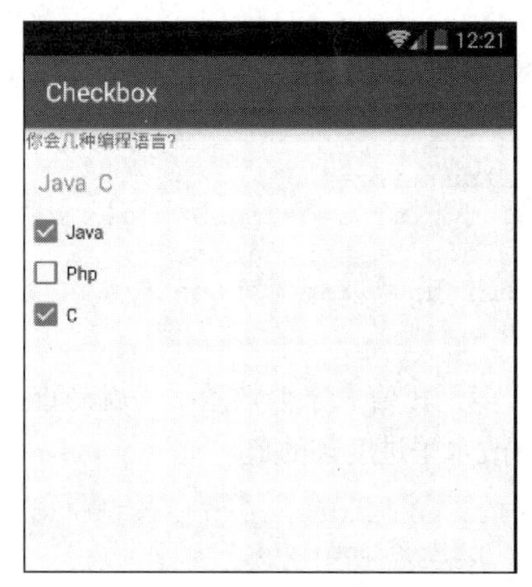

图 2-24　复选框显示效果

2.4.7 ProgressBar（进度条）

ProgressBar 在 Android 中比较常用。ProgressBar 分为确定的和不确定的两种，确定的是能明确看到进度，不确定的就是不清楚、不确定一个操作需要多长时间来完成。

本例用了水平进度条和圆形进度条，水平进度条是确定进度的，圆形进度条表示不确定进度。同时在两个进度条的上方放上两个按钮，用来操作水平进度条的进度值。

```
<?xml version="1.0" encoding="utf-8"?>
<LinearLayout xmlns:android="http://schemas.android.com/apk/res/android"
    android:layout_width="match_parent"
    android:layout_height="match_parent"
    android:orientation="vertical">

     <LinearLayout
        android:layout_width="match_parent"
        android:layout_height="wrap_content"
        android:orientation="horizontal">
        <Button
            android:id="@+id/btn_add"
            android:layout_width="wrap_content"
            android:layout_height="wrap_content"
            android:text="进度值+"/>

        <Button
            android:id="@+id/btn_reduce"
            android:layout_width="wrap_content"
            android:layout_height="wrap_content"
            android:text="进度值-"/>
     </LinearLayout>

    <ProgressBar
       android:id="@+id/pb_horizontal"
```

```
        style="?android:attr/progressBarStyleHorizontal"
        android:layout_width="match_parent"
        android:layout_height="wrap_content"
        android:max="100"
        android:progress="0"/>

    <ProgressBar
        android:id="@+id/pb_large"
        style="?android:attr/progressBarStyleLarge"
        android:layout_width="match_parent"
        android:layout_height="wrap_content"/>
</LinearLayout>
```

这种逻辑代码与之前的差不多，相信大家很熟悉了，就是查找控件。接下来给两个按钮设置点击事件，在点击回调方法中修改水平进度条的值。

```
public class MainActivity extends AppCompatActivity {
    private ProgressBar pbHorizontal;

    @Override
    protected void onCreate(Bundle savedInstanceState) {
        super.onCreate(savedInstanceState);
        setContentView(R.layout.activity_main);

        pbHorizontal= (ProgressBar) findViewById(R.id.pb_horizontal);

        findViewById(R.id.btn_add).setOnClickListener(onClickListener);
        findViewById(R.id.btn_reduce).setOnClickListener(onClickListener);
    }

    private View.OnClickListener onClickListener=new View.OnClickListener() {
        @Override
        public void onClick(View v) {
            switch (v.getId()){
                case R.id.btn_add:
                    pbHorizontal.setProgress(pbHorizontal.getProgress()+10);
                    break;
                case R.id.btn_reduce:
                    pbHorizontal.setProgress(pbHorizontal.getProgress()-10);
                    break;
            }
        }
    };
}
```

在真实的企业开发中，一般用确定进度条表示下载文件进度，用不明确的进度条表示正在访问网络。这里只是为了演示 ProgressBar 用法，所以就用两个按钮的点击来修改进度条的值。最后运行效果如图 2-25 所示。

图 2-25 两种不同的进度条

2.4.8 ProgressDialog（进度对话框）

ProgressDialog 经常用于一些费时的操作，需要用户进行等待。例如，加载网页内容，这时需要一个提示来告诉用户程序正在运行，并没有假死或者真死，而 ProgressDialog 就是专门干这项工作的。

一般使用它的步骤为：在执行耗时间的操作之前弹出 ProgressDialog 提示用户，然后开一个新线程。在新线程中执行耗时的操作，运行完毕之后通知主程序将 ProgressDialog 结束。

新建项目，首先修改 activity_main.xml 文件，增加一个按钮，布局文件比较简单，就不贴出来了，直接看 MainActivity 如何实现：

```
    public class MainActivity extends AppCompatActivity {
        private ProgressDialog staticDialog;

        @Override
        protected void onCreate(Bundle savedInstanceState) {
            super.onCreate(savedInstanceState);
            setContentView(R.layout.activity_main);

            //创建对象，调用 Dialog 的 show 方法显示
            //ProgressDialog dialog = new ProgressDialog(this);
            //dialog.setProgressStyle(ProgressDialog.STYLE_HORIZONTAL);//水平
            //dialog.incrementProgressBy(20);//设置进度值
            //dialog.setCanceledOnTouchOutside(false);
                                  // 设置在点击 Dialog 外是否取消 Dialog 进度条
            // dialog.show();//显示

            // 调用 ProgressDialog 的静态方法显示 5 秒后关闭。模拟访问网络过程
            findViewById(R.id.btn_show).setOnClickListener(new
View.OnClickListener() {
                @Override
                public void onClick(View v) {
                    staticDialog = ProgressDialog.show(MainActivity.this,"这是标题
","这是内容");
                    /* 开启一个新线程，在新线程里执行耗时的方法 */
```

```
                new Thread(new Runnable() {
                    @Override
                    public void run() {
                        try {
                            Thread.sleep(5000);//延迟 5 秒
                        } catch (InterruptedException e) {
                            e.printStackTrace();
                        }
                        handler.sendEmptyMessage(0);//延迟 5 秒之后发送消息给 handler
                    }

                }).start();
            }
        });
    }

    Handler handler = new Handler() {
        @Override
        public void handleMessage(Message msg){//handler 接收到消息后就会执行此方法
            staticDialog.dismiss();// 关闭 ProgressDialog
        }
    };
}
```

调用 ProgressDialog 的静态方法显示，开启一个新的线程，延迟 5 秒，然后给 handler 发送一个消息，在 handler 的 handleMessage 方法中关闭 ProgressDialog。程序运行结果如图 2-26 所示。

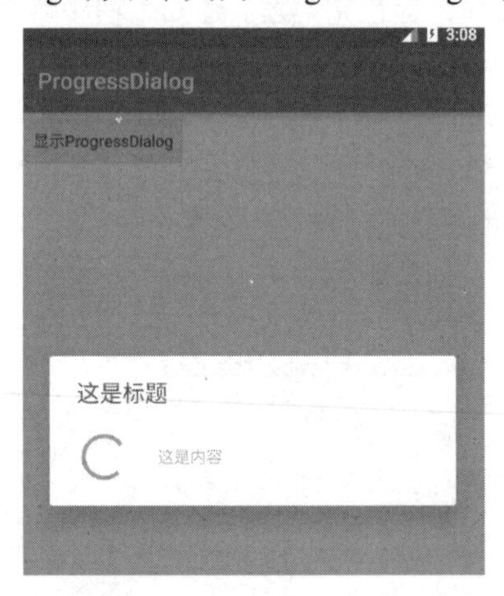

图 2-26　显示进度对话框

2.4.9　AlertDialog（简单对话框）

在 Android 开发中，经常需要在 Android 界面上弹出一些对话框，例如询问用户或者让用户选择（如删除提示对话框、警告对话框等），这些功能用 AlertDialog 对话框来实现。

```
public class MainActivity extends AppCompatActivity {
    @Override
    protected void onCreate(Bundle savedInstanceState) {
        super.onCreate(savedInstanceState);
```

```
        setContentView(R.layout.activity_main);

        findViewById(R.id.btn_show_dialog).setOnClickListener(new View.
          OnClickListener() {
            @Override
            public void onClick(View v) {
                showDialog();
            }
        });
    }

    //显示对话框
    protected void showDialog() {
        AlertDialog.Builder builder = new AlertDialog.Builder(this);
        builder.setTitle("提示");//设置标题
        builder.setMessage("确认退出吗？");//设置消息
        builder.setIcon(R.mipmap.ic_launcher);//设置 icon
        builder.setPositiveButton("确认",new DialogInterface.OnClickListener(){

            @Override
            public void onClick(DialogInterface dialog, int which) {
                dialog.dismiss();
                MainActivity.this.finish();//结束当前 Activity
            }
        });
        builder.setNegativeButton("取消",new DialogInterface.OnClickListener(){

            @Override
            public void onClick(DialogInterface dialog, int which) {
                dialog.dismiss();
            }
        });
        builder.create().show();
    }
}
```

点击按钮，弹出一个对话框，给对话框设置了标题、内容、图片、两个按钮监听事件。这段代码相信大家很好理解，运行效果图如图 2-27 所示。

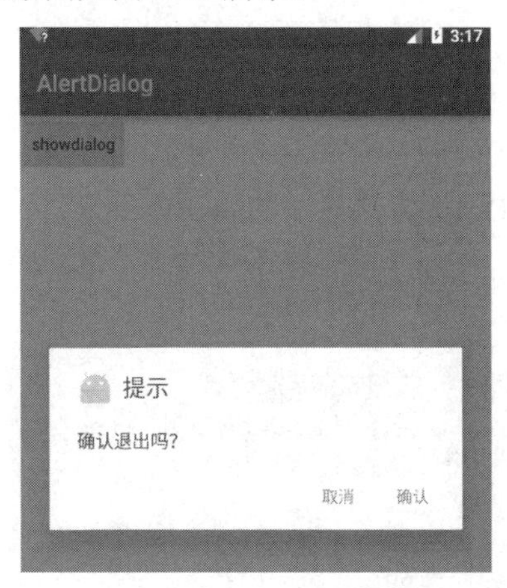

图 2-27　弹出简单对话框

2.4.10 PopupWindow（弹出式窗口）

PopupWindow 弹出一个浮动的窗口，可以显示在屏幕任意的位置，比 Dialog 对话框更加灵活（默认只能在屏幕的中间）。我们还可以通过 setAnimationStyle 方法设置 PopupWindow 的显示或隐藏动画。

本例中，PopupWindow 显示在某个控件之下。在 activity_main.xml 中放置两个按钮，从上向下显示。

```
<?xml version="1.0" encoding="utf-8"?>
<LinearLayout xmlns:android="http://schemas.android.com/apk/res/android"
    android:layout_width="match_parent"
    android:layout_height="match_parent"
    android:orientation="vertical">

    <Button
        android:id="@+id/btn_show_popupwindow"
        android:layout_width="wrap_content"
        android:layout_height="wrap_content"
        android:text="在当前位置下面弹出 PopupWindow" />

    <Button
        android:id="@+id/btn_bottom_popupwindow"
        android:layout_width="wrap_content"
        android:layout_height="wrap_content"
        android:text="从底部弹出 PopupWindow" />
</LinearLayout>
```

在活动页面中根据 id 查找第一个按钮并且设置点击事件：

```
btnShowPopupwindow = (Button) findViewById(R.id.btn_show_popupwindow);
btnShowPopupwindow.setOnClickListener(this);
```

在点击事件中调用 showAsDropDown 方法：

```
@Override
public void onClick(View v) {
    if(v.getId()==R.id.btn_show_popupwindow){//点击第一个按钮
        showAsDropDown();
    }
}
```

接下来，通过 showAsDropDown 方法显示 PopupWindow：

```
private void showAsDropDown(){
    View popView =
LayoutInflater.from(this).inflate(R.layout.popup_drop_down,null);
    //设置 PopupWindow View,宽度，高度
    PopupWindow popupWindow=new PopupWindow(popView,
   LinearLayout.LayoutParams.WRAP_CONTENT,
LinearLayout.LayoutParams.WRAP_CONTENT);
    //设置允许在外点击消失,必须要给 popupWindow 设置背景才会有效
    popupWindow.setOutsideTouchable(true);
    popupWindow.setBackgroundDrawable(new BitmapDrawable());
```

```
    //显示在 btnShowPopupwindow 按钮下面，x 位置偏移 100px 就是偏移屏幕左边 100px
    popupWindow.showAsDropDown(btnShowPopupwindow,100,0);
}
```

PopupWindow 加载的布局文件 popup_drop_down.xml 如下：

```
<?xml version="1.0" encoding="utf-8"?>
<LinearLayout xmlns:android="http://schemas.android.com/apk/res/android"
    android:layout_width="wrap_content"
    android:layout_height="wrap_content"
    android:background="@color/colorAccent">

    <TextView
        android:layout_width="wrap_content"
        android:layout_height="wrap_content"
        android:padding="10dp"
        android:text="我是点击上面那个按钮弹出的哦" />
</LinearLayout>
```

运行以上代码，其效果如图 2-28 所示。

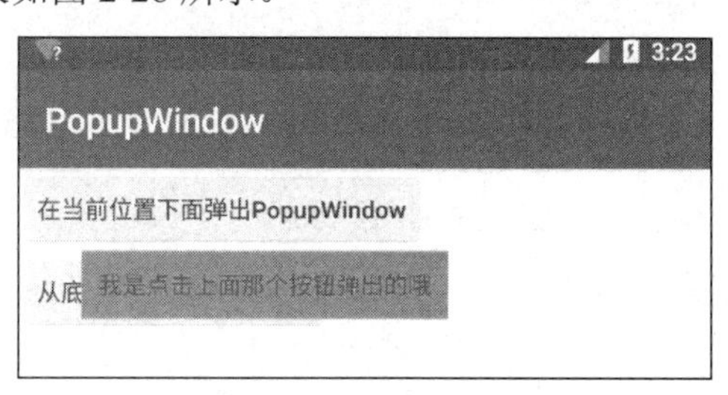

图 2-28 弹出式窗口

接下来的实例将 PopupWindow 显示在指定位置，从下往上弹出。

我们给第二个按钮设置点击事件，调用 showBottomPopupwindow 方法显示 PopupWindow。设置一个动画效果，从下往上弹出。

```
    private void showBottomPopupwindow(){
        View popView =
LayoutInflater.from(this).inflate(R.layout.popup_bottom,null);

        final PopupWindow popupWindow=new PopupWindow(popView,
LinearLayout.LayoutParams.MATCH_PARENT,
                LinearLayout.LayoutParams.WRAP_CONTENT);
        popupWindow.setOutsideTouchable(true);// 设置允许在外点击消失
        popupWindow.setBackgroundDrawable(new ColorDrawable(0x30000000));//设置背
景颜色
        popupWindow.setAnimationStyle(R.style.Animation_Bottom_Dialog);//设置动画
        View.OnClickListener onClickListener=new View.OnClickListener() {
            @Override
            public void onClick(View v) {
                if(v.getId()==R.id.btn_camera_album){
                    Toast.makeText(MainActivity.this,"点击拍照按钮",
Toast.LENGTH_SHORT).show();
                }else if(v.getId()==R.id.btn_camera_cancel){
                    Toast.makeText(MainActivity.this,"点击了取消按钮",
Toast.LENGTH_SHORT).show();
```

```
            }
            popupWindow.dismiss();
        }
    };
    popView.findViewById(R.id.btn_camera_album).setOnClickListener
        (onClickListener);
    popView.findViewById(R.id.btn_camera_cancel).setOnClickListener
        (onClickListener);
    //参数1:根视图，整个Window界面的最顶层View  参数2:显示位置
    popupWindow.showAtLocation(getWindow().getDecorView(),Gravity.BOTTOM,0,0);
}
```

底部 PopupWidnow 显示的布局文件 popup_bottom.xml 如下：

```
<?xml version="1.0" encoding="utf-8"?>
<LinearLayout xmlns:android="http://schemas.android.com/apk/res/android"
    android:layout_width="match_parent"
    android:layout_height="wrap_content"
    android:padding="10dp"
    android:orientation="vertical">

    <Button
        android:id="@+id/btn_camera_album"
        android:layout_width="match_parent"
        android:layout_height="wrap_content"
        android:background="@color/colorAccent"
        android:text="拍照"
        android:textSize="18sp" />

    <Button
        android:id="@+id/btn_camera_cancel"
        android:layout_width="match_parent"
        android:layout_height="wrap_content"
        android:layout_marginTop="10dp"
        android:background="@color/colorPrimaryDark"
        android:text="取消"
        android:textSize="18sp" />
</LinearLayout>
```

在上面的方法中，还通过 setAnimationStyle 方法设置了动画，这是因为在 style.xml 中增加了一个 style，android:windowEnterAnimation 是 PopupWindow 显示动画，android:windowExitAnimation 是 PopupWindow 消失动画。

```
<style name="Animation_Bottom_Dialog">
    <item name="android:windowEnterAnimation">@anim/bottom_dialog_enter</item>
    <item name="android:windowExitAnimation">@anim/bottom_dialog_exit</item>
</style>
```

在上面的 style 中引用了 res/anim 下的两个动画文件。

【bottom_dialog_enter.xml】

```
<?xml version="1.0" encoding="utf-8"?>
<translate xmlns:android="http://schemas.android.com/apk/res/android"
    android:duration="225"
    android:fromYDelta="100%"
```

```
    android:interpolator="@android:anim/decelerate_interpolator"
    android:toYDelta="0%"/>
```

解释上面这个xml中各个属性的作用。首先说明这是一个位置转移动画。

- android:duration：动画的运行时间，毫秒为单位。
- android:fromYDelta：动画起始时，Y坐标上的位置。
- android:toYDelta：动画结束时，Y坐标上的位置。
- android:interpolator：用来修饰动画效果，定义动画的变化率，可以使动画效果accelerated（加速）、decelerated（减速）、repeated（重复）、bounced（弹跳）等。

bottom_dialog_exit.xml中用的属性与上面的一样，就不解释了。

```
<?xml version="1.0" encoding="utf-8"?>
<translate xmlns:android="http://schemas.android.com/apk/res/android"
    android:duration="225"
    android:fromYDelta="0%"
    android:interpolator="@android:anim/accelerate_interpolator"
    android:toYDelta="100%"/>
```

运行效果如图2-29所示，显示与隐藏的动画效果无法截图，大家自己运行源代码。

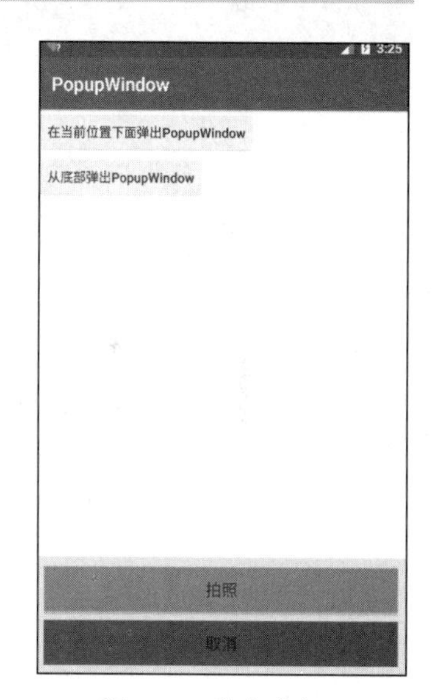

图2-29 弹出式窗口

2.4.11 DialogFragment

DialogFragment是在Android 3.0版本中被引入的，是一种基于Fragment的Dialog，可以用来创建基本对话框、警告对话框，以替代Dialog。

实现DialogFragment，需要重写DialogFragment并且实现onCreateView（LayoutInflater、ViewGroup、Bundle）方法获取对话框显示内容，或者重写onCreateDialog（Bundle）来创建一个完全自定义的对话框。

使用DialogFragment的好处如下：

- 因为继承自Fragment，所以具有Fragment的所有特性，可以更好地管理生命周期，手机配置发生变化时，我们能够进行监听。
- 在活动页面中启动对话框，要写一大堆逻辑代码、监听等，但是通过DialogFragment控制对话框，通过调用API来实现何时显示、隐藏、销毁，就能够很方便地管理对话框。

实例：重写onCreateView方法，加载布局文件

继承DialogFragment，重写onCreateView方法：

```
public class MyDialogFragment extends DialogFragment {
    static MyDialogFragment newInstance() {
        return new MyDialogFragment();
    }
```

```
    @Override
    public View onCreateView(LayoutInflater inflater, ViewGroup container,
Bundle savedInstanceState) {
        View v = inflater.inflate(R.layout.hello_world, container, false);
        TextView tv = (TextView) v.findViewById(R.id.textview);
        tv.setText("This is an instance of MyDialogFragment");
        return v;
    }
}
```

在 onCreateView 方法中加载一个布局文件 hello_world.xml，布局文件内容比较简单，最外层是 FrameLayout，里面是一个 TextView：

```
<?xml version="1.0" encoding="utf-8"?>
<FrameLayout xmlns:android="http://schemas.android.com/apk/res/android"
    android:layout_width="match_parent"
    android:layout_height="match_parent">

    <TextView
        android:id="@+id/textview"
        android:layout_width="match_parent"
        android:layout_height="match_parent"
        android:gravity="center"
        android:text="Hello World"
        android:textSize="20sp"/>
</FrameLayout>
```

这样就封装了一个 DialogFragment 对话框，我们需要在哪个 activity 中使用就调用它显示一下，一行代码即可轻松搞定。

```
MyDialogFragment.newInstance().show(getSupportFragmentManager(),"MyDialogF
ragment");
```

运行代码，效果如图 2-30 所示。

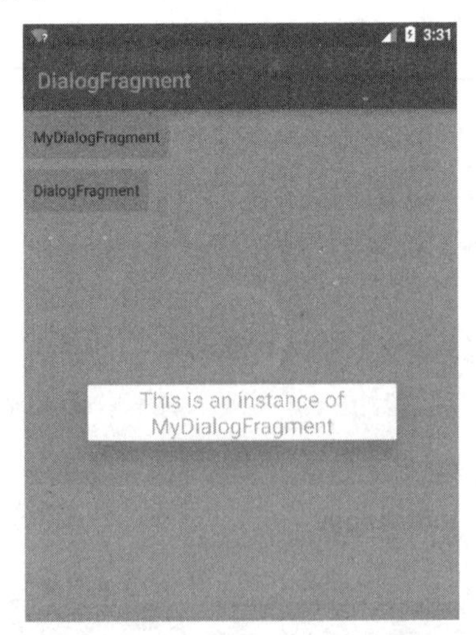

图 2-30　DialogFragment 对话框

实例：重写 onCreateView 方法，显示 AlertDialog 对话框

从如下代码中看到，在 onCreateDialog 方法中创建了一个 AlertDialog 对象返回，为对话框的两个按钮设置点击事件。

因为要处理按钮的点击事件，所以增加了 setOnClickListener 方法，调用时可以注入对话框两个按钮点击事件监听的实现。

```
    public class MyAlertDialogFragment extends DialogFragment{
       private DialogInterface.OnClickListener onClickListener;

       public static MyAlertDialogFragment newInstance() {
          return new MyAlertDialogFragment();
       }

       @Override
       public Dialog onCreateDialog(Bundle savedInstanceState) {
          AlertDialog.Builder builder=new AlertDialog.Builder(getActivity())
                .setIcon(R.mipmap.ic_launcher)
                .setTitle(R.string.app_name);
          if(onClickListener!=null){//设置对话框 ok&&取消按钮的点击事件

builder.setPositiveButton(R.string.alert_dialog_ok,onClickListener);

builder.setNegativeButton(R.string.alert_dialog_cancel,onClickListener);
          }
          return builder.create();
       }

       public void setOnClickListener(DialogInterface.OnClickListener
onClickListener) {
          this.onClickListener = onClickListener;
       }
    }
```

通过以上代码，我们可以看到设置点击监听就封装在自定义的 DialogFragment 中。

接下来我们看如何在 Activity 中调用，代码不多，几行而已。

```
     MyAlertDialogFragment
myAlertDialogFragment=MyAlertDialogFragment.newInstance();
    myAlertDialogFragment.setOnClickListener(onClickListener);
    myAlertDialogFragment.show(getSupportFragmentManager(),"MyAlertDialogFragm
ent");
```

点击对话框按钮时回调监听实现：

```
    private DialogInterface.OnClickListener onClickListener=new
DialogInterface.OnClickListener() {
       @Override
       public void onClick(DialogInterface dialog, int which) {
          switch (which){
             case DialogInterface.BUTTON_POSITIVE://ok
                Log.i("ansen","ok");
                break;
             case DialogInterface.BUTTON_NEGATIVE://cancel
                Log.i("ansen","cancel");
```

```
                break;
            }
        }
    };
```

运行效果如图 2-31 所示。

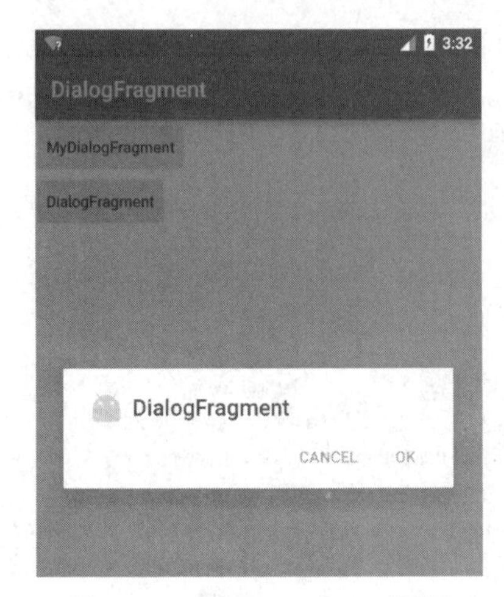

图 2-31 DialogFragment 对话框

2.5 Android 高级控件的使用

2.5.1 ListView（列表视图）

在 Android 开发中，ListView 是很常用的控件，以列表的形式展示具体内容，并且能够根据数据的长度自适应显示。

列表的显示需要三个元素：

- ListVeiw：用来展示列表的 View。
- 适配器：用来把数据映射到 ListView 上的中介。
- 数据源：ListVeiw 中的每一行 View 对应数据源的一条数据。

1. 简单使用

首先查看布局文件，比较简单。

```
<?xml version="1.0" encoding="utf-8"?>
<RelativeLayout xmlns:android="http://schemas.android.com/apk/res/android"
   android:layout_width="match_parent"
   android:layout_height="match_parent">
   <ListView
       android:id="@+id/listview"
       android:layout_width="match_parent"
       android:layout_height="wrap_content"/>
</RelativeLayout>
```

接下来查看代码，先初始化数据，循环 20 条数据放到一个集合中，这个 items 集合就是数据源，然后通过 id 获取 ListView 对象，给 ListView 设置一个适配器。

```
public class MainActivity extends AppCompatActivity {
    private ListView listView;
    private ListViewAdapter adapter;
    private List<String> items;

    @Override
    protected void onCreate(Bundle savedInstanceState) {
        super.onCreate(savedInstanceState);
        setContentView(R.layout.activity_main);

        initData();

        listView= (ListView) findViewById(R.id.listview);
        listView.setAdapter(adapter=new ListViewAdapter(this,items));
    }

    //初始化数据
    private void initData(){
        items=new ArrayList<>();
        for(int i=0;i<20;i++){
            items.add("item:"+(i+1));
        }
    }
}
```

自己写一个 ListViewAdapter 类，继承 BaseAdapter，然后重写 4 个方法。

```
public class ListViewAdapter extends BaseAdapter{
    private List<String> data;
    private LayoutInflater inflater;

    public ListViewAdapter(Context context,List<String> data){
        inflater=LayoutInflater.from(context);
        this.data=data;
    }

    //数据源长度
    @Override
    public int getCount() {
        return data.size();
    }

    //每一行的绑定数据源
    @Override
    public Object getItem(int position) {
        return data.get(position);
    }

    @Override
    public long getItemId(int position) {
        return position;
    }
```

```
        //获取每一行的 View
        @Override
        public View getView(int position, View convertView, ViewGroup parent) {
            ViewHolder viewHolder;
            if(convertView==null){
                viewHolder=new ViewHolder();
                //xml 文件加载成 View
                convertView = inflater.inflate(android.R.layout.simple_
                            list_item_1, parent, false);
                viewHolder.text1= (TextView) convertView.findViewById
                                (android.R.id.text1);

                convertView.setTag(viewHolder);
            }else{
                viewHolder= (ViewHolder) convertView.getTag();
            }

            viewHolder.text1.setText(data.get(position));
            return convertView;
        }

        private class ViewHolder{
            private TextView text1;
        }
    }
```

对于上面的代码，详细解释一下，ListView 在开始绘制时，系统首先调用 getCount()函数，根据它的返回值得到 listView 的长度，然后根据这个长度，调用 getView()逐一绘制每一行。如果 getCount()返回值是 0，列表将不显示；如果返回值是 1，就仅显示一行。

这个适配器的写法是目前为止比较标准的固定写法，getView 方法中用到了 ViewHolder 类，这是因为 ListView 有 RecycleBin 机制，列表滚动时复用 ItemView，这样做的好处就是，不管列表滑动了几千条还是上万条的数据，ListView 滚动的过程中永远只会创建一屏的 View。如图 2-32 所示，整个屏幕显示 12 条 View，说明这个列表就算一直往下滚动，也永远只会创建 12 个 View。

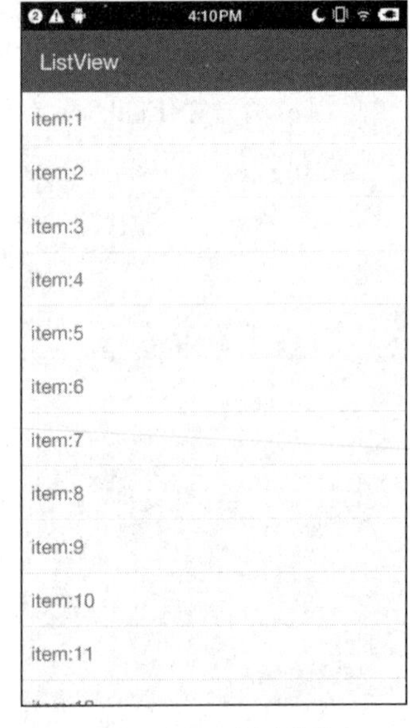

图 2-32　列表视图

2. 每一行点击监听

不需要给 ListView 每一行的 View 设置点击事件，ListView 源码已经封装了 setOnItemClickListener 方法。下面代码设置了 ListView 行点击事件，点击之后显示一个 Toast。

```
    listView.setOnItemClickListener(new AdapterView. OnItemClickListener(){
        @Override
        public void onItemClick(AdapterView<?> parent, View view, int position, long
id) {
            Toast.makeText(MainActivity.this,"点击 Item 位置:"+position,
Toast.LENGTH_SHORT).show();
        }
    });
```

3. 设置分隔线

ListView 的每一行可以通过分隔线来区分，只需要在布局中为 ListView 增加两个属性即可。

```
android:divider="@android:color/holo_red_light"  //分隔线颜色
android:dividerHeight="5dp" //分隔线高度
```

添加了分隔线之后，运行效果如图 2-33 所示。

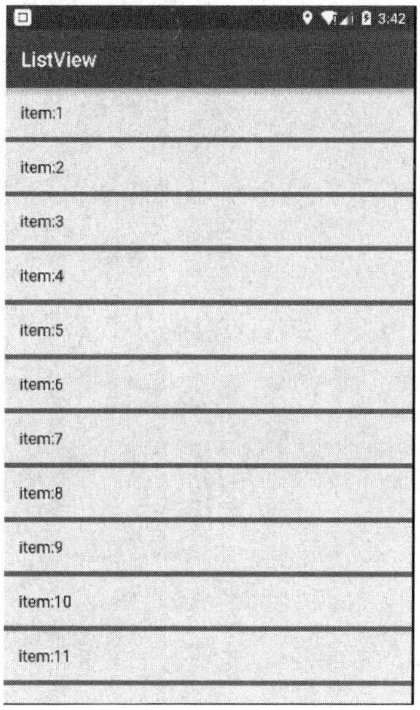

图 2-33　添加分隔线后的列表

如果不想要分隔线，可以设置分隔线为空，或者将分隔线颜色设为透明：

```
android:divider="@null"或者 android:divider="@android:color/transparent"
```

4. 添加 header 和 footer

用 LayoutInflater 类的 from 静态方法获取一个 LayoutInflater 对象，调用 inflate 方法将布局文件转化成 View，给 View 设置一个点击监听，调用 ListView 的 addHeaderView 方法把头布局添加进去。

```
View header=LayoutInflater.from(this).
inflate(R.layout.activity_listview_header,null);
header.setOnClickListener(onClickListener);//给头布局设置一个点击事件
listView.addHeaderView(header);//添加头部 View

View footer=LayoutInflater.from(this).
inflate(R.layout.activity_listview_footer,null);
footer.setOnClickListener(onClickListener);//给头布局设置一个点击事件
listView.addFooterView(footer);//添加尾部 View
```

activity_listview_header.xml 文件比较简单，就放置一个 TextView。（activity_listview_footer 代码与此几乎一样，就不贴出来了，只是布局的 id 不一样而已，用来判断点击事件。）

```
<?xml version="1.0" encoding="utf-8"?>
```

```
<LinearLayout xmlns:android="http://schemas.android.com/apk/res/android"
    android:id="@+id/ll_header"
    android:layout_width="match_parent"
    android:layout_height="wrap_content"
    android:orientation="vertical">

    <TextView
        android:id="@+id/textView"
        android:layout_width="match_parent"
        android:layout_height="wrap_content"
        android:gravity="center_horizontal"
        android:paddingTop="10dp"
        android:paddingBottom="10dp"
        android:text="这是 ListView 头部布局"
        android:background="@android:color/holo_orange_light"/>
</LinearLayout>
```

在头 View 点击监听函数中就显示了一个 Toast，大家对这样的代码应该很熟悉了。

```
private View.OnClickListener onClickListener=new View.OnClickListener() {
    @Override
    public void onClick(View v) {
        switch (v.getId()){
            case R.id.ll_header:
                Toast.makeText(MainActivity.this,"点击 ListView 头布局",
                    Toast.LENGTH_SHORT).show();
                break;
        }
    }
};
```

5. 动态修改 item

动态改变 ListView 很简单，只需要修改数据源，然后调用 adapter 的 notifyDataSetChanged 方法更新适配器即可，其他的底层已经封装好了。用代码来举个例子：点击尾部 View 为 ListView 添加一行数据。

```
private View.OnClickListener onClickListener=new View.OnClickListener() {
    @Override
    public void onClick(View v) {
        switch (v.getId()){
            case R.id.ll_footer://点击底部
                items.add("点击底部添加的 item");
                adapter.notifyDataSetChanged();
                break;
        }
    }
};
```

在底部点击监听方法中，给数据源（也就是集合）添加一个字符串，然后刷新适配器，两行代码就能为 ListView 增加一行数据。如果想删除一行也是一样的，删除源数据，然后更新适配器。有兴趣的读者可以花点时间自己去实现。

6. 设置显示位置

```
listView.setSelection(items.size()-1); //显示列表最后一条
```

运行结果如图2-34所示。

7. 实现聊天界面

一般情况下使用的ListView中每一行的View都是固定的，但是一些特殊界面会有多个布局类型，例如微信、陌陌、QQ等聊天App，在聊天详情界面就有发送与接收两个布局，其效果如图2-35所示。

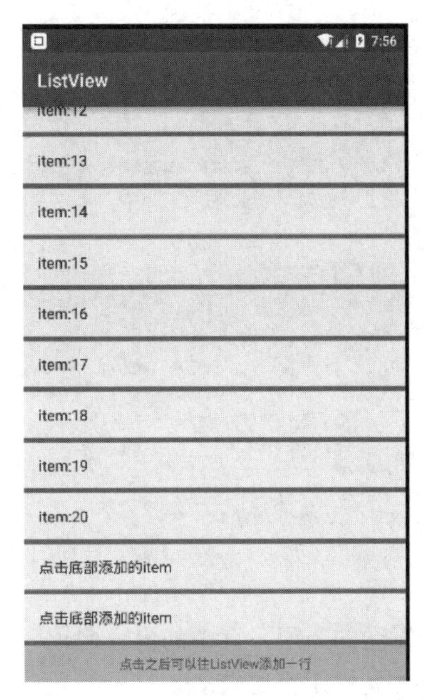

图2-34 ListView设置显示位置

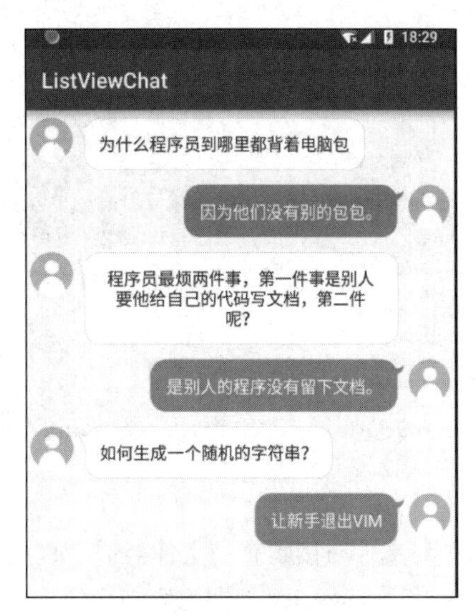

图2-35 聊天界面

想要实现图2-35中的效果，需要给Adapter加载不同的item布局，重写Adapter的getViewTypeCount和getItemViewType方法。

```
//item类型数量
@Override
public int getViewTypeCount() {
    return TYPE_ACCEPT+1;
}

//每个类型对应的int类型的值必须从0开始
@Override
public int getItemViewType(int position) {
    if(messages.get(position).isSended()){
        return TYPE_SEND;//发送类型
    }
    return TYPE_ACCEPT;//接收类型
}
```

为每种类型定义一个常量，写在ListViewAdapter类中。

```
private final int TYPE_SEND=0;//消息发送
```

```
    private final int TYPE_ACCEPT=1;//消息接收
```

当前，在 getView 中也需要处理一下，根据不同的类型加载不同的布局文件。

```
    @Override
    public View getView(int position, View convertView, ViewGroup parent) {
        int type = getItemViewType(position);
        Message message=messages.get(position);

        ViewHolder viewHolder;
        if(convertView==null){
            viewHolder=new ViewHolder();
            if(type==TYPE_SEND){//发送的消息
                convertView = inflater.inflate(R.layout.item_message_chat_send,
null);
            }else if(type==TYPE_ACCEPT){//接收
                convertView = inflater.inflate(R.layout.item_message_chat_accept,
null);
            }

            viewHolder.tvContent= (TextView)
convertView.findViewById(R.id.tv_content);
            convertView.setTag(viewHolder);
        }else{
            viewHolder= (ViewHolder) convertView.getTag();
        }

        viewHolder.tvContent.setText(message.getContent());
        return convertView;
    }
```

发送布局文件与接收布局文件类似，这里贴出一个发送的布局，其中的内容比较简单，仅一个 TextView 和 ImageView，即内容与头像。不过，控件的属性用得比较多。

```
    <?xml version="1.0" encoding="utf-8"?>
    <RelativeLayout xmlns:android="http://schemas.android.com/apk/res/android"
    android:layout_width="wrap_content"
    android:layout_height="wrap_content"
    android:paddingBottom="7dip"
    android:paddingTop="7dip">

    <TextView
        android:id="@+id/tv_content"
        android:layout_width="wrap_content"
        android:layout_height="wrap_content"
        android:layout_marginLeft="55dp"
        android:layout_toLeftOf="@+id/iv_message_from_head_image"
        android:background="@mipmap/icon_message_from"
        android:gravity="center"
        android:paddingRight="20dip"
        android:text="我已经吃过了"
        android:textColor="@color/white_normal"
        android:textSize="16dip" />

    <!--        -->
```

```
<ImageView
    android:id="@+id/iv_message_from_head_image"
    android:layout_width="40dp"
    android:layout_height="40dp"
    android:layout_alignParentRight="true"
    android:layout_marginLeft="5dp"
    android:layout_marginRight="5dp"
    android:src="@mipmap/slide_left_avatar_default" />
</RelativeLayout>
```

如果仔细阅读了前面的内容，就会发现每一行绑定的数据源不是 String 类型，而是 Message 类型了。

```
public class Message {
    private String content;// 消息内容
    private boolean sended;// 是否发送

    public Message(){
    }

    public Message(String content,boolean sended){
        this.content=content;
        this.sended=sended;
    }

    public String getContent() {
        return content;
    }

    public void setContent(String content) {
        this.content = content;
    }

    public boolean isSended() {
        return sended;
    }

    public void setSended(boolean sended) {
        this.sended = sended;
    }
}
```

如果学会了以上内容，ListView 的基本使用就没问题了。这里在代码中加了注释，并在代码后面做了详细的解释，希望能够帮助大家更好地理解。

2.5.2 GridView（网格视图）

GridView 是按照行列的方式来显示内容的，一般用于显示图片列表，比如九宫格列表，使用 GridView 实现起来很简单。GridView 的用法与 ListView 类似，首先看图 2-36，效果图中显示的两张图片是网上找的。

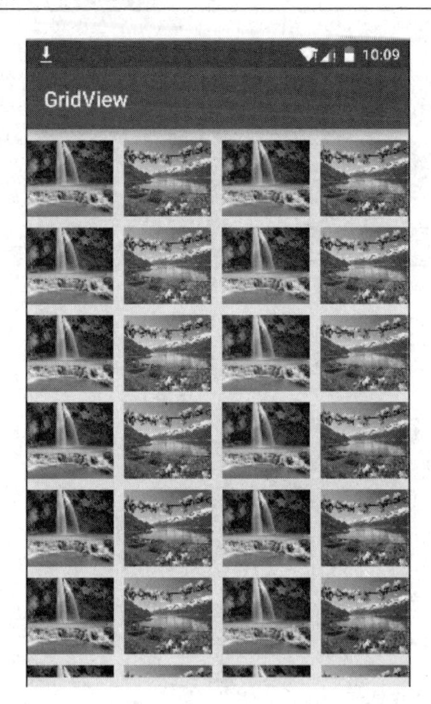

图 2-36　网格视图

新建项目，首先修改布局文件 activity_main.xml。

```
<?xml version="1.0" encoding="utf-8"?>
<RelativeLayout xmlns:android="http://schemas.android.com/apk/res/android"
    android:layout_width="match_parent"
    android:layout_height="match_parent">
    <GridView
        android:id="@+id/gridview"
        android:numColumns="4"
        android:scrollbars="none"
        android:layout_marginBottom="10dp"
        android:verticalSpacing="10dp"
        android:horizontalSpacing="10dp"
        android:layout_width="match_parent"
        android:layout_height="wrap_content"/>
</RelativeLayout>
```

GridView 控件中的几个属性作用如下：

- android:numColumns="4"：一行显示 4 列。
- android:scrollbars="none"：去掉滚动条。
- android:verticalSpacing="10dp"：两行之间的间距。
- android:horizontalSpacing="10dp"：两列之间的间距。

Activity 的代码比较简单，初始化数据、设置适配器、设置点击事件。

```
public class MainActivity extends AppCompatActivity {
    private GridView gridview;
    private List<Integer> images;
    private GridAdapter gridAdapter;
```

```
    @Override
    protected void onCreate(Bundle savedInstanceState) {
        super.onCreate(savedInstanceState);
        setContentView(R.layout.activity_main);

        initData();
        gridview= (GridView) findViewById(R.id.gridview);
        gridview.setAdapter(gridAdapter=new GridAdapter(this,images));

        //item 设置点击事件
        gridview.setOnItemClickListener(onItemClickListener);
    }

    private AdapterView.OnItemClickListener onItemClickListener=new
AdapterView.OnItemClickListener() {
        @Override
        public void onItemClick(AdapterView<?> parent, View view, int position,
long id) {
            Toast.makeText(MainActivity.this,"当前选中了" +
                    ":"+position,Toast.LENGTH_SHORT).show();
        }
    };

    //初始化数据源
    private void initData(){
        images=new ArrayList<>();
        for(int i=0;i<100;i++){
            if(i%2==1){//对 2 取余数，结果为 1
                images.add(R.mipmap.test_one);
            }else{
                images.add(R.mipmap.test_two);
            }
        }
    }
}
```

GridView 适配器与 ListView 适配器类似，GridAdapter.java 代码如下：（因为类似所以代码不做详细解释）

```
public class GridAdapter extends BaseAdapter{
    private LayoutInflater inflater;
    private List<Integer> images;

    public GridAdapter(Context context,List<Integer> images){
        inflater=LayoutInflater.from(context);
        this.images=images;
    }

    @Override
    public int getCount() {
        return images.size();
    }

    @Override
```

```
        public Object getItem(int position) {
            return images.get(position);
        }

        @Override
        public long getItemId(int position) {
            return position;
        }

        @Override
        public View getView(int position, View convertView, ViewGroup parent) {
            ViewHolder viewHolder;
            if(convertView==null){
                viewHolder=new ViewHolder();
                convertView = inflater.inflate(R.layout.activity_grid_item, parent,
false);
                viewHolder.imageview = (ImageView) convertView.findViewById
(R.id.imageview);
                convertView.setTag(viewHolder);
            }else{
                viewHolder= (ViewHolder) convertView.getTag();
            }

            viewHolder.imageview.setImageResource(images.get(position));
            return convertView;
        }

        private class ViewHolder{
            private ImageView imageview;
        }
    }
```

item 的布局文件：activity_grid_item.xml。

```
    <?xml version="1.0" encoding="utf-8"?>
    <LinearLayout xmlns:android="http://schemas.android.com/apk/res/android"
        android:layout_width="match_parent"
        android:layout_height="match_parent"
        android:orientation="vertical">
        <ImageView
            android:id="@+id/imageview"
            android:layout_gravity="center_horizontal"
            android:layout_width="match_parent"
            android:layout_height="70dp"
            android:scaleType="centerCrop"
            android:src="@mipmap/test_one" />
    </LinearLayout>
```

android:scaleType="centerCrop" 均衡地缩放图像（保持图像原始比例），使图片的宽高都大于等于 View 宽高。就本例而言，ImageView 的高度是 70dp，宽度是包裹内容（随内容多少而改变），设置了这个属性之后，图片会按照比例缩放到宽高都是 70dp 为止。

2.5.3 RecyclerView（循环视图）

RecylerView 是 support-v7 包中的新组件，是一个强大的滑动组件，与经典的 ListView 相比，同样拥有 item 回收复用的功能，这一点从它的名字 RecylerView（循环视图）也可以看出。官方对于它的介绍是：RecyclerView 是 ListView 的升级版本，更加先进和灵活。RecyclerView 通过设置 LayoutManager、ItemDecoration、ItemAnimator 可实现更多效果。

- 使用 LayoutManager 来确定每一个 item 的排列方式。
- 使用 ItemDecoration 自己绘制分隔线，更加灵活。
- 使用 ItemAnimator 为增加或删除一行设置动画效果。

注　意

新建完项目，需要在 app/build.gradle 增加 RecylerView 依赖，不然找不到 RecyclerView 类：

```
compile 'com.android.support:recyclerview-v7:23.1.0'
```

1. RecylerView 简单的 Demo

我们来看 Activity 代码，与 ListView 写法其实差不多，这里只是多设置了一个布局管理器。

```
    public class LinearLayoutActivity extends AppCompatActivity {
        private RecyclerView recyclerView;
        private RecyclerViewAdapter adapter;
        private List<String> datas;

        @Override
        public void onCreate(Bundle savedInstanceState) {
            super.onCreate(savedInstanceState);
            setContentView(R.layout.recycler_main);

            initData();

            recyclerView= (RecyclerView) findViewById(R.id.recyclerview);
            recyclerView.setLayoutManager(new LinearLayoutManager(this));
            //设置布局管理器
            recyclerView.addItemDecoration(new DividerItemDecoration(this));
            recyclerView.setAdapter(adapter=new
RecyclerViewAdapter(this,datas));
        }

        private void initData(){
            datas=new ArrayList<>();
            for(int i=0;i<100;i++){
                datas.add("item:"+i);
            }
        }
    }
```

Activity 对应的布局文件：recycler_main.xml。因为 RecyclerView 是 v7 包才有的控件，所以需要在布局文件中指定包名+类名。

```
<?xml version="1.0" encoding="utf-8"?>
<RelativeLayout xmlns:android="http://schemas.android.com/apk/res/android"
   android:layout_width="match_parent"
   android:layout_height="match_parent">

   <android.support.v7.widget.RecyclerView
      android:id="@+id/recyclerview"
      android:layout_width="match_parent"
      android:layout_height="match_parent"/>
</RelativeLayout>
```

Adapter 相对 ListView 的 Adapter 来说变化比较大。RecyclerViewAdapter.java 代码如下：

```
public class RecyclerViewAdapter extends
RecyclerView.Adapter<RecyclerViewAdapter.MyViewHolder>{
   private List<String> datas;
   private LayoutInflater inflater;

   public  RecyclerViewAdapter(Context context,List<String> datas){
      inflater=LayoutInflater.from(context);
      this.datas=datas;
   }

   //创建每一行的 View 用 RecyclerView.ViewHolder 包装
   @Override
   public RecyclerViewAdapter.MyViewHolder onCreateViewHolder(ViewGroup parent,
int viewType) {
      View itemView=inflater.inflate(R.layout.recycler_item,null);
      return new MyViewHolder(itemView);
   }

   //给每一行 View 填充数据
   @Override
   public void onBindViewHolder(RecyclerViewAdapter.MyViewHolder holder, int
position) {
      holder.textview.setText(datas.get(position));
   }

   //数据源的数量
   @Override
   public int getItemCount() {
      return datas.size();
   }

   class MyViewHolder extends RecyclerView.ViewHolder{
      private TextView textview;

      public MyViewHolder(View itemView) {
         super(itemView);
         textview= (TextView) itemView.findViewById(R.id.textview);
      }
   }
}
```

从上面的代码中看到需要继承 RecyclerView.Adapter，重写三个方法（onCreateViewHolder、 onBindViewHolder、getItemCount），同时MyViewHolder也需要继承RecyclerView.ViewHolder。

运行代码，效果如图2-37所示。

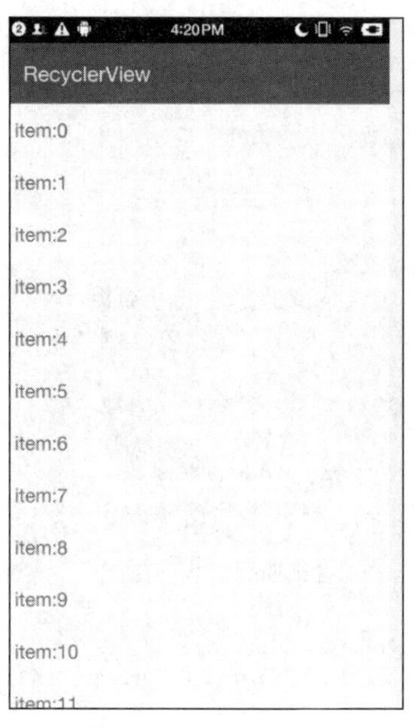

图2-37 RecyclerView

2. RecyclerView增加分隔线

RecyclerView是没有android:divider与android:dividerHeight属性的，如果需要分隔线，只能自己动手去实现。

- 需要继承ItemDecoration类，实现onDraw与getItemOffsets方法。
- 调用RecyclerView的addItemDecoration方法。

先编写一个DividerItemDecoration类，继承RecyclerView.ItemDecoration，在getItemOffsets留出item之间的间隔，然后调用onDraw方法绘制（onDraw的绘制优先于每一行的绘制）。

```
public class DividerItemDecoration extends RecyclerView.ItemDecoration{
    /*
    * RecyclerView的布局方向，默认先赋值为纵向布局
    * RecyclerView 布局可横向，也可纵向
    * 横向和纵向对应的分隔线画法不一样
    * */
    private int mOrientation = LinearLayoutManager.VERTICAL;

    private int mItemSize = 1;//item之间分隔线的size，默认为1

    private Paint mPaint;//绘制item分隔线的画笔，并设置其属性

    public DividerItemDecoration(Context context) {
        this(context,LinearLayoutManager.VERTICAL,R.color.colorAccent);
    }

    public DividerItemDecoration(Context context, int orientation) {
```

```
        this(context,orientation, R.color.colorAccent);
    }

    public DividerItemDecoration(Context context, int orientation, int
dividerColor){
        this(context,orientation,dividerColor,1);
    }

    /**
     * @param context
     * @param orientation 绘制方向
     * @param dividerColor 分隔线颜色，颜色资源 id
     * @param mItemSize 分隔线宽度，传入 dp 值就行
     */
    public DividerItemDecoration(Context context, int orientation, int
dividerColor, int mItemSize){
        this.mOrientation = orientation;
        if(orientation != LinearLayoutManager.VERTICAL && orientation !=
LinearLayoutManager.HORIZONTAL){
            throw new IllegalArgumentException("请传入正确的参数") ;
        }
        //把 dp 值换算成 px
        this.mItemSize = (int) TypedValue.applyDimension(TypedValue.COMPLEX_
            UNIT_DIP,mItemSize,context.getResources().getDisplayMetrics());
        mPaint = new Paint(Paint.ANTI_ALIAS_FLAG);

        mPaint.setColor(context.getResources().getColor(dividerColor));
    }


    @Override
    public void onDraw(Canvas c, RecyclerView parent, RecyclerView.State state) {
        if(mOrientation == LinearLayoutManager.VERTICAL){
            drawVertical(c,parent) ;
        }else {
            drawHorizontal(c,parent) ;
        }
    }

    /**
     * 绘制纵向 item 分隔线
     * @param canvas
     * @param parent
     */
    private void drawVertical(Canvas canvas,RecyclerView parent){
        final int left = parent.getPaddingLeft() ;
        final int right = parent.getMeasuredWidth() - parent.getPaddingRight();
        final int childSize = parent.getChildCount() ;
        for(int i = 0 ; i < childSize ; i ++){
            final View child = parent.getChildAt( i ) ;
            RecyclerView.LayoutParams layoutParams =
(RecyclerView.LayoutParams) child.getLayoutParams();
            final int top = child.getBottom() + layoutParams.bottomMargin ;
            final int bottom = top + mItemSize ;
            canvas.drawRect(left,top,right,bottom,mPaint);
```

```
            }
        }

        /**
         * 绘制横向 item 分隔线
         * @param canvas
         * @param parent
         */
        private void drawHorizontal(Canvas canvas,RecyclerView parent){
            final int top = parent.getPaddingTop() ;
            final int bottom = parent.getMeasuredHeight() -
parent.getPaddingBottom() ;
            final int childSize = parent.getChildCount() ;
            for(int i = 0 ; i < childSize ; i ++){
                final View child = parent.getChildAt( i ) ;
                RecyclerView.LayoutParams layoutParams =
(RecyclerView.LayoutParams) child.getLayoutParams();
                final int left = child.getRight() + layoutParams.rightMargin ;
                final int right = left + mItemSize ;
                canvas.drawRect(left,top,right,bottom,mPaint);
            }
        }

        /**
         * 设置 item 分隔线的 size
         * @param outRect
         * @param view
         * @param parent
         * @param state
         */
        @Override
        public void getItemOffsets(Rect outRect, View view, RecyclerView parent,
RecyclerView.State state) {
            if(mOrientation ==
LinearLayoutManager.VERTICAL){
                outRect.set(0,0,0,mItemSize);
                 //垂直排列,底部偏移
            }else {
                outRect.set(0,0,mItemSize,0);
                 //水平排列,右边偏移
            }
        }
    }
```

不要忘记调用 addItemDecoration 方法哦！

```
    recyclerView.addItemDecoration(new
DividerItemDecoration(this));//添加分隔线
```

重新运行，效果如图 2-38 所示。

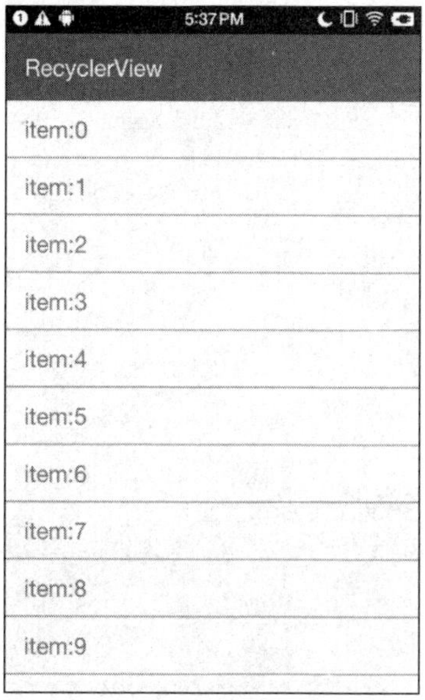

大家读到这里肯定会有一个疑问，这个比 ListView 麻烦多了，但是 Google 官方为什么要说是 ListView 的升级版呢？接下来开始放大招。

5:37 PM
RecyclerView
item:0
item:1
item:2
item:3
item:4
item:5
item:6
item:7
item:8
item:9

图 2-38　添加分隔线

3. GridLayoutManager

在 RecyclerView 中实现不同的列表，只需切换不同的 LayoutManager（布局管理器）即可。RecyclerView.LayoutManager 与 RecyclerView.ItemDecoration 一样，都是 RecyclerView 静态抽象内部类，但是 LayoutManager 有三个官方写好的实现类。

- LinearLayoutManager（线性布局管理器）：与 ListView 功能相似。
- GridLayoutManager（网格布局管理器）：与 GridView 功能相似。
- StaggeredGridLayoutManager（瀑布流布局管理器）。

刚刚用的是 LinearLayoutManager，现在切换到 GridLayoutManager。看到下面这句代码，有没有感觉到很轻松呢？

```
    recyclerView.setLayoutManager(new GridLayoutManager(this,2));
```

如果要显示多列或者要纵向显示，只需新建不同的构造方法即可。以下代码纵向显示 4 列。当然，如果还需要反方向显示，把 false 改成 true 即可。

```
    recyclerView.setLayoutManager(new
GridLayoutManager(this,4,GridLayoutManager.HORIZONTAL,false));
```

因为用的是网格布局，所以绘制分隔线的代码需要重新修改一下。网格布局一行可以有多列，并且最后一列与最后一行不需要绘制，所以需要重新创建一个类：DividerGridItemDecoration.java。

```
    public class DividerGridItemDecoration extends RecyclerView.ItemDecoration {
        /*
        * RecyclerView 的布局方向，默认先赋值为纵向布局
        * RecyclerView 的布局可横向，也可纵向
        * 横向和纵向对应的分隔线画法不一样
        * */
        private int mOrientation = LinearLayoutManager.VERTICAL;

        private int mItemSize = 1;//item 之间分隔线的 size，默认为 1

        private Paint mPaint;//绘制 item 分隔线的画笔并设置其属性

        public DividerGridItemDecoration(Context context) {
            this(context,LinearLayoutManager.VERTICAL,R.color.colorAccent);
        }

        public DividerGridItemDecoration(Context context, int orientation) {
            this(context,orientation, R.color.colorAccent);
        }

        public DividerGridItemDecoration(Context context, int orientation, int
dividerColor){
            this(context,orientation,dividerColor,1);
        }

        /**
         * @param context
         * @param orientation 绘制方向
         * @param dividerColor 分隔线颜色，颜色资源 id
         * @param mItemSize 分隔线宽度，传入 dp 值就行
```

```
     */
    public DividerGridItemDecoration(Context context, int orientation, int
dividerColor, int mItemSize){
        this.mOrientation = orientation;
        if(orientation != LinearLayoutManager.VERTICAL && orientation !=
LinearLayoutManager.HORIZONTAL){
            throw new IllegalArgumentException("请传入正确的参数") ;
        }
        //把dp值换算成px
        this.mItemSize = (int) TypedValue.applyDimension(TypedValue.
COMPLEX_UNIT_DIP, mItemSize,context.getResources().getDisplayMetrics());
        mPaint = new Paint(Paint.ANTI_ALIAS_FLAG);

        mPaint.setColor(context.getResources().getColor(dividerColor));
    }

    @Override
    public void onDraw(Canvas c, RecyclerView parent, RecyclerView.State state) {
        drawHorizontal(c, parent);
        drawVertical(c, parent);
    }

    private int getSpanCount(RecyclerView parent) {
        // 列数
        int spanCount = -1;
        RecyclerView.LayoutManager layoutManager = parent.getLayoutManager();
        if (layoutManager instanceof GridLayoutManager) {
            spanCount = ((GridLayoutManager) layoutManager).getSpanCount();
        } else if (layoutManager instanceof StaggeredGridLayoutManager) {
            spanCount=((StaggeredGridLayoutManager)layoutManager).getSpanCount();
        }
        return spanCount;
    }

    public void drawHorizontal(Canvas canvas, RecyclerView parent) {
        int childCount = parent.getChildCount();
        for (int i = 0; i < childCount; i++) {
            final View child = parent.getChildAt(i);
            final RecyclerView.LayoutParams params = (RecyclerView.LayoutParams)
child.getLayoutParams();
            final int left = child.getLeft() - params.leftMargin;
            final int right = child.getRight() + params.rightMargin + mItemSize;
            final int top = child.getBottom() + params.bottomMargin;
            final int bottom = top + mItemSize;
            canvas.drawRect(left,top,right,bottom,mPaint);
        }
    }

    public void drawVertical(Canvas canvas, RecyclerView parent) {
        final int childCount = parent.getChildCount();
        for (int i = 0; i < childCount; i++) {
            final View child = parent.getChildAt(i);

            final RecyclerView.LayoutParams params = (RecyclerView.LayoutParams)
child.getLayoutParams();
```

```
            final int top = child.getTop() - params.topMargin;
            final int bottom = child.getBottom() + params.bottomMargin;
            final int left = child.getRight() + params.rightMargin;
            final int right = left + mItemSize;
            canvas.drawRect(left,top,right,bottom,mPaint);
        }
    }

    @Override
    public void getItemOffsets(Rect outRect, int itemPosition,RecyclerView
parent) {
        int spanCount = getSpanCount(parent);
        int childCount = parent.getAdapter().getItemCount();

        if (isLastRow(parent, itemPosition, spanCount, childCount)){
         //如果是最后一行，不需要绘制底部
            outRect.set(0, 0, mItemSize, 0);
        } else if (isLastColum(parent, itemPosition, spanCount, childCount)){
          // 如果是最后一列，不需要绘制右边
            outRect.set(0, 0, 0, mItemSize);
        } else {
            outRect.set(0, 0, mItemSize,mItemSize);
        }
    }

    private boolean isLastColum(RecyclerView parent, int pos, int spanCount,
int childCount) {
        RecyclerView.LayoutManager layoutManager = parent.getLayoutManager();
        if (layoutManager instanceof GridLayoutManager) {
            if ((pos + 1) % spanCount == 0){// 如果是最后一列，不需要绘制右边
                return true;
            }
        } else if (layoutManager instanceof StaggeredGridLayoutManager) {
            int orientation = ((StaggeredGridLayoutManager)
layoutManager).getOrientation();
            if (orientation == StaggeredGridLayoutManager.VERTICAL) {
                if ((pos + 1) % spanCount == 0){// 如果是最后一列，不需要绘制右边
                    return true;
                }
            } else {
                childCount = childCount - childCount % spanCount;
                if (pos >= childCount)// 如果是最后一列，不需要绘制右边
                    return true;
            }
        }
        return false;
    }

    private boolean isLastRow(RecyclerView parent, int pos, int spanCount, int
childCount) {
        RecyclerView.LayoutManager layoutManager = parent.getLayoutManager();
        if (layoutManager instanceof GridLayoutManager) {
            childCount = childCount - childCount % spanCount;
            if (pos >= childCount)//最后一行
                return true;
```

```
        } else if (layoutManager instanceof StaggeredGridLayoutManager) {
            int orientation = ((StaggeredGridLayoutManager) layoutManager).
getOrientation();
            if (orientation == StaggeredGridLayoutManager.VERTICAL){//纵向
                childCount = childCount - childCount % spanCount;
                if (pos >= childCount)//最后一行
                    return true;
            } else{ //横向
                if ((pos + 1) % spanCount == 0) {//最后一行
                    return true;
                }
            }
        }
        return false;
    }
}
```

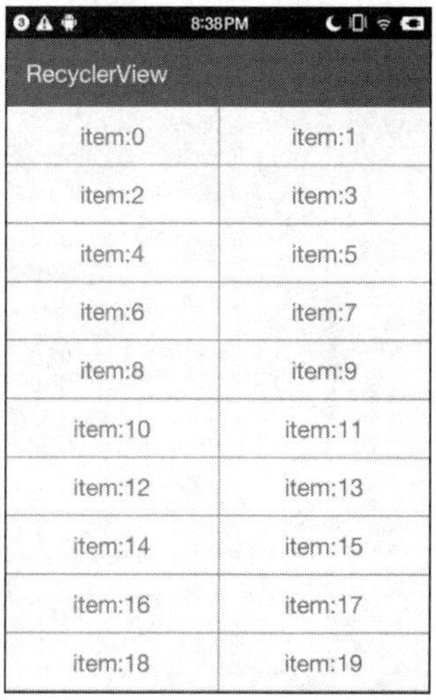

写了这两个画分隔线的类和主流的布局，线性列表与网格列表都能展示了。运行代码，结果如图 2-39 所示。

图 2-39　线性列表与网格列表

4. StaggeredGridLayoutManager

在 Activity 页面中修改布局管理器，大家应该很熟悉了。

```
recyclerView.setLayoutManager(new
StaggeredGridLayoutManager(3,StaggeredGridLayoutManag
er.VERTICAL));
```

一般瀑布流列表的列高度是不一致的，为了模拟不同的宽高，把 String 类型改成对象，然后初始化时随机一个高度即可。

```
public class ItemData {
    private String content;//item 内容
    private int height;//item 高度

    public ItemData() {
    }

    public ItemData(String content, int height) {
        this.content = content;
        this.height = height;
    }

    public String getContent() {
        return content;
    }

    public void setContent(String content) {
        this.content = content;
    }

    public int getHeight() {
        return height;
    }

    public void setHeight(int height) {
```

```
            this.height = height;
        }
    }
```

对应的适配器代码 StaggeredGridAdapter.java：

```
    public class StaggeredGridAdapter extends
RecyclerView.Adapter<StaggeredGridAdapter.MyViewHolder>{
        private List<ItemData> datas;
        private LayoutInflater inflater;

        public StaggeredGridAdapter(Context context, List<ItemData> datas){
            inflater=LayoutInflater.from(context);
            this.datas=datas;
        }

        //创建每一行的 View，用 RecyclerView.ViewHolder 包装
        @Override
        public StaggeredGridAdapter.MyViewHolder onCreateViewHolder(ViewGroup
parent, int viewType) {
            View itemView=inflater.inflate(R.layout.recycler_staggered_item,
parent,false);
            return new MyViewHolder(itemView);
        }

        //给每一行 View 填充数据
        @Override
        public void onBindViewHolder(StaggeredGridAdapter.MyViewHolder holder,
int position) {
            ItemData itemData=datas.get(position);
            holder.textview.setText(itemData.getContent());
            //手动更改高度，不同位置的高度有所不同
            holder.textview.setHeight(itemData.getHeight());
        }

        //数据源的数量
        @Override
        public int getItemCount() {
            return datas.size();
        }

        class MyViewHolder extends RecyclerView.ViewHolder{
            private TextView textview;

            public MyViewHolder(View itemView) {
                super(itemView);
                textview= (TextView) itemView.findViewById(R.id.textview);
            }
        }
    }
```

需要注意的是，我们在适配器的 onBindViewHolder 方法中给 item 布局中的 TextView 设置了一个高度，这个高度是创建实体类的时候随机生成的。

item 对应的布局文件：recycler_staggered_item.xml。

```
<?xml version="1.0" encoding="utf-8"?>
<FrameLayout xmlns:android="http://schemas.android.com/apk/res/android"
    android:padding="5dp"
    android:layout_width="wrap_content"
    android:layout_height="match_parent">

    <TextView
        android:id="@+id/textview"
        android:background="@color/colorAccent"
        android:layout_width="100dp"
        android:layout_height="wrap_content"
        android:gravity="center"
        android:text="122"
        android:textSize="20sp"/>
</FrameLayout>
```

瀑布流列表没有添加分割线，为了展示效果，给item布局设置了android:padding属性。

是不是感觉很容易？赶紧运行代码，查看效果，整个页面看起来变得生动活泼了，如图2-40所示。

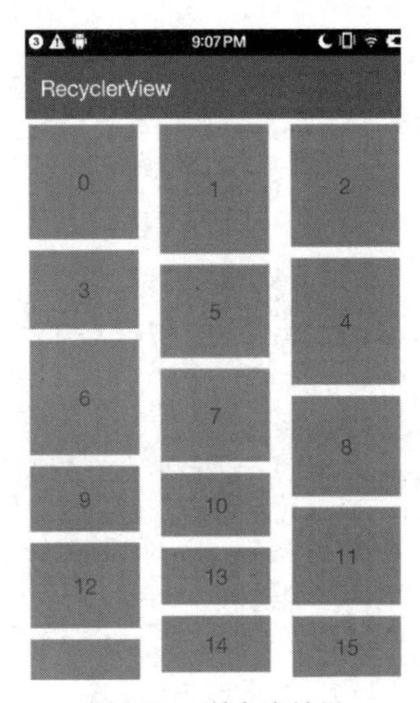

图2-40　瀑布流效果

5. 添加header和footer

RecyclerView添加头部与底部是没有对应的api的，但是很多需求都会用到，于是只能自己想办法实现了。可以通过适配器的getItemViewType方法来实现这个功能。

修改后的适配器代码：RecyclerHeadFootViewAdapter.java。

```
    public class RecyclerHeadFootViewAdapter extends
RecyclerView.Adapter<RecyclerView.ViewHolder>{
        private List<String> datas;
        private LayoutInflater inflater;

        public static final int TYPE_HEADER=1;//header类型
        public static final int TYPE_FOOTER=2;//footer类型
        private View header=null;//头View
```

```
        private View footer=null;//脚 View

        public  RecyclerHeadFootViewAdapter(Context context, List<String> datas){
            inflater=LayoutInflater.from(context);
            this.datas=datas;
        }

        //创建每一行的 View，用 RecyclerView.ViewHolder 包装
        @Override
        public RecyclerView.ViewHolder onCreateViewHolder(ViewGroup parent, int
viewType) {
            if(viewType==TYPE_HEADER){
                return new RecyclerView.ViewHolder(header){};
            }else if(viewType==TYPE_FOOTER){
                return new RecyclerView.ViewHolder(footer){};
            }
            View itemView=inflater.inflate(R.layout.recycler_item,null);
            return new MyViewHolder(itemView);
        }

        //给每一行 View 填充数据
        @Override
        public void onBindViewHolder(RecyclerView.ViewHolder holder, int
position){
            if(getItemViewType(position)==TYPE_HEADER||getItemViewType(position)
                ==TYPE_FOOTER){
                return;
            }
            MyViewHolder myholder= (MyViewHolder) holder;
            myholder.textview.setText(datas.get(getRealPosition(position)));
        }

        //如果有头部，position 的位置是从 1 开始的，所以需要-1
        public int getRealPosition(int position){
            return header==null?position:position-1;
        }

        //数据源的数量
        @Override
        public int getItemCount() {
            if(header == null && footer == null){//没有 head 跟 foot
                return datas.size();
            }else if(header == null && footer != null){//head 为空&&foot 不为空
                return datas.size() + 1;
            }else if (header != null && footer == null){//head 不为空&&foot 为空
                return datas.size() + 1;
            }else {
                return datas.size() + 2;//head 不为空&&foot 不为空
            }
        }

        @Override
        public int getItemViewType(int position){
            //如果头布局不为空&&位置是第一个那就是 head 类型
            if(header!=null&&position==0){
```

```
            return TYPE_HEADER;
        }else if(footer!=null&&position==getItemCount()-1){
                        //如果 footer 不为空&&最后一个
            return TYPE_FOOTER;
        }
        return super.getItemViewType(position);
    }

    public void setHeader(View header) {
        this.header = header;

        notifyItemInserted(0);//在位置 0 处插入一条数据，然后刷新
    }

    public void setFooter(View footer) {
        this.footer = footer;
        notifyItemInserted(datas.size()-1);//在尾部插入一条数据，然后刷新
    }

    class MyViewHolder extends RecyclerView.ViewHolder{
        private TextView textview;

        public MyViewHolder(View itemView) {
            super(itemView);
            textview= (TextView) itemView.findViewById(R.id.textview);
        }
    }
}
```

- getItemCount：有 header 和 footer 时，需要在源数据长度基础上进行增加。
- getItemViewType：通过 getItemViewType 判断不同的类型。
- onCreateViewHolder：通过不同的类型创建 item 的 View。
- onBindViewHolder：如果是 header 与 footer 类型是不需要绑定数据的，header 与 footer 的 View 一般在 Activity 页面中创建，不需要处理，所以这两种类型就不往下执行。如果有头布局，position==0 的位置就会被 header 占用，但是数据源也就是集合的下标是从 0 开始的，所以这里需要改为-1。
- setHeader：设置头布局，在第一行插入一条数据，然后刷新。注意，这个方法调用后会有插入的动画，这个动画可以使用默认的，也可以自己定义。
- setFooter：设置尾部布局，在尾部插入一条数据，然后刷新。

添加 header 与 footer 的方法终于封装好了，在 Activity 页面中只需要两行代码就能添加 header，与 ListView 调用 addHeader 方法一样简单。这里需要注意的是，初始化 View 时，inflate 方法需要三个参数。

- resource：布局文件资源 id。
- root：父 View。
- attachToRoot：需要传入一个 boolean 类型的值。如果传入 true，布局文件将转化为 View 并绑定到 root，然后返回 root 作为根节点的整个 View。如果传入 false，布局文件转化为 View 但不绑定到 root，返回以布局文件根节点为根节点的 View。

```
    //添加 header
    View header=LayoutInflater.from(this).inflate(R.layout.recycler_header,
recyclerView,false);
    adapter.setHeader(header);

    //添加 footer
    View footer=LayoutInflater.from(this).inflate(R.layout.recycler_footer,
recyclerView,false);
    adapter.setFooter(footer);
```

recycler_header 和 recycler_footer 布局文件只有一个 TextView，直接看效果，如图 2-41 所示。

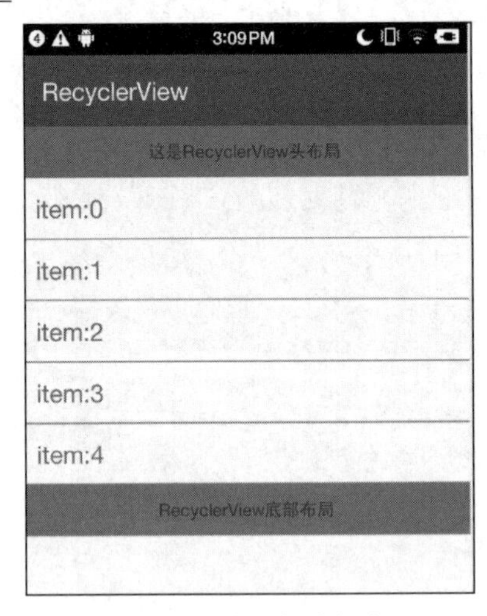

图 2-41　头布局与底部布局

6. item 点击事件&&增加或删除带动画效果

当调用 RecyclerView 的 setOnItemClickListener 方法时，发现居然没有该方法。用 RecyclerView 要习惯什么东西都自己封装。

首先从 adapter 着手，内部写一个接口、一个实例变量，提供一个公共方法，设置监听。

```
    private RecyclerViewItemClick recyclerViewItemClick;

    public void setRecyclerViewItemClick(RecyclerViewItemClick
recyclerViewItemClick) {
        this.recyclerViewItemClick = recyclerViewItemClick;
    }

    public interface RecyclerViewItemClick{
        /**
         * item 点击
         * @param realPosition 数据源 position
         * @param position view position
         */
        void onItemClick(int realPosition,int position);
    }
```

在 onBindViewHolder 方法中为 item 监听点击事件：

```
    if(recyclerViewItemClick!=null) {
```

```
    myholder.itemView.setOnClickListener(new View.OnClickListener() {
        @Override
        public void onClick(View v) {
            recyclerViewItemClick.onItemClick(getRealPosition(position),
             position);
        }
    });
}
```

在 Activity 页面的 onCreate 方法中进行监听，顺便设置 item 增加或删除动画，用的是 SDK 自带的默认动画。

```
adapter.setRecyclerViewItemClick(recyclerViewItemClick);
recyclerView.setItemAnimator(new DefaultItemAnimator());
private RecyclerHeadFootViewAdapter.RecyclerViewItemClick
recyclerViewItemClick=new RecyclerHeadFootViewAdapter.RecyclerViewItemClick() {
    @Override
    public void onItemClick(int realPosition, int position) {
        Log.i("ansen","删除数据:"+realPosition+" view位置:"+position);
        Log.i("ansen","当前位置:"+position+" 更新item数量:
"+(adapter.getItemCount()-position-1));

        datas.remove(realPosition);//删除数据源
        adapter.notifyItemRemoved(position);//item移除动画
        //更新position至adapter.getItemCount()-1的数据
        adapter.notifyItemRangeChanged(position,adapter.getItemCount()
          -position-1);
    }
};
```

2.5.4 SwipeRefreshLayout（下拉刷新）

SwipeRefreshLayout 是一种下拉刷新控件，在 Version 19.1 之后被放到 support v4 中，继承自 ViewGroup。SwipeRefreshLayout 控件只允许有一个子元素，子元素一般是 ListView 或者 RecyclerView。

1. SwipeRefreshLayou 下拉刷新

上一节学习了 RecyclerView，这里就在 RecyclerView 简单 Demo 的基础上加上下拉刷新。

首先看布局文件 activity_main.xml，把最外层的布局控件换成 SwipeRefreshLayout。

```
<?xml version="1.0" encoding="utf-8"?>
<android.support.v4.widget.SwipeRefreshLayout
    xmlns:android="http://schemas.android.com/apk/res/android"
    android:id="@+id/swipeRefreshLayout"
    android:layout_width="match_parent"
    android:layout_height="match_parent">

    <RecyclerView
        android:id="@+id/recyclerview"
        android:layout_width="match_parent"
        android:layout_height="match_parent"/>
</android.support.v4.widget.SwipeRefreshLayout>
```

接下来，在 Activity 页面中设置刷新监听，设置刷新箭头的颜色：

```
swipeRefreshLayout=(SwipeRefreshLayout)findViewById(R.id.swipeRefreshLayou
t;
//监听刷新状态
swipeRefreshLayout.setOnRefreshListener(this);
//设置下拉刷新的箭头颜色(可以设置多个颜色)
swipeRefreshLayout.setColorSchemeResources(R.color.colorAccent,
      android.R.color.holo_green_light,R.color.colorPrimary);
```

刷新监听回调方法，这里为了模拟访问网络请求，用 Handle 来达到延时的效果。有关 Handle 的使用方法，后续的章节会详细介绍。

```
@Override
public void onRefresh() {
   //延迟 3000 毫秒,发送空消息跟 handle, handle 的 handleMessage 方法会接收到
   handler.sendEmptyMessageDelayed(PULL_TO_REFRESH,3000);
}
```

PULL_TO_REFRESH 是一个常量，在 Activity 中定义如下：

```
public static final int PULL_TO_REFRESH=1;//下拉刷新
```

下面来看 Handle 是怎么实现的，可以通过 msg 对象的 what 来判断是什么消息。

```
private Handler handler=new Handler(){
   @Override
   public void handleMessage(Message msg) {
      switch (msg.what){
         case PULL_TO_REFRESH://下拉刷新
            if(datas.size()>0){
               datas.remove(0);//删除第一条
               adapter.notifyDataSetChanged();//更新第一条记录
               swipeRefreshLayout.setRefreshing(false);
                  //false, 刷新完成; true, 正在刷新
            }
            break;
      }
   }
};
```

通过上面为数不多的代码，就实现了 RecyclerView 下拉刷新的效果。运行代码，正在刷新中的效果如图 2-42 所示。

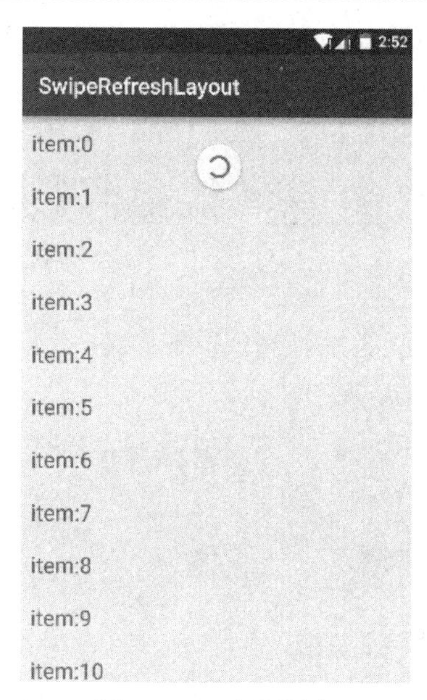

图2-42 正在刷新中

2. RecyclerView 加载更多

既然有了下拉刷新功能，肯定还需要加载更多，在 SwipeRefreshLayout 控件的 api 中并没有这个功能，只能从 RecyclerView 入手。

首先修改 item 布局文件 recycler_item.xml，增加 ProgressBar 进度条。

```
<?xml version="1.0" encoding="utf-8"?>
<LinearLayout xmlns:android="http://schemas.android.com/apk/res/android"
    android:layout_width="match_parent"
    android:layout_height="wrap_content"
    android:orientation="vertical">

    <TextView
        android:id="@+id/textview"
        android:layout_width="match_parent"
        android:layout_height="wrap_content"
        android:padding="10dp"
        android:text="122"
        android:textSize="20sp"/>

    <ProgressBar
        android:id="@+id/progressBar"
        style="@style/Widget.AppCompat.ProgressBar"
        android:layout_width="wrap_content"
        android:layout_height="wrap_content"
        android:layout_gravity="center_horizontal"
        android:padding="3dp" />
</LinearLayout>
```

在 adapter 内部类 MyViewHolder 中查找这个控件：

```
class MyViewHolder extends RecyclerView.ViewHolder{
    private TextView textview;
```

```
        private ProgressBar progressBar;

        public MyViewHolder(View itemView) {
            super(itemView);
            textview= (TextView) itemView.findViewById(R.id.textview);
            progressBar= (ProgressBar) itemView.findViewById(R.id.progressBar);
        }
    }
```

在 adapter 中的 onBindViewHolder 方法增加以下代码：（如果列表有 5 条记录并且是最后一条记录，将显示进度条，否则隐藏进度条。）

```
    if(position>5 && position == datas.size()-1){
        holder.progressBar.setVisibility(View.VISIBLE);
    }else{
        holder.progressBar.setVisibility(View.GONE);
    }
```

在 Activity 页面中监听 recyclerView 滚动：

```
    recyclerView.addOnScrollListener(onScrollListener);
```

下面用一个实例变量控制加载状态：

```
    private RecyclerView.OnScrollListener onScrollListener=new
RecyclerView.OnScrollListener() {
            @Override
            public void onScrolled(RecyclerView recyclerView, int dx, int dy) {
                super.onScrolled(recyclerView, dx, dy);
                RecyclerView.LayoutManager mLayoutManager = recyclerView.
getLayoutManager();
                int lastVisibleItem = ((LinearLayoutManager) mLayoutManager).
findLastVisibleItemPosition();
                int totalItemCount = mLayoutManager.getItemCount();
                //最后一项显示&&下滑状态的时候回调加载更多
                if (lastVisibleItem >= totalItemCount-1 && dy > 0) {
                    if(!isLoadMore){
                        loadMore();//加载更多
                        isLoadMore=true;
                    }
                }
            }
        };
```

加载更多的方法，模拟访问网络延迟 1000 毫秒给 handle 发送一个消息：

```
    public void loadMore() {
        handler.sendEmptyMessageDelayed(UP_TO_REFRESH,1000);
    }
```

为在 handle 中判断 what 的 switch 增加一个 case，这里要加载更多，为列表增加三行数据。在企业开发中，基本都是请求下一页数据，然后更新列表。

```
    case UP_TO_REFRESH://上拉加载更多
        for(int i=0;i<3;i++){
            datas.add("load more item:"+i);
        }
```

```
        adapter.notifyDataSetChanged();//更新列表
        isLoadMore=false;//加载更多完成
        break;
```

因为书中能展示的代码有限，所以都是展示要修改的代码，如果大家看起来有困难，可以对照源代码学习。运行代码，有一个波纹效果，这是 5.0 操作系统自带的，如图 2-43 所示。

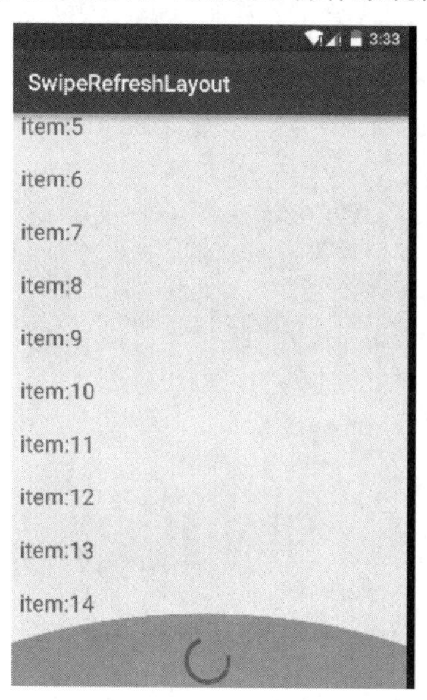

图 2-43　RecyclerView 的效果

2.5.5　ViewPager（翻页视图）

ViewPager 是 Android 扩展包 v4 中的类，可以让用户左右切换 View。ViewPager 的用法与 ListView 类似，需要设置 PagerAdapter 来完成页面和数据的绑定，这个 PagerAdapter 是一个基类适配器，经常用来实现 App 引导图。它的子类有 FragmentPagerAdapter 和 FragmentStatePagerAdapter，和 Fragment 一起使用。在 Android 应用中，它们就像 ListView 一样出现得较频繁。

本节实例将用 ViewPager 来实现 Fragment 的滑动。

首先看布局文件 activity_main.xml，外层是 RelativeLayout，里面包含一个 ViewPager 控件：

```
<?xml version="1.0" encoding="utf-8"?>
<RelativeLayout xmlns:android="http://schemas.android.com/apk/res/android"
    android:layout_width="match_parent"
    android:layout_height="match_parent">

    <android.support.v4.view.ViewPager
        android:id="@+id/viewPager"
        android:layout_width="match_parent"
        android:layout_height="match_parent"/>
</RelativeLayout>
```

接下来看布局文件对应的 Activity 如何实现，首先查找 Viewpager 控件，设置缓存页数，设置当前显示第几个 Fragment，然后初始化 FragmentAdapter 适配器，这个适配器是我们自己写的，集成自 FragmentStatePagerAdapter，接下来调用适配器的 addFragment 方法将要显示的 Fragment 添加进去。最后调用 ViewPager 的 setAdapter 方法将适配器设置进去。当想知道 ViewPager 滑动到哪个 Fragment 时，可以通过 addOnPageChangeListener 方法来设置监听。

```
    public class MainActivity extends AppCompatActivity {
        @Override
        protected void onCreate(Bundle savedInstanceState) {
            super.onCreate(savedInstanceState);
            setContentView(R.layout.activity_main);

            ViewPager vPager = (ViewPager) findViewById(R.id.viewPager);
            vPager.setOffscreenPageLimit(2);//设置缓存页数
            vPager.setCurrentItem(0);//设置当前显示的 item 0 表示显示第一个

            FragmentAdapter pagerAdapter = new FragmentAdapter
(getSupportFragmentManager());

            pagerAdapter.addFragment(new FragmentTest("页面 1",android.R.color.
holo_red_dark));
            pagerAdapter.addFragment(new FragmentTest("页面 2",android.R.color.
holo_green_dark));
            pagerAdapter.addFragment(new FragmentTest("页面 3",android.R.color.
holo_red_dark));
            pagerAdapter.addFragment(new FragmentTest("页面 4",android.R.color.
holo_green_dark));

            //给 ViewPager 设置适配器
            vPager.setAdapter(pagerAdapter);
            //页面改变监听
            vPager.addOnPageChangeListener(onPageChangeListener);
        }

        private ViewPager.OnPageChangeListener onPageChangeListener=new ViewPager.
OnPageChangeListener() {
            //页面滑动
            @Override
            public void onPageScrolled(int position, float positionOffset, int
positionOffsetPixels) {}

            //页面选择
            @Override
            public void onPageSelected(int position) {
                Log.i("MainActivity","选中了页面"+(position+1));
            }

            //页面滑动状态改变
            @Override
            public void onPageScrollStateChanged(int state) {}
        };
    }
```

FragmentAdapter.java 继承 FragmentStatePagerAdapter 类，与 BaseAdapter 有点类似。相对于 BaseAdapter 适配器，FragmentStatePagerAdapter 不需要重写 getView 方法。

```
public class FragmentAdapter extends FragmentStatePagerAdapter {
    private final List<Fragment> fragmentList = new ArrayList<Fragment>();

    public FragmentAdapter(FragmentManager fm) {
        super(fm);
    }

    @Override
    public Fragment getItem(int arg0) {
        return fragmentList.get(arg0);
    }

    @Override
    public int getCount() {
        return fragmentList.size();
    }

    public void addFragment(Fragment fragment) {
        fragmentList.add(fragment);
    }
}
```

FragmentTest.xml 在构造方法中传入两个参数，参数 1 是内容，参数 2 是页面背景颜色。fragment_test.xml 只有一个 TextView，非常简单，代码就不贴出来了。

```
@SuppressLint("ValidFragment")
public class FragmentTest extends Fragment {
    private String content;
    private int backgroundResourceId;

    public FragmentTest(String content,int backgroundResourceId){
        this.content=content;
        this.backgroundResourceId=backgroundResourceId;
    }

    @Override
    public View onCreateView(LayoutInflater inflater, ViewGroup container,
Bundle savedInstanceState){
        View rootView=LayoutInflater.from(getActivity()).inflate(R.layout.
                    fragment_test, null);
        TextView tvContent= (TextView) rootView.findViewById
                    (R.id.tv_content);
        tvContent.setText(content);

        rootView.setBackgroundResource(backgroundResourceId);
        return rootView;
    }
}
```

效果如图 2-44 所示。这是 ViewPager 滑动到一半的截图，页面一和页面二都显示一半。

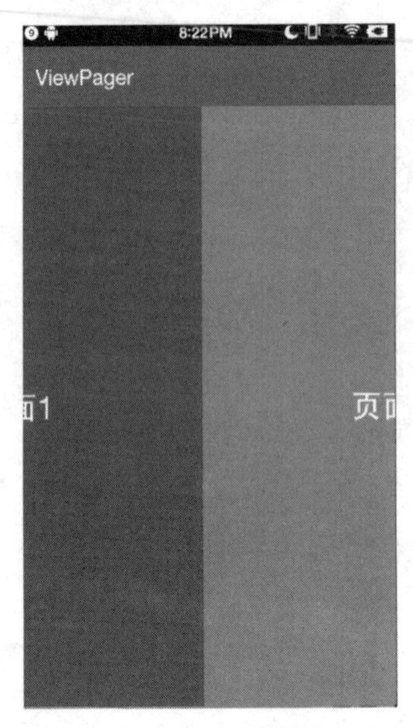

图 2-44　翻页视图

2.6　通过 xml 文件修饰 View

2.6.1　shapes（设置圆角、边框、填充色、渐变色）

shape 用来控制控件（View）的几何形状，有 6 个子标签：

- corners：圆角。
- solid：填充颜色。
- stroke：描边。
- padding：内边距。
- size：宽高。
- gradient：渐变。

在 res/drawable 目录下新建一个文件 tv_shape.xml：

```
<?xml version="1.0" encoding="utf-8"?>
<shape xmlns:android="http://schemas.android.com/apk/res/android">
    <!--圆角-->
    <corners android:radius="10dp"/>
    <!--填充颜色-->
    <solid android:color="@android:color/white"/>
    <!-- 描边 -->
    <stroke
        android:width="5dp"
        android:color="@color/colorAccent" />
    <!--内边距-->
    <padding android:top="10dp" android:bottom="10dp" android:left="10dp"
```

```
android:right="10dp"/>
    </shape>
```

在布局文件中为 TextView 设置 background，引用刚刚写的 shape。

```
    <TextView
        android:layout_width="wrap_content"
        android:layout_height="wrap_content"
        android:background="@drawable/tv_shape"
        android:text="Hello World!" />
```

效果如图 2-45 所示。

图 2-45　设置圆角图形

因为渐变色和填充色不能同时使用，所以接着为渐变色新建一个 shape 文件 tv_shape_two.xml，顺便增加一个 size 标签，指定该 View 的宽高都是 100dp。

```
    <?xml version="1.0" encoding="utf-8"?>
    <shape xmlns:android="http://schemas.android.com/apk/res/android">
        <!--宽高-->
        <size
            android:width="100dp"
            android:height="100dp" />
        <!-- 渐变 -->
        <gradient
            android:endColor="#ff8c00"
            android:gradientRadius="50"
            android:startColor="#FFFFFF"
            android:type="linear" />
    </shape>
```

解释上面的渐变标签，android:startColor 和 android:endColor 分别为起始和结束颜色，android:angle 是渐变角度，必须为 45 的整数倍。另外，渐变默认的模式为 android:type="linear"，即线性渐变。可以指定渐变为径向渐变，android:type="radial"，径向渐变需要指定半径 android:gradientRadius="50"。

效果如图 2-46 所示。

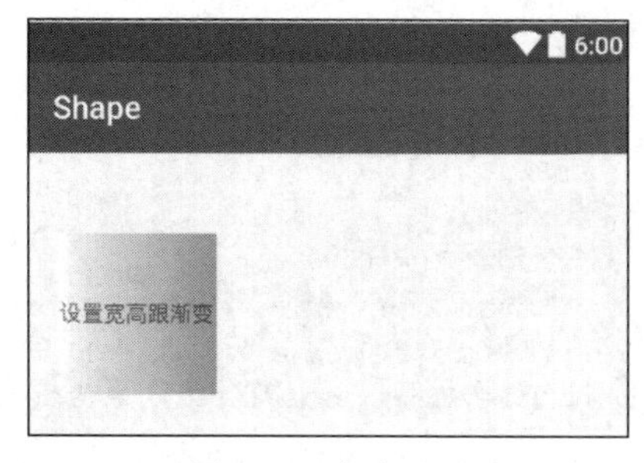

图 2-46　渐变效果

2.6.2 selector（设置点击、选中点击效果）

当用户点击界面的某个按钮或者图片的时候，需要告诉用户：你的点击我收到了。例如，大部分 App 注册时需要选择用户性别，都可能会用到 selector。下面介绍 selector 标签中 item 标签的常用属性。

1. android:state_pressed 属性

在 drawable 目录下新建一个 selector_button_pressed.xml 文件：

```
    <?xml version="1.0" encoding="utf-8"?>
    <selector xmlns:android="http://schemas.android.com/apk/res/android">
        <!--按下背景颜色-->
        <item android:drawable="@color/colorAccent"
android:state_pressed="true"/>
        <!--默认背景颜色-->
        <item android:drawable="@color/colorPrimary"/>
    </selector>
```

给按钮设置一个背景颜色：

```
    android:background="@drawable/selector_button_pressed"
```

左边是默认效果，右边是按下效果，如图 2-47 所示。

图 2-47 为按钮设置背景颜色

2. android:state_checked

一般 Checkbox（复选框）使用这个属性比较多。在 drawable 目录下新建一个 selector_check_state.xml 文件，用 android:drawable 引用图片，是之前准备好的两张图片，表示勾选状态和未勾选状态。

```
    <?xml version="1.0" encoding="utf-8"?>
    <selector xmlns:android="http://schemas.android.com/apk/res/android">
        <!--选中时显示图片-->
        <item android:drawable="@mipmap/icon_checkbox_selector"
android:state_checked="true" />
        <!--未选中时显示图片-->
        <item android:drawable="@mipmap/icon_checkbox_normal"
android:state_checked="false" />
    </selector>
```

用 Checkbox 控件的 android:button 引用这个文件：

```
    <CheckBox
        android:layout_marginTop="20dp"
        android:layout_width="wrap_content"
        android:layout_height="wrap_content"
        android:button="@drawable/selector_check_state"/>
```

未勾选状态跟勾选状态的效果如图 2-48 所示。

图 2-48 复选框的两种状态

3. android:state_selected

选中状态和未选中状态其实与 android:state_checked 类似。这个属性大部分控件都能使用，一般需要自己在代码中设置选中和未选中状态。

新建一个文件 selector_imageview.xml，引用的图片与上一个属性图片一样。

```
<?xml version="1.0" encoding="utf-8"?>
<selector xmlns:android="http://schemas.android.com/apk/res/android">
    <!--选中时显示图片-->
    <item android:drawable="@mipmap/icon_checkbox_selector"
android:state_selected="true" />
    <!--未选中时显示图片-->
    <item android:drawable="@mipmap/icon_checkbox_normal"
android:state_selected="false" />
</selector>
```

给 ImageView 控件设置 android:background：

```
android:background="@drawable/selector_imageview"
```

在 MainActivity 中设置 ImageView 点击事件，在点击事件设置选中状态。

```
@Override
public void onClick(View v) {
    //获取状态，取反，设置取反后的状态
    imageView.setSelected(!imageView.isSelected());
}
```

效果图和 CheckBox 的勾选效果一样，只不过用的控件以及 selector 的属性不一样，如图 2-49 所示。

图 2-49 复选框的两种状态

2.6.3 layer-list（把 item 按照顺序层叠显示）

使用 layer-list 可以将多个 drawable 按照顺序层叠在一起显示。最先定义的在最下方显示，后面的依次往上叠放。

在 drawable 目录下新建 layer_list_textview.xml 文件：

```
<?xml version="1.0" encoding="utf-8"?>
<layer-list xmlns:android="http://schemas.android.com/apk/res/android">
    <item >
        <shape android:shape="rectangle" >
            <solid android:color="#0000ff"/>
        </shape>
    </item>

    <item android:bottom="25dp" android:top="25dp" android:left="25dp"
android:right="25dp">
        <shape android:shape="rectangle" >
            <solid android:color="#00ff00"/>
        </shape>
    </item>

    <item android:bottom="50dp" android:top="50dp" android:left="50dp"
android:right="50dp">
        <shape android:shape="rectangle" >
            <solid android:color="#ff0000" />
        </shape>
    </item>
</layer-list>
```

为 TextView 设置 background 属性引用这个文件，并且设置 padding 属性。

```
<TextView
    android:layout_width="wrap_content"
    android:layout_height="wrap_content"
    android:background="@drawable/layer_list_textview"
    android:padding="50dp"
    android:text="Hello World!"/>
```

如图 2-50 所示，蓝色在最底部（定义的第一个 item），红色在最上面（定义的最后一个 item）。

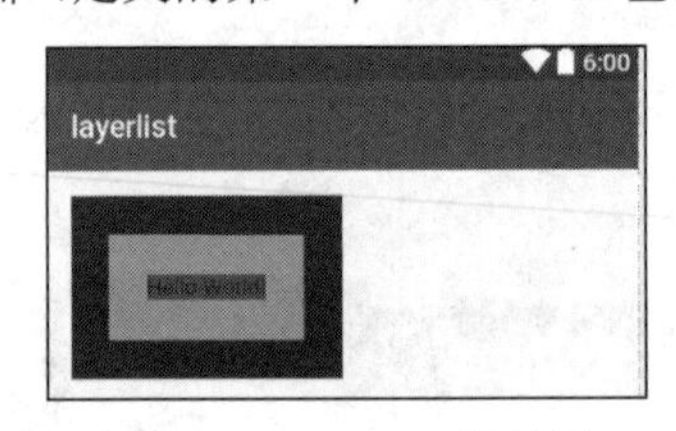

图 2-50　layer-list 的使用

2.7　本章小结

现在市场上的应用程序五花八门，每一个类型都有几十个应用程序相互竞争，熟练使用控件才能做出更好的 App。本章主要介绍了使用较多的三种布局以及控件的使用，每个控件都提供了案例，让读者加深对组件的理解。通过本章的学习，相信大家在开发过程中也能灵活应用，在不同的 UI 界面上知道应该使用哪种控件最合适。

第 3 章

Android 四大组件

Android 四大核心组件指的是 Activity、Service、Broadcast Receiver 和 ContentProvider。四大组件由 Android 系统进行管理和维护，一般都需要在清单文件中注册或者在代码中动态注册。

- Activity：代表一个页面（窗口）。
- Service：在后台默默做一些耗时操作。
- Broadcast Receiver：对感兴趣的外部事件进行监听，例如监听系统短信、手机网络状态改变等，当然也可以监听自己发送的广播。
- ContentProvider：多个应用程序之间数据共享。

3.1 Activity（活动）

Activity（活动）是用得最多、最基本的组件，是一种可以包含界面的组件。Activity 代表一个页面，开发者可以通过 setContentView（View）把界面（UI）放到该页面上。实现 Activity 需要进行以下两步：

- 重写 onCreate 方法，Activity 创建的时候会调用这个方法。
- 在 onCreate 中调用 setContentView（int）设置界面布局，并且可以使用 findViewById（int）查找界面上的某个控件。

3.1.1 Activity 的生命周期

Activity 是由 Activity 任务栈（任务栈后面介绍）进行管理的。假设我们现在已经开启了一个 Activity A，当再次开启一个新的 Activity B 时，Activity B 就会被放到栈的顶部，也会标示成运行状态。Activity A 同时放到栈的下面，进入后台状态。如果新开启的 Activity B 被销毁，Activity A 会继续回到栈的顶部。

图 3-1 表示了 Activity 的生命周期，正方形表示 Activity 在状态之间切换要回调的方法，彩色椭圆形表示 Activity 处于主要状态。所谓的生命周期就是 Activity 从创建到销毁的生命过程。

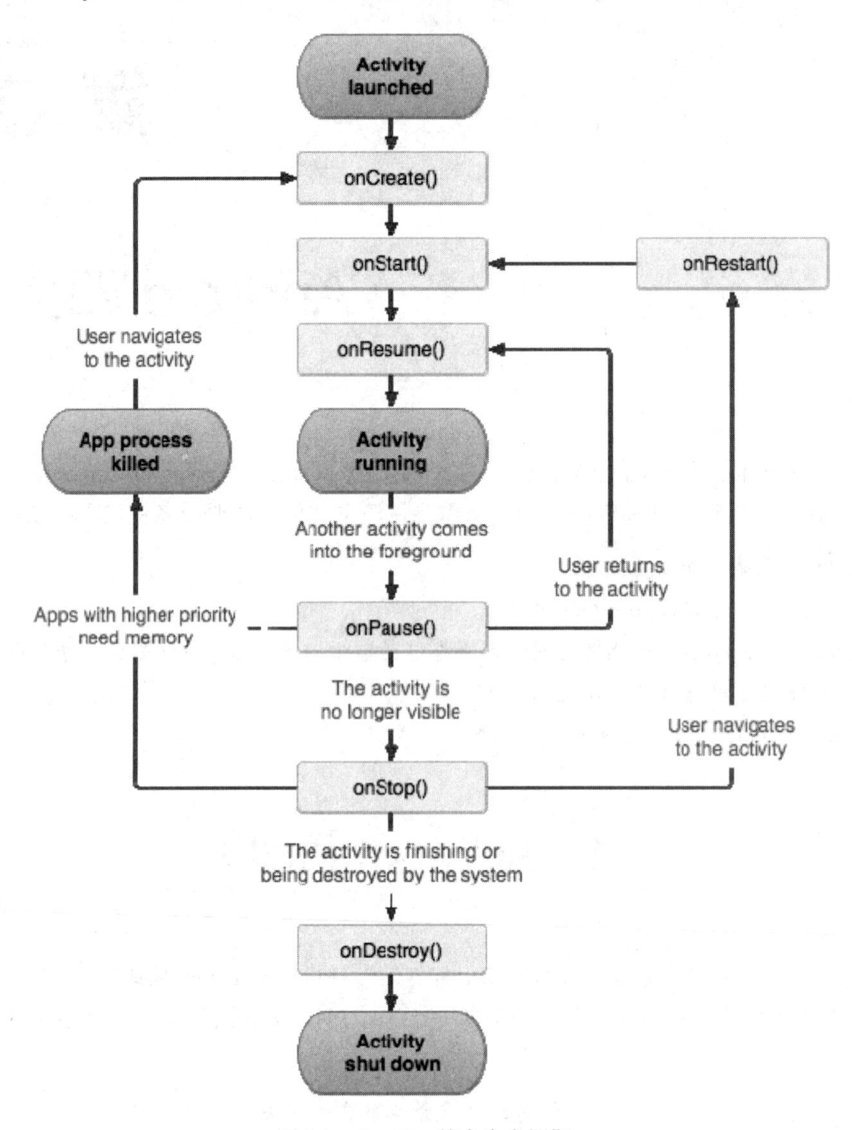

图 3-1 Activity 整个生命周期

我们首先介绍一下正方形中各方法的调用顺序。

- onCreate: 该方法在 Activity 创建时调用，是整个生命周期中第一个调用的方法，我们在创建一个 Activity 时一般都要重写这个方法，在该方法中做一些初始化的操作，例如通过 setContentView 设置布局资源、查找布局文件中的控件等。
- onStart: 此方法回调时表示 Activity 正在启动。此时 Activity 已处于可见状态，只是还没有在前台显示，也无法与用户进行交互。

- onResume：此方法回调时说明 Activity 已经在前台可见、可以跟用户进行交互了。
- onPause：此方法调用时表示 Activity 正在停止。
- onStop：表示 Activity 即将停止，此时 Activity 不可见，仅在后台运行。
- onRestart：表示 Activity 正在重启，当 Activity 由不可见变为可见状态时，该方法被回调。一般是用户打开了一个新的 Activity，当前的 Activity 会被暂停（onPause 和 onStop 被执行了），接着又回到当前 Activity 页面，onRestart 方法就会被回调。
- onDestroy：此方法调用时表示 Activity 即将被销毁，是 Activity 生命周期的最后一个方法。在这个方法中可以做一些回收工作和最终的资源释放（例如广播、Service 等）。

接下来通过代码来验证一下。在 MainActivity 中重写图 3-1 正方形中的所有方法，并且在每个方法中打印 Log。运行之后，看这些方法都在什么情况下调用。重写生命周期中的方法第一行要调用 super。

```
public class MainActivity extends AppCompatActivity{
    private String TAG="ansen";

    @Override
    protected void onCreate(Bundle savedInstanceState) {
        super.onCreate(savedInstanceState);
        setContentView(R.layout.activity_main);
        Log.i(TAG,"onCreate");
    }

    @Override
    protected void onStart() {
        super.onStart();
        Log.i(TAG,"onStart");
    }

    @Override
    protected void onResume() {
        super.onResume();
        Log.i(TAG,"onResume");
    }

    @Override
    protected void onPause() {
        super.onPause();
        Log.i(TAG,"onPause");
    }

    @Override
    protected void onStop() {
        super.onStop();
        Log.i(TAG,"onStop");
    }

    @Override
    protected void onDestroy() {
        super.onDestroy();
        Log.i(TAG,"onDestroy");
```

```
    }

    @Override
    protected void onRestart() {
        super.onRestart();
        Log.i(TAG,"onRestart");
    }
}
```

直接运行代码，打印 log 如下，回调方法过程是 onCreate→onStart→onResume。

```
04-26 20:45:37.655 8655-8655/com.ansen.activity I/ansen: onCreate
04-26 20:45:37.657 8655-8655/com.ansen.activity I/ansen: onStart
04-26 20:45:37.662 8655-8655/com.ansen.activity I/ansen: onResume
```

在当前页面按住系统 Home 键，这时会回到首页。打印 log 如下，回调方法过程是 onPause→onStop。

```
04-26 20:49:23.549 8655-8655/com.ansen.activity I/ansen: onPause
04-26 20:49:23.585 8655-8655/com.ansen.activity I/ansen: onStop
```

从手机首页找到 App，点击之后进入 App 首页，重新显示 MainActivity。打印 log 如下，回调方法过程是 onRestart→onStart→onResume。

```
04-26 20:51:09.737 8655-8655/com.ansen.activity I/ansen: onRestart
04-26 20:51:09.748 8655-8655/com.ansen.activity I/ansen: onStart
04-26 20:51:09.751 8655-8655/com.ansen.activity I/ansen: onResume
```

按住系统返回键回到手机首页，这时当前 Activity 销毁，应用程序退出。打印 log 如下，回调方法过程是 onPause→onStop→onDestroy。

```
04-26 20:52:39.398 8655-8655/com.ansen.activity I/ansen: onPause
04-26 20:52:39.746 8655-8655/com.ansen.activity I/ansen: onStop
04-26 20:52:39.747 8655-8655/com.ansen.activity I/ansen: onDestroy
```

通过以上的 log 分析再回头看看图 3-1，相信读者对 Activity 的生命周期有了更深刻的认识。

3.1.2 启动 Activity 的两种方式

启动一个 Activity，有显式启动和隐式启动两种方式。

- 显式 Intent：一般用于启动同一个应用中的 Activity，效率高。
- 隐式 Intent：一般用于启动不同应用中的 Activity，通过 Intent Filter 来实现。因为没有明确指出目标 Activity，所以效率相对低一些。

1. 显式 Intent 启动 Activity

新建一个 SecondActivity 类继承 AppCompatActivity：

```
public class SecondActivity extends AppCompatActivity {
    @Override
    protected void onCreate(Bundle savedInstanceState){
        super.onCreate(savedInstanceState);
        setContentView(R.layout.activity_second);
```

```
    }
}
```

如果写新的 Activity，一定要在 AndroidManifest.xml 中注册：

```
<activity android:name="com.ansen.activity.SecondActivity"/>
```

然后在 MainActivity 中启动我们刚创建的 SecondActivity：

```
Intent intent=new Intent(this,SecondActivity.class);
startActivity(intent);
```

2. 隐式 Intent 启动 Activity

新建一个 ThreeActivity 类继承 AppCompatActivity：

```
public class ThreeActivity extends AppCompatActivity {
    @Override
    protected void onCreate(Bundle savedInstanceState){
        super.onCreate(savedInstanceState);
        setContentView(R.layout.activity_three);
    }
}
```

同样需要在 Activity 中注册，因为 ThreeActivity 是隐式启动，所以需要增加 intent-filter 过滤器。在过滤器中增加标签，指定 name 的值（是一个字符串），隐式启动时就是根据这个值来找到 Activity。标签是固定的写法。

```
<activity android:name="com.ansen.activity.ThreeActivity" >
    <intent-filter>
        <action android:name="com.ansen.activity.ThreeActivity"/>
        <category android:name="android.intent.category.DEFAULT" />
    </intent-filter>
</activity>
```

在 MainActivity 中隐式启动 ThreeActivity，用上一步标签 name 的值给 Intent 的构造方法即可。

```
Intent threeIntent=new Intent("com.ansen.activity.ThreeActivity");
startActivity(threeIntent);
```

3.1.3　在 Activity 中使用 Toast

Toast 是 Android 系统的一种提示方式，显示某一段话来提示用户，显示一段时间自动消息，不占用屏幕空间。

Toast 与其他组件一样，都属于 UI 界面中的内容，因此在子线程中无法使用 Toast 弹出提示内容。如果强行在子线程中添加 Toast，就会导致错误。

Toast 使用起来很简单，官方 SDK 已经帮助封装好了，只需要一行代码即可。最简单的写法如下：

```
Toast.makeText(this,"提示",Toast.LENGTH_LONG).show();
```

调用 Toast 类的 makeText 静态方法，传入三个参数：Context、提示内容、显示时长。makeText 方法会返回 Toast 对象，调用 show 方法显示出来。运行效果如图 3-2 所示。

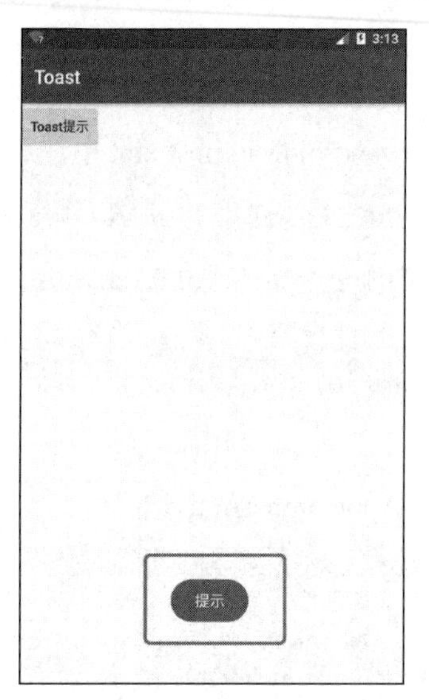

图 3-2　Toast 显示效果

系统 Toast 基本都是半透明背景、黑色的字体，不太好看，并且不同版本的 Android 操作系统显示的效果还不一样。不像 iOS 开发，能够统一规范管理。Google 早给我们考虑到了，只需要调用 Toast 对象的 setView 方法就能设置显示我们自定义的 View。

我们自己封装一个 CustomToast 类，通过单例模式（学过 Java 的读者对这个设计模式应该都很熟悉，不知道的请求助于网页搜索）获取这个类的对象。

```
    public class CustomToast {
        private static CustomToast _instance = null;
        private Toast toast = null;

        private final int MARGIN_DP = 50;

        private CustomToast() {
        }

        public static CustomToast getInstance() {
            if (_instance == null) {
                _instance = new CustomToast();
            }
            return _instance;
        }

        public void cancel() {
            if (toast != null) {
                toast.cancel();
                toast = null;
            }
        }

        public void showToastCustom(Context ctx, String msg, int gravity) {
            showToastCustom(ctx, msg,R.layout.toast_msg, R.id.txt_toast_message,
gravity);
        }
```

```
    /**
     * 显示自定义布局
     * @param ctx 上下文对象
     * @param msg 显示内容
     * @param layoutResId 布局文件资源 id
     * @param txtResId 布局文件中 TextView 控件 id
     * @param gravity 显示位置
     */
    public void showToastCustom(Context ctx, String msg, int layoutResId, int
txtResId, int gravity) {
        cancel();//显示之前取消上次显示，这样每次都能显示最新的
        try {
            if (TextUtils.isEmpty(msg)) {
                return;
            }
            View layout = View.inflate(ctx, layoutResId, null);
            TextView txtMsg = layout.findViewById(txtResId);
            txtMsg.setText(msg);
            toast = new Toast(ctx);
            toast.setDuration(Toast.LENGTH_SHORT);
            if (gravity == Gravity.TOP) {
                int marginVertical = (int) dip2px(ctx, MARGIN_DP);
                toast.setGravity(gravity, 0,marginVertical);
            } else if (gravity == Gravity.BOTTOM) {
                int marginVertical = (int) dip2px(ctx, MARGIN_DP);
                toast.setGravity(gravity, 0,marginVertical);
            } else {
                toast.setGravity(gravity, 0, 0);
            }
            toast.setView(layout);
            toast.show();
        } catch (Exception e) {
            e.printStackTrace();
        }
    }

    /**
     * dp 转 px
     * @param context
     * @param dpValue
     * @return
     */
    public static float dip2px(Context context, float dpValue) {
        final float scale = context.getResources().getDisplayMetrics().density;
        float result = dpValue * scale + 0.5f;
        return result;
    }
}
```

- getInstance：通过这个方法获取对象。
- cancel：取消显示。
- showToastCustom：有两个方法名一样的方法，只是参数不一样，这是方法重载，我们主要查看下面那个参数多的方法。这个方法中第一行代码调用 cancel 方法，如果之前已有显示，那

么 Toast 就取消之前的；如果显示内容为空，则直接返回，调用 View.inflate 方法将布局资源 id 变成 View 对象，从布局文件中查找 TextView，将显示内容赋值给 TextView。新建一个 Toast 对象，设置显示时间，判断 Toast 显示位置，如果显示在顶部或者底部，则需要设置 yOffset 参数，不然 Toast 紧挨着顶部或者底部效果肯定不好看。调用 Toast 对象的 setView 方法将 View 赋值进去，最后调用 show 方法进行显示。

- dip2px：将 dip 的值转换成 px，根据不同屏幕的分辨率转成 px 的值是不一样的，这样做的目的是为了兼容不同的分辨率。

Toast 显示的布局文件：toast_msg.xml。

```
<LinearLayout xmlns:android="http://schemas.android.com/apk/res/android"
    android:id="@+id/layout_custom_toast"
    android:layout_width="wrap_content"
    android:layout_height="wrap_content"
    android:background="@drawable/toast_bg"
    android:orientation="vertical">

    <TextView
        android:id="@+id/txt_toast_message"
        android:layout_margin="5dp"
        android:layout_width="match_parent"
        android:layout_height="wrap_content"
        android:paddingLeft="10dip"
        android:paddingRight="10dp"
        android:text=""
        android:textColor="#FFFFFF"/>
</LinearLayout>
```

最外层是 LinearLayout，里面一个 TextView 用来显示内容。最外层的 LinearLayout 设置了 background 属性，指向了一个 shape 文件 toast_bg.xml，内容如下：

```
<?xml version="1.0" encoding="utf-8"?>
<shape xmlns:android="http://schemas.android.com/apk/res/android"
    android:shape="rectangle">

    <gradient
        android:angle="-90"
        android:endColor="@color/colorAccent"
        android:startColor="@color/mainColor" />

    <corners android:radius="5dip" />
    <stroke
        android:width="1dip"
        android:color="#EEEEEE" />
</shape>
```

在 shape 文件中给 Toast 的 View 设置渐变色、弧度和边框。

这样 CustomToast 类就封装好了，接下来在 MainActivity 中进行调用，查看效果。

调用自定义 Toast 显示在 Activity 顶部，如图 3-3 所示。

```
    CustomToast.getInstance().showToastCustom(MainActivity.this,"显示顶部",
Gravity.TOP);
```

调用自定义 Toast 显示在 Activity 中间，效果如图 3-4 所示。

```
    CustomToast.getInstance().showToastCustom(MainActivity.this,"显示中间",
Gravity.CENTER);
```

调用自定义 Toast 显示在 Activity 底部，效果如图 3-5 所示。

```
    CustomToast.getInstance().showToastCustom(MainActivity.this,"显示底部",
Gravity.BOTTOM);
```

自定义的 Toast 比 Android 自带的 Toast 好看，并且想显示什么效果时只需修改布局文件即可。这样产品经理更改需求时，程序员就不用慌了。

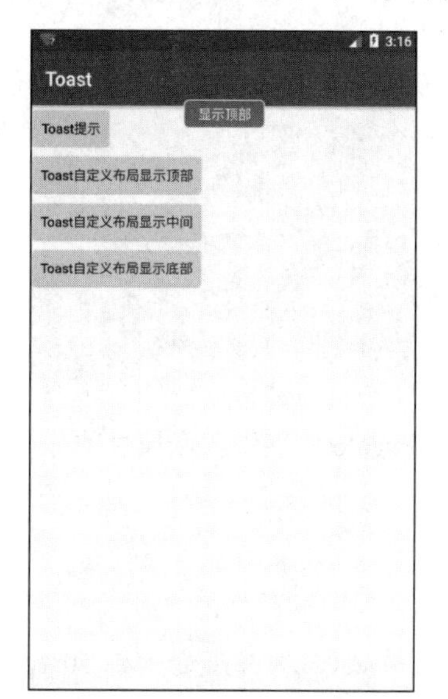

图 3-3　自定义 Toast 显示顶部

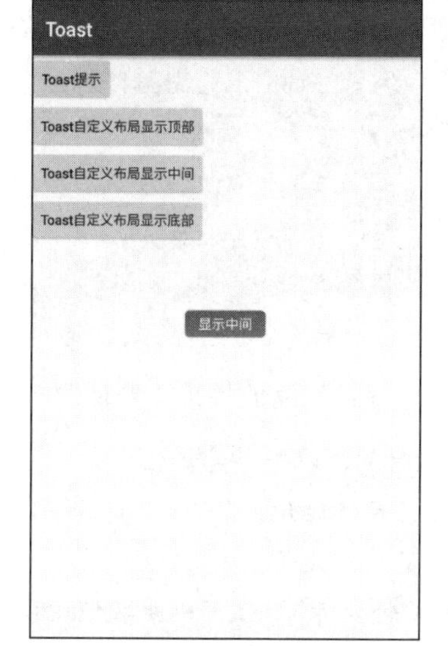

图 3-4　自定义 Toast 显示中间

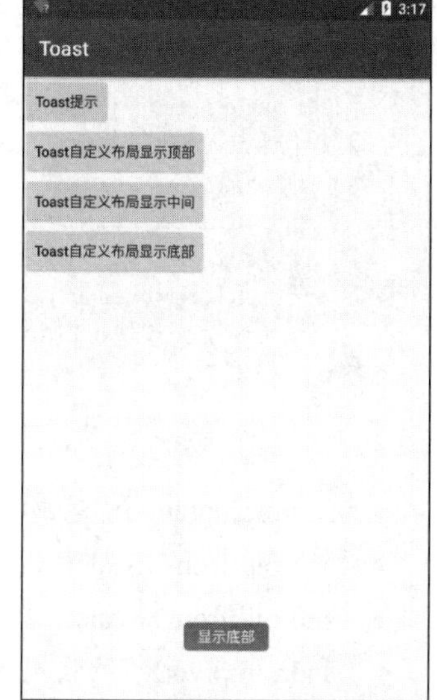

图 3-5　自定义 Toast 显示底部

3.1.4　Activity 启动与退出动画

在 App 中打开一个新的页面时，Activity 之间的切换是不可避免的，那么怎么才能让 Acitivity 的切换更优雅呢？在 Android 中提供了一个方法来解决这个问题，即 Activity 的 overridePendingTransition（enterAnim，exitAnim）方法，它有两个参数。

- 参数 enterAnim 表示的是从 Activity A 跳转到 Activity B，进入 B 时的动画效果。
- 参数 exitAnim 表示的是从 Activity A 跳转到 Activity B，离开 A 时的动画效果。

使用这个方法有两个注意事项：

- overridePendingTransition 方法需要在 startAtivity 方法或者 finish 方法调用之后立即执行。
- 若进入 B 或者离开 A 时不需要动画效果，则可以传值为 0。

Android 本身也自带了一些切换 Activity 的动画 xml 文件，例如从 Activity A 跳转到 Activity B 时，Activity B 从屏幕左边切入。

```
    Intent intent = new Intent(MainActivity.this, SecondActivity.class);
```

```
    startActivity(intent);
    //SecondActivity 从屏幕左边滑进来，使用自带的动画 xml
    overridePendingTransition(android.R.anim.slide_in_left,android.R.anim.slid
e_out_right);
```

当然，官方自带的动画效果总是有限的，Google 也不可能在 SDK 中集成很多的 Activity 切换动画效果，这时就需要写动画 xml 文件了，例如实现 iPhone 手机的进入和退出时的效果。

首先在 res 文件夹下新建 anim 文件夹，在 anim 文件夹下新建 zoomin.xml 文件，具体内容如下。

```
<?xml version="1.0" encoding="utf-8"?>
<set xmlns:android="http://schemas.android.com/apk/res/android"
   android:interpolator="@android:anim/decelerate_interpolator">
   <scale
      android:duration="@android:integer/config_mediumAnimTime"
      android:fromXScale="2.0"
      android:fromYScale="2.0"
      android:pivotX="50%p"
      android:pivotY="50%p"
      android:toXScale="1.0"
      android:toYScale="1.0"/>
</set>
```

其中有一个 scale 标签，这个标签起到缩放效果。顺便解释一下 scale 标签中各个属性的作用。

- android:duration：动画持续时间，以毫秒为单位。
- android:fromXScale：起始 X 尺寸比例。
- android:fromYScale：起始 Y 尺寸比例。
- android:pivotX：缩放起点 X 轴坐标，取值可以是数值（50）、百分数（50%）、百分数 p（50%p），当取值为数值时，缩放起点为 View 左上角坐标加上具体数值像素；当取值为百分数时，表示在当前 View 左上角坐标加上 View 宽度的具体百分比；当取值为百分数 p 时，表示在 View 左上角坐标加上父控件宽度的具体百分比。
- android:pivotY：同 android:pivotX。
- android:toXScale：最终 X 尺寸比例。
- android:toYScale：最终 Y 尺寸比例。

在 anim 文件夹下继续新建一个文件 zoomout.xml，内容如下：

```
<?xml version="1.0" encoding="utf-8"?>
<set xmlns:android="http://schemas.android.com/apk/res/android"
   android:interpolator="@android:anim/decelerate_interpolator"
   android:zAdjustment="top">

   <scale
      android:duration="@android:integer/config_mediumAnimTime"
      android:fromXScale="1.0"
      android:fromYScale="1.0"
      android:pivotX="50%p"
      android:pivotY="50%p"
      android:toXScale=".5"
      android:toYScale=".5" />
```

```
    <alpha
        android:duration="@android:integer/config_mediumAnimTime"
        android:fromAlpha="1.0"
        android:toAlpha="0"/>
</set>
```

除了 scale 标签之外，还增加了一个 alpha 标签，这个标签用于改变透明图。下面解释 alpha 标签各个属性的作用：

- android:duration：动画持续时间，以毫秒为单位。
- android:fromAlpha：动画开始的透明度，取值为 0.0~1.0，0.0 表示完全透明，1.0 表示保持原有状态不变。
- 动画最终的透明度取值和 android:fromAlpha 一样。

两个动画 xml 文件写完了，开启一个新的 Activity 时进行调用：

```
intent = new Intent(MainActivity.this, ThreeActivity.class);
startActivity(intent);
overridePendingTransition(R.anim.zoomin,R.anim.zoomout);
```

overridePendingTransition 方法第一次参数传入的是 zoomin，第二个参数传入的是 zoomout，也就是从 MainActivity 跳转到 ThreeActivity 时，ThreeActivity 进行放大。MainActivity 缩小并且改变透明度为 0，就是不可见。这样切换页面就是类似 iPhone 的进入和退出时的效果。

运行代码效果如图 3-6 与图 3-7 所示。图 3-6 是 Activity 切换时动画运行时，图 3-7 是 Activity 切换完成，也就是动画结束之后的效果。如果想看清切换动画的过程，则将动画时间延长一些，例如改成 5000 毫秒。

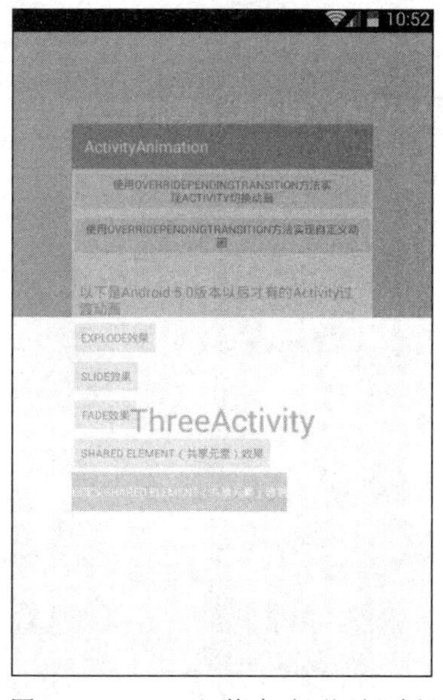

图 3-6　Activity 切换中动画运行过程

图 3-7　Activity 切换动画结束

Android 5.0 以后 Activity 之间切换动画

上面讲解通过 overridePendingTransition 方法，基本上可以满足日常中对 Activity 跳转动画的需求了。不过，炫酷的 MD（Material Design）风格出来之后，overridePendingTransition 这种老旧、生硬的方式不能适合 MD（Material Design）风格的 App 了。

好在 Google 在新的 SDK 中提供了另一种 Activity 的过渡动画——ActivityOptions，并且提供了兼容包 ActivityOptionsCompat。ActivityOptionsCompat 是一个静态类，提供了相应的 Activity 跳转动画效果，通过其可以实现不少炫酷的动画效果。

内置的 Activity 之间切换动画有以下几种：

- Explode 效果。
- Slide 效果。
- Fade 效果。
- Shared Element 效果。

（1）Explode 效果

设置 Explode 爆炸效果，从场景的中心移入或移出。实现起来也很简单：在项目的 res 文件夹下新建 transition 文件夹，并在 transition 文件夹下新建 explode.xml 文件，内容如下：

```
<explode xmlns:android="http://schemas.android.com/apk/res/android"
    android:duration="1000"/>
```

最外层是 explode 标签，仅 android:duration 一个属性，该属性用来设置动画的执行时间。

新建 ExplodeActivity 类，继承 AppCompatActivity。

```
public class ExplodeActivity extends AppCompatActivity {
    @Override
    protected void onCreate(Bundle savedInstanceState) {
        super.onCreate(savedInstanceState);

        if (Build.VERSION.SDK_INT>Build.VERSION_CODES.KITKAT_WATCH){
            //版本号大于 4.4
            Transition explode = TransitionInflater.from(this).
inflateTransition (R.transition.explode);
            getWindow().setEnterTransition(explode);//第一次进入时动画
        }

        setContentView(R.layout.activity_explode);

        findViewById(R.id.ll_root).setOnClickListener(new
View.OnClickListener() {
            @Override
            public void onClick(View v) {
                finish();
            }
        });
    }
}
```

主要了解 setContentView 前面几行代码，首先判断版本号是不是 5.0 以上，然后用 TransitionInflater 类初始化写好的 xml 文件，返回一个 Transition 对象，调用 Window 的 setEnterTransition 方法设置进去。

启动带爆炸效果的 Activity，跟我们之前启动 Activity 的方式稍微有点区别。例如，启动

ExplodeActivity，修改之后如下：

```
intent = new Intent(MainActivity.this,ExplodeActivity.class);
startActivity(intent,ActivityOptions.
makeSceneTransitionAnimation(MainActivity.this).toBundle());
```

startActivity 方法需要传入两个参数，第一个参数是 intent 对象，第二个参数是 Bundle 对象，用 ActivityOptions 类的 makeSceneTransitionAnimation 方法创建 ActivityOptions 对象，调用 ActivityOptions 对象的 toBundle 方法，将返回的 Bundle 对象传入进去。

运行代码，效果如图 3-8、图 3-9、图 3-10 所示。

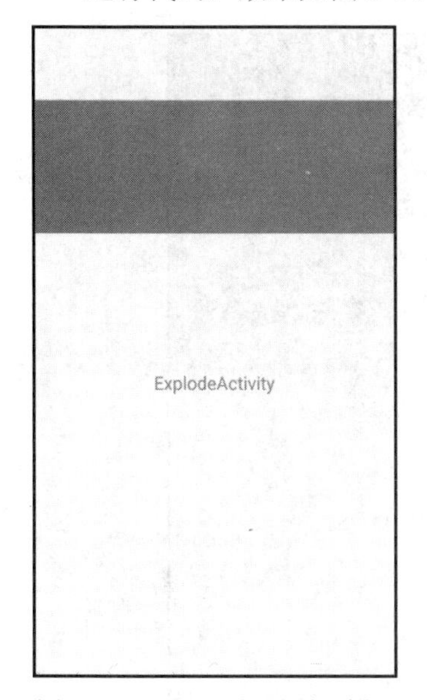

图 3-8　Explode 动画运行过程 1

图 3-9　Explode 动画运行过程 2

图 3-10　Explode 动画结束

（2）Slide 效果

Slide 使用与 Explode 相同的步骤，在 res/transition 文件夹下新建 slide.xml，内容如下：

```
<slide xmlns:android="http://schemas.android.com/apk/res/android"
    android:interpolator="@android:interpolator/decelerate_cubic"
    android:slideEdge="end"
    android:duration="1000"/>
```

- android:interpolators 属性设置插值器，就是用来控制滑动速度。这里使用 SDK 自带的 decelerate_cubic。
- android:slideEdge 有 4 个值，表示不同的方向滑动效果：end 表示右侧，start 表示左侧，top 表示顶部，bottom 表示底部。
- android:duration 设置动画执行时间。

如果不希望顶部的状态栏一起执行动画，可以在 xml 中根据控件 id 进行忽略，修改的 slide.xml 如下：

```
<slide xmlns:android="http://schemas.android.com/apk/res/android"
    android:duration="1000"
    android:interpolator="@android:interpolator/decelerate_cubic"
```

```
    android:slideEdge="end">
    <targets>
        <target android:excludeId="@android:id/statusBarBackground" />
    </targets>
</slide>
```

从 xml 中可以看到在 slide 标签中增加了 targets 标签，在 targets 标签中增加 target 标签，设置 android:excludeId 属性，属性的值就是忽略的控件 id。如果需要忽略多个 View，则一直增加 target 标签。

运行代码，其效果如图 3-11 与图 3-12 所示。

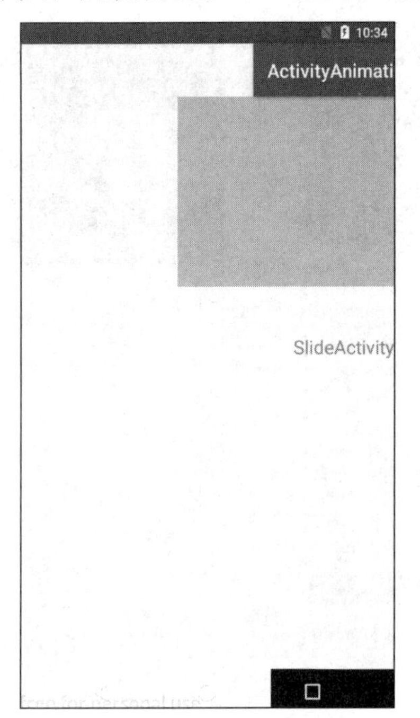

图 3-11　slide 动画运行过程

图 3-12　slide 动画结束

（3）Fade 效果

Fade 淡化效果使用的方法与前面两种效果类似。新建 fade.xml 文件，内容如下：

```
<fade xmlns:android="http://schemas.android.com/apk/res/android"
    android:duration="1000" />
```

Fade 效果就是将 View 从无到有，将透明度从 0 慢慢变成 1 的过程。这里就不贴效果图了，大家自己运行源码查看效果。

（4）Shared Element 效果

Shared Element 共享元素效果与前面几种效果不同，共享元素就是从 Activity A 过渡到 Activity B，可以指定元素过渡，给 View 增加 android:transitionName 属性。先查看运行效果图，如图 3-13、图 3-14、图 3-15 所示。

从效果图可以看到 Activity 切换过程中，顶部的两个 View 慢慢移动到底部，并且在移动的过程中会改变 View 的宽度。有了这种切换动画，用户可能感觉不到开启了一个新的界面，接下来看看是如何实现的。

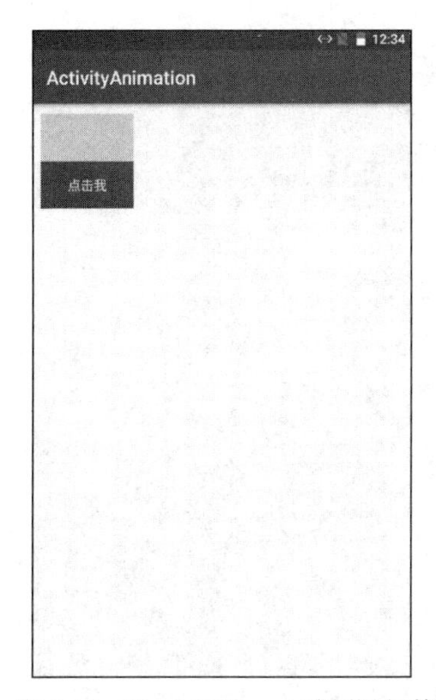

图 3-13　Shared Element 动画运行前　图 3-14　Shared Element 动画运行中　图 3-15　Shared Element 动画运行结束

首先查看 Activity A 的布局文件，也就是动画开始前显示的界面。这个文件很简单，即一个 View 和 TextView，需要注意的是给这两个 View 设置了 android:transitionName 属性，属性的值是一个字符串。

```
<?xml version="1.0" encoding="utf-8"?>
<LinearLayout xmlns:android="http://schemas.android.com/apk/res/android"
    android:layout_width="match_parent"
    android:layout_height="match_parent"
    android:id="@+id/rl_root"
    android:orientation="vertical"
    android:padding="10dp">

    <View
        android:id="@+id/view_share_one"
        android:layout_width="100dp"
        android:layout_height="50dp"
        android:background="@color/primary_light"
        android:transitionName="share_one"/>

    <TextView
        android:id="@+id/view_share_two"
        android:layout_width="100dp"
        android:layout_height="50dp"
        android:background="@color/colorPrimary"
        android:text="点击我"
        android:gravity="center"
        android:textColor="#FFFFFF"
        android:transitionName="share_two"/>
</LinearLayout>
```

SharedElementActivity.java（第一个 Activity）的内容如下：

```
public class SharedElementActivity extends AppCompatActivity implements
```

```
View.OnClickListener{
        private View viewShareOne;
        private TextView viewShareTwo;

        @Override
        protected void onCreate(Bundle savedInstanceState) {
            super.onCreate(savedInstanceState);
            setContentView(R.layout.activity_shared_element);

            viewShareOne=findViewById(R.id.view_share_one);
            viewShareTwo=findViewById(R.id.view_share_two);

            findViewById(R.id.rl_root).setOnClickListener(this);
            viewShareTwo.setOnClickListener(this);
        }

        @Override
        public void onClick(View view) {
            if(view.getId()==R.id.rl_root){
                finish();
            }else if(view.getId()==R.id.view_share_two){//开启过渡效果
                Intent intent = new Intent(this, SecondShareElemActivity.class);

                Pair onePair = new Pair<>(viewShareOne,
                 ViewCompat.getTransitionName(viewShareOne));
                Pair twoPair = new Pair<>(viewShareTwo,
                 ViewCompat.getTransitionName(viewShareTwo));

                ActivityOptionsCompat transitionActivityOptions =
                        ActivityOptionsCompat.makeSceneTransitionAnimation(
                                this, onePair, twoPair);
                ActivityCompat.startActivity(this,intent,
                  transitionActivityOptions.toBundle());
            }
        }
    }
```

这里需要注意的是，将两个 View 对象构建成两个 Pair 对象，通过 ActivityOptionsCompat 类的 makeSceneTransitionAnimation 静态方法得到 ActivityOptionsCompat 对象，这个方法的第二个参数是一个数组，可以传入多个 Pair。最后调用 ActivityCompat.startActivity 开启 Activity。

activity_second_share_elem.xml 是 Activity B 显示的布局文件，最外层是 RelativeLayout，第一个 View 是 Button，底部还有一个 LinearLayout，LinearLayout 中的两个 View 与 Activity A 的 View 差不多，改变了控件宽度。需要注意的是，这两个 View 同样设置了 android:transitionName 属性，这个值需要与 Activity A 显示的布局文件中的值进行对应，就是通过这个属性的值来判断从哪个 View 过渡到哪个 View。

```
<?xml version="1.0" encoding="utf-8"?>
<RelativeLayout xmlns:android="http://schemas.android.com/apk/res/android"
    android:id="@+id/rl_root"
    android:layout_width="match_parent"
    android:layout_height="match_parent"
    android:padding="10dp">
```

```
    <Button
        android:id="@+id/btn_close"
        android:layout_width="wrap_content"
        android:layout_height="wrap_content"
        android:layout_gravity="center_horizontal"
        android:text="点击我关闭当前界面"
        android:layout_centerHorizontal="true"
        android:layout_marginTop="100dp"
        android:textColor="@color/accent"
        android:textSize="25sp"/>

    <LinearLayout
        android:layout_width="match_parent"
        android:layout_height="wrap_content"
        android:layout_alignParentBottom="true"
        android:orientation="vertical">

        <View
            android:layout_width="match_parent"
            android:layout_height="100dp"
            android:background="@color/primary_light"
            android:transitionName="share_one" />

        <TextView
            android:layout_width="match_parent"
            android:layout_height="100dp"
            android:background="@color/colorPrimary"
            android:gravity="center"
            android:textColor="#FFFFFF"
            android:transitionName="share_two" />
    </LinearLayout>
</RelativeLayout>
```

SecondShareElemActivity.java（第二个 Activity）文件的 Activity 代码很少，需要注意的是，点击按钮时调用的是 finishAfterTransition 来关闭当前的 Activity，这样的目的是产生一个退出动画效果。

```
public class SecondShareElemActivity extends AppCompatActivity {
    @Override
    protected void onCreate(Bundle savedInstanceState) {
        super.onCreate(savedInstanceState);
        setContentView(R.layout.activity_second_share_elem);

        findViewById(R.id.btn_close).setOnClickListener(new
         View.OnClickListener() {
            @Override
            public void onClick(View v) {
                finishAfterTransition();//进行退出动画
            }
        });

        findViewById(R.id.rl_root).setOnClickListener(new
          View.OnClickListener() {
           @Override
```

```
            public void onClick(View view) {
                finish();
            }
        });
    }
}
```

3.1.5 Activity 销毁

每一个 Activity 都有自己的生命周期，打开了就要及时关闭，从而释放内存，也可以避免程序出现未知错误。

要关闭当前的 Activity，直接调用 finish 方法即可：

```
finish();
```

如果当前打开了多个 Activity，这时想退出程序怎么办呢？finish 只能退出当前的一个 Activity。还有两种方案，能够直接终止程序进程。

方法一：

```
System.exit(0);//使用系统的方法，强制退出，终止程序
```

方法二：

```
android.os.Process.killProcess(android.os.Process.myPid());//直接终止当前进程
```

这两个方法确实能把程序进程终止，但是会重新启动 App，并重新创建一个新的进程。因为 Android 系统认为 App 是被意外终止的（如内存不足），系统底层有监听服务，App 被意外终止会自动重启。

销毁所有的 Activity

当打开多个 Activity 时，系统提供销毁所有 Activity 的方法都会有一些问题，于是我们只能自己想办法去实现，解决方案就是把开启的 Activity 保存到栈中，退出软件时把栈中所有的 Activity 一一销毁。

首先新建一个 MyApplication 类，继承系统的 Application 类，其中增加几个方法，每个方法都有注释，就不解释了。重点了解 finishAllActivity 方法，因为每次打开一个新的 Activity 时，就会把 Activity 保存到栈中，因此退出程序时只需要将栈中的所有 Activity 退出来，调用 finish 方法。

```
public class MyApplication extends Application{
    private static Stack<Activity> activityStack;//activity 栈

    /**
     * 添加 Activity 到堆栈
     */
    public void addActivity(Activity activity) {
        if (activityStack == null) {
            activityStack = new Stack<>();
        }

        if(!activityStack.contains(activity)){
            Log.i("ansen","添加 Activity:"+activity.getLocalClassName());
```

```
            activityStack.add(activity);
        }
    }

    /**
     * 获取当前 Activity（堆栈中最后一个压入的）
     */
    public Activity currentActivity() {
        Activity activity = activityStack.lastElement();
        return activity;
    }

    public void removeActivity(Activity activity) {
        if (activity != null&&activityStack.contains(activity)) {
            Log.i("ansen","删除 Activity:"+activity.getLocalClassName());
            activityStack.remove(activity);
        }
    }

    /**
     * 结束所有 Activity
     */
    public void finishAllActivity() {
        for (int i = 0, size = activityStack.size(); i < size; i++) {
            if (null != activityStack.get(i)) {
                activityStack.get(i).finish();
            }
        }
        activityStack.clear();
        Log.i("ansen","结束所有 Activity");
    }
}
```

为什么要将保存 Activity 的栈放到 Application 中呢？因为 Application 的生命周期与应用程序是绑定的，这样能够最大地保证 Activity 栈静态变量不会被系统回收。

因为重写了 Application，需要在 AndroidManifest.xml 文件的 application 标签的 android:name 属性中指定重写的 Application。

```
<application
    ...
    android:name=".MyApplication">
    ...
</application>
```

合理地利用 Activity 的生命周期，Activity 创建（onCreate）与销毁（onDestroy）方法调用时将当前 Activity 从 Activity 栈中加入或者移除。我们不可能每次新建一个 Activity 就去重写 onCreate 和 onDestroy 方法，可以封装一个 BaseActivity，所有的 Activity 都继承自 BaseActivity 即可。然后增加 finishAllActivity 方法，销毁所有的 Activity。

```
public class BaseActivity extends AppCompatActivity {
    private MyApplication myApplication;

    @Override
```

```
    protected void onCreate(Bundle savedInstanceState) {
        super.onCreate(savedInstanceState);

        if (myApplication == null) {
            // 得到 Application 对象
            myApplication = (MyApplication) getApplication();
        }
        myApplication.addActivity(this);
    }

    //销毁所有 Activity 方法
    public void finishAllActivity() {
        myApplication.finishAllActivity();//调用 myApplication 销毁所有 Activity 方法
    }

    @Override
    protected void onDestroy() {
        super.onDestroy();

        myApplication.removeActivity(this);
    }
}
```

如果想在哪个 Activity 退出程序时销毁所有 Activity，只需要调用父类 BaseActivity 的 finishAllActivity 方法即可。

3.1.6 Activity 与 Activity 之间传递数据

1. 传递参数

打开一个新的页面时，可能需要把一些值传递过去，称为 Activity 传递参数。Activity 与 Activity 之间基本采用 Intent 传递参数，是通过 Bundle 来实现的。当调用 intent.putExtra 方法时，系统会创建一个 Bundle 对象，Bundle 对象有一个 ArrayMap，用来存储数据。

上一节通过显式 Intent 启动了 SecondActivity，现在要传递一个值给 SecondActivity，就要在原来的基础上增加一行代码。这里我们传递了一个字符串：

```
Intent intent=new Intent(this,SecondActivity.class);
//参数 1:key  参数 2:value
intent.putExtra("parameter","SecondActivity parameter");
startActivity(intent);
```

参数通过 Intent 传递过来了，现在需要在 SecondActivity 中获取这个值：

```
public class SecondActivity extends AppCompatActivity {
    @Override
    protected void onCreate(Bundle savedInstanceState){
        super.onCreate(savedInstanceState);
        setContentView(R.layout.activity_second);
        String value=getIntent().getStringExtra("parameter");
        Log.i("ansen",value);
    }
}
```

2. Activity 的传值与回传值

在 MainActivity 启动 FourActivity，将通过 startActivityForResult 方法来启动 Activity，并且包含第二个参数 requestCode，接下来会用到。

```
Intent fourIntent=new Intent(this,FourActivity.class);
fourIntent.putExtra("parameter","FourActivity parameter");
startActivityForResult(fourIntent,REQUEST_FOURACTIVITY_CODE);
```

还需要重写 MainActivity 的 onActivityResult 方法，通过 requestCode 来判断是不是自己启动的，然后通过 Intent 获取回传值。

```
@Override
protected void onActivityResult(int requestCode, int resultCode, Intent data) {
    if(requestCode==REQUEST_FOURACTIVITY_CODE&&resultCode==RESULT_OK){
        String resultStr=data.getStringExtra("result");
        Log.i("ansen",resultStr);
    }
    super.onActivityResult(requestCode, resultCode, data);
}
```

接下来查看 FourActivity 是如何写的，在 onCreate 中获取请求参数，重写 onKeyDown 方法。如果按了系统返回键，通过 setResult 方法将 Intent 回传给上一个 Activity。

```
public class FourActivity extends AppCompatActivity{
    @Override
    protected void onCreate(Bundle savedInstanceState){
        super.onCreate(savedInstanceState);
        setContentView(R.layout.activity_four);

        String value=getIntent().getStringExtra("parameter");
        Log.i("ansen",value);
    }

    @Override
    public boolean onKeyDown(int keyCode, KeyEvent event) {
        //系统返回键
        if (keyCode == KeyEvent.KEYCODE_BACK && event.getAction() ==
KeyEvent.ACTION_DOWN) {
            Intent intent = new Intent();
            intent.putExtra("result","FourActivityResultValue");//封装回传值
            setResult(RESULT_OK,intent);
            finish();//结束当前 Activity
        }
        return super.onKeyDown(keyCode, event);
    }
}
```

先启动 SecondActivity 并返回，再启动 FourActivity 并返回，打印 Log 截图，如图 3-16 所示。

```
01-17 18:16:41.874 29834-29834/com.ansen.activity I/ansen: onPause
01-17 18:16:41.894 29834-29834/com.ansen.activity I/ansen: SecondActivity parameter
01-17 18:16:42.394 29834-29834/com.ansen.activity I/ansen: onStop
01-17 18:16:44.604 29834-29834/com.ansen.activity I/ansen: onRestart
01-17 18:16:44.604 29834-29834/com.ansen.activity I/ansen: onStart
01-17 18:16:44.604 29834-29834/com.ansen.activity I/ansen: onResume
01-17 18:16:47.134 29834-29834/com.ansen.activity I/ansen: onPause
01-17 18:16:47.154 29834-29834/com.ansen.activity I/ansen: FourActivity parameter
01-17 18:16:47.534 29834-29834/com.ansen.activity I/ansen: onStop
01-17 18:16:49.234 29834-29834/com.ansen.activity I/ansen: FourActivityResultValue
01-17 18:16:49.234 29834-29834/com.ansen.activity I/ansen: onRestart
01-17 18:16:49.234 29834-29834/com.ansen.activity I/ansen: onStart
01-17 18:16:49.234 29834-29834/com.ansen.activity I/ansen: onResume
```

图 3-16　Activity 的传值与回传值

3.1.7　Activity 的软键盘弹出方式

在 AndroidManifest.xml 中给 Activiy 设置 android:windowSoftInputMode 属性，可以避免输入法面板遮挡输入框的问题。

android:windowSoftInputMode 属性有以下值：

- stateUnspecified：软键盘的状态并没有指定，系统将选择一个合适的状态或依赖于主题的设置。
- stateUnchanged：当这个 Activity 出现时，软键盘将一直保持在上一个 Activity 中的状态，无论是隐藏还是显示。
- stateHidden：用户选择 Activity 时，软键盘总是被隐藏。
- stateAlwaysHidden：当该 Activity 主窗口获取焦点时，软键盘也总是被隐藏的。
- stateVisible：软键盘通常是可见的。
- stateAlwaysVisible：用户选择 Activity 时，软键盘总是显示的状态。
- adjustUnspecified：默认设置，通常由系统自行决定是隐藏或显示。
- adjustResize：该 Activity 总是调整屏幕的大小，以便留出软键盘的空间。
- adjustPan：当前窗口的内容将自动移动，以便当前焦点从不被键盘覆盖和用户总是能够看到输入内容的部分。

可以设置一个值或者多个值，多个用“|”分割，例如：

```
android:windowSoftInputMode="stateVisible|adjustResize"
```

下面举一个实例，在布局文件的底部放置一个 EditText 控件，软键盘弹出时的效果图如图 3-17 所示。从效果图中可以看到软键盘弹出时标题栏被顶上去看不到了。

这种情况下就需要用到 android:windowSoftInputMode 了。设置 adjustResize，软键盘弹出，并调整屏幕的大小。

```
<activity android:name=".MainActivity"android:windowSoftInputMode=
"adjustResize"/>
```

运行代码，从效果图中可以看到软键盘弹出，标题栏正常显示，如图 3-18 所示。

图 3-17 设置软键盘的弹出方式

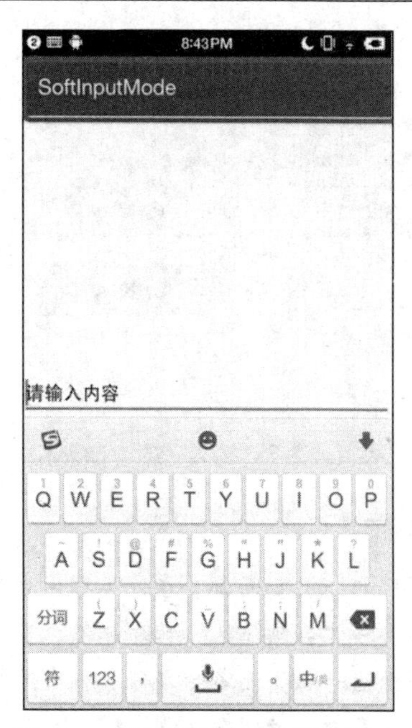

图 3-18 软键盘弹出时标题正常显示

还有其他的值，大家在实际开发中根据需求进行设置即可。

3.1.8 Activity 任务栈

Android 任务栈有以下特点：

- Android 任务栈又称为 Task，是一个栈结构，具有后进先出的特性，用于存放我们的 Activity 组件。
- 我们每次打开一个新的 Activity 或者销毁 Activity 都会在任务栈中增加一条记录或者减少一条记录，任务栈保存 Activity 集合。
- 任务栈可以移动到后台，保留每一个 Activity 的状态，有序地给用户列出它们的任务，并且不丢失状态信息。
- 当把所有任务栈中的 Activity 清除出栈时，任务栈会被销毁，程序退出。

Android 系统通过任务栈有序地管理每个 Activity，并决定用哪个 Activity 跟用户交互，只有任务栈顶的 Activity 才能跟用户进行交互。

图 3-19 显示了 Activity 任务栈的执行过程，从中可以看到任务栈中只有 Activity 1，然后开启了 Activity 2，之后又开启了 Activity3，这时点击系统返回键，把 Activity 3 从任务栈中移除，于是留下了 Activity2 跟 Activity1。

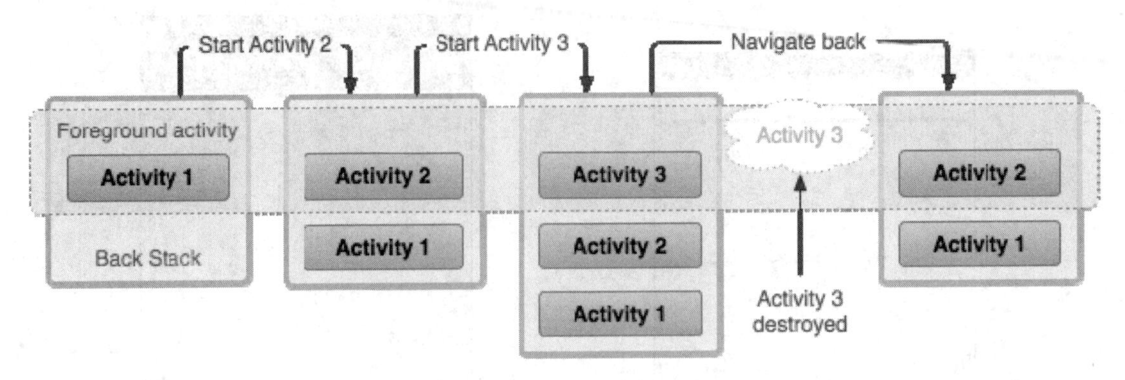

图 3-19 Activity 任务栈执行过程

任务栈的缺点：

- 每开启一次页面都会在任务栈中添加一个 Activity，只有任务栈中的 Activity 全部清除出栈时，任务栈才会被销毁，程序才算真正退出，需要点击多次返回键才能退出程序。
- 每开启一次页面都会在任务栈中添加一个 Activity，还会造成数据冗余，重复数据太多，会导致内存溢出的问题（OOM）。

为了解决任务栈的缺点，我们引入了启动模式。

3.1.9 Activity 四种启动模式

启动模式（launchMode）在多个 Activity 跳转的过程中扮演着重要的角色，可以决定是否生成新的 Activity 实例，是否重用已存在的 Activity 实例，是否和其他 Activity 实例共用一个 task。这里简单介绍一下 task 的概念。task 是一个具有栈结构的对象，一个 task 可以管理多个 Activity，启动一个应用，也就创建一个与之对应的 task。

Activity 一共有四种 launchMode：

- standard：系统默认的启动模式，即标准模式。
- singleTop：栈顶复用模式。
- singleTask：栈内复用模式。
- singleInstance：全局唯一模式。

怎么使用呢？很简单，只需要在 AndroidManifest.xml 文件中给 activity 设置 android:launchMode 属性即可。参考如下代码：

```
<activity android:name=".MainActivity" android:launchMode="standard">
    ....
</activity>
```

这四种启动模式对应不同的跳转模式。接下来详细介绍一下。

1. standard（标准模式）

standard 是 Activity 默认的启动方式，如果你需要这种启动方式可以不需要设置

android:launchMode 属性。这种模式是每次启动一个 Activiy 都会创建一个新的实例，不管这个例之前是否存在，这种模式下，谁启动了 Activity，该 Activity 就属于启动该 Activity 的任务栈。

stardard 启动模式效果如图 3-20 所示。

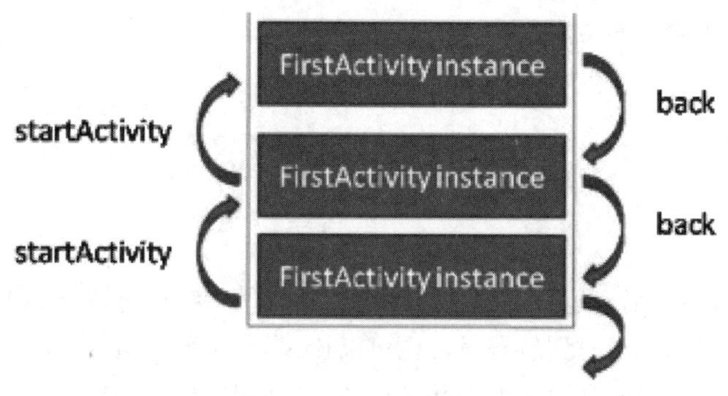

图 3-20 stardard 模式

2. singleTop（栈顶复用模式）

这种模式下，如果新打开的 Activity 已经在栈顶了，那就不会重新创建 Activity 实例，只会调用 onNewIntent 方法，如果新打开的 Activity 不在栈顶，而是在栈底或者栈中间，还是会创建一个新的实例。

3. singleTask（栈内复用模式）

singleTask 模式下，如果打开一个新的 Activity，这个 Activity 在栈中存在，就会把这个 Activity 之上的 Activity 都销毁，然后这个 Activity 就会置顶。

假设现在有三个 Activity，即 Activity1、Activity2、Activity3，给 Activity1 设置成 singleTask 模式。Activity1 启动 Activity2，Activity2 启动 Activity3。这时任务栈效果如图 3-21 左边的情况，里面有三个 Activity，栈顶是 Activity3，这个时候我们开启 Activity1，运行之后效果如图 3-21 右边所示，任务栈只有 Activity1 还会存在。

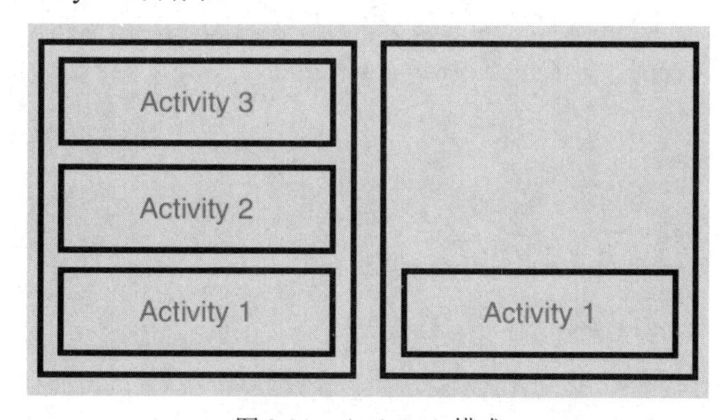

图 3-21 singleTask 模式

使用 singleTask 模式有以下两点需要注意：

- 如果是其他 App 以 singleTask 模式启动 Activity1，将会创建一个新的任务栈。
- 如果以 singleTask 模式启动的 Activity1 已经在后台的一个任务栈中，那么启动后，后台的任务栈一起切换到前台。

4．singleInstance（全局唯一模式）

singleInstance 模式比较特殊，这种模式下的 Activity 会单独占用一个栈，这个栈在整个系统中是唯一的。不同的 App 去打开 singleInstance 模式的 Activity，如果这个实例存在，不会重新创建。系统中永远只会有一个这样的实例。

3.2 Service（服务）

Service 是在 Android 后台运行的组件，能够在后台做一些耗时较长的事情，这样就不会影响应用体验了。另外，当程序退出到后台时，也可以让 Service 继续运行。例如，首先打开了一个音乐播放器，这时又想打开微信刷朋友圈，并且让音乐播放器继续播放音乐，就可以用 Service 在后台播放音乐了。

3.2.1 Activity 中启动 Service 以及销毁 Service

首先新建一个 LocalService 类，继承 Service，重写 onCreate、onStartCommand、onDestroy 方法。在每个方法中都打印 Log。

```
public class LocalService extends Service{
    @Override
    public void onCreate() {
        Log.i("LocalService","onCreate");
        super.onCreate();
    }

    @Override
    public int onStartCommand(Intent intent, int flags, int startId) {
        Log.i("LocalService","onStartCommand start id " + startId);
        return super.onStartCommand(intent, flags, startId);
    }

    @Override
    public void onDestroy() {
        super.onDestroy();
        Log.i("LocalService","onDestroy");
    }

    @Override
    public IBinder onBind(Intent intent) {
        return null;
    }
}
```

Service 组件和 Activity 一样，一定要记住在 AndroidManifest.xml 文件中注册：

```
<service android:name="com.ansen.service.LocalService"/>
```

接下来，在 activity_main.xml 文件中添加两个按钮：

```
<?xml version="1.0" encoding="utf-8"?>
<LinearLayout xmlns:android="http://schemas.android.com/apk/res/android"
    android:layout_width="match_parent"
    android:layout_height="match_parent"
    android:orientation="vertical">

    <Button
        android:id="@+id/btn_start_service"
        android:layout_width="match_parent"
        android:layout_height="wrap_content"
        android:text="Start Service" />

    <Button
        android:id="@+id/btn_stop_service"
        android:layout_width="match_parent"
        android:layout_height="wrap_content"
        android:text="Stop Service"/>
</LinearLayout>
```

在 MainActivity 中设置两个点击事件，这种代码前面介绍很多次了，大家应该非常熟悉。通过 startService 方法启动服务，通过 stopService 方法停止服务，两个方法都是传入一个 Intent 对象。

```
public class MainActivity extends AppCompatActivity implements
View.OnClickListener{
    @Override
    protected void onCreate(Bundle savedInstanceState) {
        super.onCreate(savedInstanceState);
        setContentView(R.layout.activity_main);

        findViewById(R.id.btn_start_service).setOnClickListener(this);
        findViewById(R.id.btn_stop_service).setOnClickListener(this);
    }

    @Override
    public void onClick(View v) {
        switch (v.getId()){
            case R.id.btn_start_service://启动 Service
                Intent startIntent=new Intent(this,LocalService.class);
                startService(startIntent);
                break;
            case R.id.btn_stop_service://停止 Service
                Intent stopIntent=new Intent(this,LocalService.class);
                stopService(stopIntent);
                break;
        }
    }
}
```

接下来是点击按钮，通过 Log 来分析 Service 的执行过程。第一次点击 Start Service 按钮时，调用了 Service 的 onCreate 和 onStartCommand 方法。

```
01-19 13:40:42.950 14814-14814/com.ansen.service I/LocalService: onCreate
01-19 13:40:42.950 14814-14814/com.ansen.service I/LocalService:
```

```
onStartCommand start id 1
```

再次点击 Start Service 按钮，仅调用了 onStartCommand 方法，startId 从之前的 1 变成了 2。从这个打印的 Log 中可以知道 Service 启动了就保存在后台，第二次启动只会调用 onStartCommand 方法，只要 Service 没有被销毁，不管启动多少次都不会调用 onCreate，通过 startId 可以知道启动了多少次。

```
    01-19 13:46:33.970 14814-14814/com.ansen.service I/LocalService:
onStartCommand start id 2
```

点击 Stop Service 按钮，onDestroy 运行了，Service 就销毁了。

```
    01-19 14:02:38.700 14814-14814/com.ansen.service I/LocalService: onDestroy
```

提　示

Service 与 Thread 是什么关系？

其实 Service 与 Thread 没有任何关系，之所以有不少人会将它们联系起来，主要就是因为 Service 的后台概念。Thread 用于开启一个子线程，在这里去执行一些耗时操作而不会阻塞主线程的运行。而 Service 最初理解时，总会觉得它是用来处理一些后台任务的，一些比较耗时的操作也可以放在这里运行，就会让人产生混淆了。但是，如果告诉你 Service 其实是运行在主线程中的，还会觉得它和 Thread 有什么关系吗？

改动 3.2.1 节中的示例代码，在 MainActivity 的 onCreate 方法中加一句代码打印当前线程 id：

```
    Log.i("MainActivity","onCreate ThreadId:"+android.os.Process.myTid());
```

同样，在 LocalService 的 onCreate 方法中也打印当前线程 id：

```
    Log.i("LocalService","onCreate ThreadId:"+android.os.Process.myTid());
```

重新运行程序，点击 Start Service 按钮，打印 Log 如下，可以看到线程 id 是一样的。由此证实了 Service 确实是运行在主线程中的，也就是说，如果在 Service 中编写了非常耗时的代码，程序会出现“应用程序无响应”对话框。如何解决这个问题呢？可以在 Service 中新建一个线程，执行耗时比较大的工作，例如音乐播放器下载歌曲，就可以在 Service 中开启一个线程去做这件事情。

```
    01-19 14:10:02.160 24514-24514/com.ansen.service I/MainActivity: onCreate
ThreadId:24514
    01-19 14:10:04.050 24514-24514/com.ansen.service I/LocalService: onCreate
ThreadId:24514
```

提　示

为什么还需要 Service?

我们都知道在主线程可以用 Thread 开启一个线程去执行一些耗时的操作，这样不会阻塞 UI，但是如果当前的 Activity 销毁了，就没有办法获取 Thread 的实例，也就不能再去操作那个 Thread。这样 Thread 的生命周期和 Activity 就绑定在一起了。Service 不一样，只要启动了没有销毁，就一直存在后台中，多个 Activity 能和一个 Service 进行关联调用。

3.2.2 Activity 与 Service 通信

细心的读者会发现，LocalService类中还有一个onBind方法，只要重写了Service类，就必须要实现这个方法。这个方法有一个返回值，是一个IBinder对象。Activity和Service通信就是通过这个对象来实现的。

我们继续修改LocalService，其改动如下：

- 编写一个内部类LocalBinder，继承自Binder，增加一个getService方法返回当前Service对象。
- 实例化LocalBinder，在onBind方法中将LocalBinder类的实例返回。
- 增加downMusic方法模拟下载音乐，提供Activity调用。

```
public class LocalService extends Service{
    private final IBinder mBinder = new LocalBinder();

    @Override
    public void onCreate() {
        Log.i("LocalService","onCreate ThreadId:"+android.os.Process.myTid());
        super.onCreate();
    }

    @Override
    public int onStartCommand(Intent intent, int flags, int startId) {
        Log.i("LocalService","onStartCommand start id " + startId);
        return super.onStartCommand(intent, flags, startId);
    }

    @Override
    public void onDestroy() {
        super.onDestroy();
        Log.i("LocalService","onDestroy");
    }

    @Override
    public IBinder onBind(Intent intent){
        Log.i("LocalService","onBind");
        return mBinder;
    }

    //下载音乐的方法
    public void downMusic(){
        //这里我只是打印了一个Log，如果在开发中需要下载音乐，开一个线程去下载
        Log.i("LocalService","downMusic");
    }

    public class LocalBinder extends Binder{
        //返回当前的Service对象
        LocalService getService() {
            return LocalService.this;
        }
    }
}
```

继续修改 activity_main.xml 布局文件，增加 Bind Service 和 Unbind Service 两个按钮：

```
<?xml version="1.0" encoding="utf-8"?>
<LinearLayout xmlns:android="http://schemas.android.com/apk/res/android"
    android:layout_width="match_parent"
    android:layout_height="match_parent"
    android:orientation="vertical">

    <Button
        android:id="@+id/btn_start_service"
        android:layout_width="match_parent"
        android:layout_height="wrap_content"
        android:text="Start Service" />

    <Button
        android:id="@+id/btn_stop_service"
        android:layout_width="match_parent"
        android:layout_height="wrap_content"
        android:text="Stop Service"/>

    <Button
        android:id="@+id/btn_bind_service"
        android:layout_width="match_parent"
        android:layout_height="wrap_content"
        android:text="Bind Service" />

    <Button
        android:id="@+id/btn_unbind_service"
        android:layout_width="match_parent"
        android:layout_height="wrap_content"
        android:text="Unbind Service"/>
</LinearLayout>
```

接下来查看 MainActivity，如果点击了 Bind Service 按钮，通过 bindService 方法将 Activity 和 Service 绑定起来，它有三个参数：

- 参数一：Intent 实例。
- 参数二：ServiceConnection 接口实例，Activity 和 Serivice 绑定与解除绑定都会调用它。通过内部类方式实例化 ServiceConnection，其中重写了两个方法，onServiceConnected 是 Activity 和 Service 连接时调用，onServiceDisconnected 是解绑时调用。在 onServiceConnected 方法中，通过 binder 来获取连接的 Service 实例。拿到了 Service 实例之后，可以调用 Service 中的任何 public 方法。
- 参数三：一个标志位，有很多值，例如 BIND_AUTO_CREATE、BIND_DEBUG_UNBIND 和 BIND_NOT_FOREGROUND 等。其中，BIND_AUTO_CREATE 表示当收到绑定请求时，如果 Service 尚未创建，就即刻创建。

绑定成功状用 mIsBound 成员变量来标示，在 MainActivity 中可以通过这个变量来判断是否绑定了 Service。

Unbind Service 按钮的代码比较简单，通过 mIsBound 判断是否绑定。如果绑定了调用 unbindService 方法解绑，传入 ServiceConnection 实例。

我们还重写了Activity的onDestroy方法，调用unbindService方法解绑。如果忘记手动解绑了，Activity销毁时也能解绑。

如果没有重写，而且又绑定Service，MainActivity销毁时会报错“Activity has leaked ServiceConnection that was originally bound here”，所以必须要解绑。

```
    public class MainActivity extends AppCompatActivity implements
View.OnClickListener{
        private LocalService localService;//service对象
        private boolean mIsBound;//是否绑定

        ServiceConnection serviceConnection=new ServiceConnection() {
            //Activity与Service绑定时调用
            @Override
            public void onServiceConnected(ComponentName name, IBinder binder) {
                localService=((LocalService.LocalBinder)binder).getService();
                localService.downMusic();//调用下载音乐的方法
            }

            //Activity与Service解绑时调用
            @Override
            public void onServiceDisconnected(ComponentName name) {
                localService=null;
            }
        };

        @Override
        protected void onCreate(Bundle savedInstanceState) {
            super.onCreate(savedInstanceState);
            setContentView(R.layout.activity_main);

            Log.i("MainActivity","onCreate ThreadId:"+android.os.Process.myTid());

            findViewById(R.id.btn_start_service).setOnClickListener(this);
            findViewById(R.id.btn_stop_service).setOnClickListener(this);

            findViewById(R.id.btn_bind_service).setOnClickListener(this);
            findViewById(R.id.btn_unbind_service).setOnClickListener(this);
        }

        @Override
        public void onClick(View v) {
            switch (v.getId()){
                case R.id.btn_start_service://启动Service
                    Intent startIntent=new Intent(this,LocalService.class);
                    startService(startIntent);
                    break;
                case R.id.btn_stop_service://停止Service
                    Intent stopIntent=new Intent(this,LocalService.class);
                    stopService(stopIntent);
                    break;
                case R.id.btn_bind_service://绑定Service
                    Intent bindIntent=new Intent(this,LocalService.class);
                    bindService(bindIntent,serviceConnection,BIND_AUTO_CREATE);
                    mIsBound = true;
```

```
                break;
            case R.id.btn_unbind_service://取消绑定 Service
                unbindService();
                break;
        }
    }

    private void unbindService(){
        if(mIsBound){
            unbindService(serviceConnection);
            mIsBound=false;
        }
    }

    @Override
    protected void onDestroy() {
        super.onDestroy();
        unbindService();
    }
}
```

运行代码，首先点击 Bind Service 按钮，打印 log 内容如下，从 log 中看到调用了 LocalService 类的 onCreate、onBind 方法，并没有调用 onStartCommand 方法，downMusic 是 ServiceConnection 的 onServiceConnected 方法中调用的。

```
    01-23 10:46:06.261 28645-28645/com.ansen.service I/LocalService: onCreate
ThreadId:28645
    01-23 10:46:06.261 28645-28645/com.ansen.service I/LocalService: onBind
    01-23 10:46:06.261 28645-28645/com.ansen.service I/LocalService: downMusic
```

点击 Unbind Service 按钮，打印 log 如下，调用 Service 的 onDestroy 方法，说明 service 已销毁。

```
    01-23 10:49:08.061 28645-28645/com.ansen.service I/LocalService: onDestroy
```

如果同时启动 Service，又绑定 Service 会发生什么情况呢？首先点击 Start Service 按钮，再点击 Bind Service 按钮，从 Log 中可以得出一个结论，Service 的 onCreate 只会执行一次，也就是说后台只有一个 Service。

```
    01-24 17:35:18.741 4696-4696/com.ansen.service I/LocalService: onCreate
ThreadId:4696
    01-24 17:35:18.741 4696-4696/com.ansen.service I/LocalService: onStartCommand
start id 1
    01-24 17:35:23.361 4696-4696/com.ansen.service I/LocalService: onBind
    01-24 17:35:23.361 4696-4696/com.ansen.service I/LocalService: downMusic
```

继续点击 Stop Service 按钮，发现 Log 中什么日志也没有打印，然后点击 Unbind Service，打印执行 onDestroy 方法的 Log。也就是说，如果既调用了 startService 方法又调用了 bindService 方法，就必须调用 stopService 和 unbindService 才能销毁这个 Service。

```
    01-24 17:39:31.641 4696-4696/com.ansen.service I/LocalService: onDestroy
```

3.3 Broadcast Receiver（广播接收器）

Broadcast Receiver 用来发送或接收广播。Android 应用程序可以发送或接收来自 Android 系统和其他 Android 应用程序的广播消息，类似于发布订阅设计模式。例如，Android 系统在各种系统事件发生时发送广播（如系统启动或设备开始充电时）。应用程序还可以发送自定义广播，例如通知其他应用程序的内容（一些新的数据已被下载等）。应用程序可以注册接收特定的广播。当发送广播时，系统自动将广播路由到订阅该特定类型广播的应用程序。一般来说，广播可以作为跨应用程序跟应用内的消息传递系统。

3.3.1 动态注册广播

动态注册广播是一种灵活的注册方式，通过代码来注册广播或销毁广播。

首先新建 DynamicBroadcast 类，继承 BroadcastReceiver 类，用来接收广播，重写 onReceive 方法，通过 onReceive 方法 intent 参数可以获取发送广播时传入的参数。

```
public class DynamicBroadcast extends BroadcastReceiver {
    @Override
    public void onReceive(Context context, Intent intent){
        String data = intent.getStringExtra("data");
        Log.i("data",data);
    }
}
```

在 activity_main.xml 文件中增加一个按钮：

```
<?xml version="1.0" encoding="utf-8"?>
<LinearLayout xmlns:android="http://schemas.android.com/apk/res/android"
    android:layout_width="match_parent"
    android:layout_height="match_parent"
    android:orientation="vertical">
    <Button
        android:id="@+id/btn_dynamic_broadcast_send_message"
        android:layout_width="wrap_content"
        android:layout_height="wrap_content"
        android:text="给动态注册的广播发送消息"/>
</LinearLayout>
```

最后查看一下 MainActivity.java 文件。

```
public class MainActivity extends AppCompatActivity implements
View.OnClickListener{
    public static final String ACTION_DYNAMIC_BROADCAST=
        "android.intent.action.DYNAMIC_BROADCAST";
    private DynamicBroadcast dynamicBroadcast;

    @Override
    protected void onCreate(Bundle savedInstanceState) {
```

```
        super.onCreate(savedInstanceState);
        setContentView(R.layout.activity_main);

        findViewById(R.id.btn_dynamic_broadcast_send_message).
         setOnClickListener(this);

        //动态注册广播
        dynamicBroadcast=new DynamicBroadcast();
        IntentFilter intentFilter=new IntentFilter(ACTION_DYNAMIC_BROADCAST);
        registerReceiver(dynamicBroadcast,intentFilter);
    }

    @Override
    public void onClick(View v){
        switch (v.getId()){
            case R.id.btn_dynamic_broadcast_send_message:
                Intent intent = new Intent(ACTION_DYNAMIC_BROADCAST);
                intent.putExtra("data","Dynamic Broadcast Parameter");
                          //通过 intent 传递参数
                sendBroadcast(intent);//发送广播消息
                break;
        }
    }

    @Override
    protected void onDestroy() {
        super.onDestroy();
        Log.i("MainActivity onDestroy","销毁广播");
        unregisterReceiver(dynamicBroadcast);
    }
}
```

在 onCreate 中通过 registerReceiver 方法注册一个广播，需要 BroadcastReceiver 和 IntentFilter 对象两个参数。同时给“发送广播”按钮设置点击监听，点击之后通过 sendBroadcast 方法发送广播，这里需要一个 Intent 对象，构造 Intent 对象时传入 Action，这个 Action 与注册广播时的 Action 一致。

我们还重写了 onDestroy 方法，当 Activity 销毁时也销毁广播，因此本例中广播的生命周期和 Activity 是一样的。

运行代码，点击“给动态注册的广播发送消息”按钮，Log 打印如下：

```
    02-03 16:32:11.194 7095-7095/com.ansen.broadcastreceiver I/data: Dynamic
Broadcast Parameter
```

3.3.2 静态注册广播

静态注册广播是在 AndroidManifest.xml 文件中注册的，无论这个程序是否启动，都会接收到这个广播。

在动态注册广播的 Demo 上增加代码，新建 StaticBroadcast 类，继承 BroadcastReceiver，实现 onReceive 方法，和动态广播的接收器代码几乎一样。

```
public class StaticBroadcast extends BroadcastReceiver{
    @Override
    public void onReceive(Context context, Intent intent){
        String data = intent.getStringExtra("data");
        Log.i("data",data);
    }
}
```

接下来需要在AndroidManifest.xml文件中注册这个广播，通过receiver标签的name属性指定类，再增加intent-filter标签，给action标签设置name属性值，发送广播时需要用到此值。

```
<receiver android:name=".StaticBroadcast" android:exported="true">
    <intent-filter>
        <action android:name="android.intent.action.STATIC_BROADCAST"/>
    </intent-filter>
</receiver>
```

在activity_main.xml文件中增加一个"给静态注册的广播发送消息"的按钮。

```
<?xml version="1.0" encoding="utf-8"?>
<LinearLayout xmlns:android="http://schemas.android.com/apk/res/android"
    android:layout_width="match_parent"
    android:layout_height="match_parent"
    android:orientation="vertical">
    <Button
        android:id="@+id/btn_dynamic_broadcast_send_message"
        android:layout_width="wrap_content"
        android:layout_height="wrap_content"
        android:text="给动态注册的广播发送消息"/>
    <Button
        android:id="@+id/btn_static_broadcast_send_message"
        android:layout_width="wrap_content"
        android:layout_height="wrap_content"
        android:text="给静态注册的广播发送消息"/>
</LinearLayout>
```

在MainActivity中给"静态注册的广播发送消息"的按钮设置点击监听事件，在对应的onClick方法中发送广播，从如下代码中可以看到Intent的构造方法有传入一个字符串，这个值与在XML中receiver标签→intent-filter→action的name属性的值必须一致。在广播底层源码中就是通过action来区分不同的广播接收者。

```
Intent staticIntent = new Intent("android.intent.action.STATIC_BROADCAST");
staticIntent.putExtra("data","Static Broadcast Parameter");
    //通过intent传递参数
sendBroadcast(staticIntent);//发送广播消息
```

因为没有增加很多代码，MainActivity的代码就不全部贴出来了，重新运行代码，点击"给静态注册的广播发送消息"按钮，打印的Log如下：

```
02-06 14:02:38.735 10749-10749/com.ansen.broadcastreceiver I/data: Static
Broadcast Parameter
```

3.3.3 广播基本总结

1. 动态注册广播与静态注册广播的区别

前面写了一个 Demo，也介绍了动态注册广播与静态注册广播，这里再总结一下。

- 动态注册：广播的生命周期自己灵活控制，消耗资源少。
- 静态注册：广播一直存在，除非卸载软件。消耗资源稍微大一些。当然，现在的手机硬件都跟得上了，这点资源可以忽略不计。

2. 广播注意事项

我们都知道收到了广播就会执行 onReceive 方法，但是在这个方法中不能进行耗时超过 10 秒的事情，否则会弹出“应用程序无响应”（ANR，Application Not Response）对话框。如果有特殊需要，可以再启动一个 Thread 处理耗时操作。

3.3.4 应用内广播 LocalBroadcastManager

在 Android 系统中，BroadcastReceiver 设计的初衷是从全局考虑，能够方便应用程序和系统、应用程序之间、应用程序内的通信。因此，对单个应用程序而言，BroadcastReceiver 是存在安全性问题的（恶意程序脚本不断地去发送你所接收的广播）。为了解决这个问题，我们就需要使用应用内广播 LocalBroadcastManager。

LocalBroadcastManager 是 Android Support 包提供的一个工具，用来在同一个 App 内的不同组件之间发送 Broadcast，可以解决 BroadcastReceiver 的安全问题（恶意程序脚本不断地发送所接收的广播）。

使用 LocalBroadcastManager 具有以下好处：

- 发送的广播只会在自己 App 内传播，不会泄露给其他 App，确保隐私数据不会泄露。
- 其他 App 也无法向你的 App 发送该广播，不用担心其他 App 会来搞破坏。
- 比系统全局广播更加高效。

LocalBroadcastManager 的使用方法与动态注册广播类似。首先需要获取 LocalBroadcastManager 对象，通过 getInstance 方法（单例模式）获取，然后调用 registerReceiver 方法。

```
    broadcastManager = LocalBroadcastManager.getInstance(this);
    localReceiver=new LocalBroadcastReceiver();
    broadcastManager.registerReceiver(localReceiver,new
IntentFilter(ACTION_LOCAL_BROADCAST));
```

发送广播也类似。这里一定需要调用 LocalBroadcastManager 对象的 sendBroadcast 方法来发送广播，不然接收不到广播。

```
    Intent localIntent=new Intent(ACTION_LOCAL_BROADCAST);
    localIntent.putExtra("data","Local Broadcast Parameter");//通过 intent 传递参数
    LocalBroadcastManager.getInstance(this).sendBroadcast(localIntent);
```

顺便在 onDestroy 方法中对广播进行反注册。

```
@Override
protected void onDestroy(){
    super.onDestroy();
    Log.i("MainActivity onDestroy","销毁广播");
    unregisterReceiver(dynamicBroadcast);
    broadcastManager.unregisterReceiver(localReceiver);
}
```

因为和动态注册广播类似，所以这里仅贴出了关键代码。

3.4　ContentProvider（内容提供者）

ContentProvider 是 Android 中的四大组件之一，主要用于对外共享数据，也就是通过 ContentProvider 将应用中的数据共享给其他应用访问，其他应用可以通过 ContentProvider 对指定应用中的数据进行操作。

下面介绍如何使用 ContentProvider 获取手机联系人列表。联系人 App 是每个 Android 手机都会内置的应用程序，这些联系人数据保存在 sqlite 数据库中。如果你的应用程序想获取联系人数据，就可以通过 ContentProvider 方式获取。

这里需要借助 ContentResolver 类，在 Activity 中调用 getContentResolver 方法即可获取。下面查询联系人列表，所以调用 ContentResolver 类的 query 方法。

```
    Cursor query(Uri uri,String[] projection,String selection,String[]
selectionArgs,String sortOrder)
```

这个方法提供 5 个参数：

- uri：暴露给外部 App 的唯一标示，例如有提供联系人的、有提供短信的、有提供图片的，总归要区分开来。ContactsContract.CommonDataKinds.Phone.CONTENT_URI 就是提供联系人的 uri。
- projection：返回列（字段），例如联系人表中包括 id、name、phone 等字段，如果想要返回表中全部的字段直接填 null。
- selection：设置条件，相当于 SQL 语句中的 where。null 表示不进行筛选。
- selectionArgs：这个参数是配合第三个参数使用的，如果在第三个参数中有“?”，那么在 selectionArgs 写的数据就会替换为“?”。
- sortOrder：按照什么进行排序，相当于 SQL 语句中的 Order by。例如，想要结果按照 id 降序排列，可以写成 android.provider.ContactsContract.Contacts._ID + " DESC"。

这个方法返回的是 Cursor 对象（Cursor 翻译成中文为“游标”），其实就是一个结果集，可以从这个对象中遍历出查询结果。

查询手机联系人列表代码如下：

```
    private void readContacts(){
```

```
Cursor cursor=null;
try{
   cursor=getContentResolver().query(
         ContactsContract.CommonDataKinds.Phone.CONTENT_URI,null,
          null,null,null
   );
   while(cursor.moveToNext()){
      String displayName =cursor.getString(cursor.getColumnIndex
        (ContactsContract.CommonDataKinds.Phone.DISPLAY_NAME));
      String number=cursor.getString(cursor.getColumnIndex
        (ContactsContract.CommonDataKinds.Phone.NUMBER));
      Log.i("ansen","显示名字:"+displayName+" 电话号码:"+number);
   }
}catch (Exception e){
   e.printStackTrace();
}finally {
   if(cursor!=null){
      cursor.close();//在 finally 中关闭游标
   }
}
}
```

首先查找所有的联系人数据，返回 Cursor 对象，然后 while 循环 Cursor，看到 while 中的条件是 cursor.moveToNext()，这个方法首先会判断还有没有下一条记录，如果有就将游标指向下一条；如果没有下一条，就返回 false，循环结束。

用 ContentProvider 可以访问很多系统 App 的数据，借助于网页搜索，读者可以自己尝试一下。当然，也可以通过 ContentProvider 将自己 App 的数据提供给其他 App 访问。

3.5 本章小结

本章学习了 Android 中最重要的四大组件，作为一个 Android 开发者，必定离不开四大组件。在企业真实开发中，Activity 使用更频繁，所以内容也是最多的。

从 Activity 的生命周期到 Activity 四种启动模式，由浅到深系统全面地介绍了 Activity，学习了 Toast，在提示内容上又多了一个选择（需要用户交互使用 Dialog，短暂内容提示使用 Toast），还学习了 Activity 启动与退出动画、在界面跳转时添加动画效果，使 App 更加流畅，体验更好。Activity 四种启动模式的使用可以让用户更好地管理 Activity。

学习了 Service，可以让我们的 App 将一些耗时任务放到后台运行，而不影响用户体验。

接下来，学习了广播的两种注册方式：动态注册与静态注册。可以根据自己的应用场景进行选择。此外，还介绍了应用内广播，这样可以解决恶意攻击问题，让程序运行平稳。

最后简单介绍了 ContentProvider 的使用，编写了一个查找联系人的案例。

本章的内容比较零散，而涉及的知识点在开发中却经常用到，希望大家能够熟练掌握这些组件的使用。

第4章

Fragment 探索

本章将学习 Fragment。Fragment 是 Android 3.0 被引入的 API，主要是体现在大屏幕（如平板电脑）上更加动态和灵活的 UI 设计。本章首先描述了 Fragment 的优缺点，接着介绍了 Fragment 的生命周期，最后介绍了 Fragment 的简单使用。本章的内容虽不多，但需要读者熟练掌握，Fragment 比 Activity 的开销要低很多，合理地使用 Fragment 可以对 APP 性能有较大提升，使你的 APP 如丝般顺滑，这种提升在低端机上尤为明显，响应速度甚至能快好几倍。

4.1 Fragment 简介

Fragment（片段）表示 Activity 中的某部分界面或者行为，必须与 Activity 配合使用。

Fragment 有自己的生命周期，可以加载自己的布局文件，但是它的生命周期依赖于 Activity，如果 Activity 停止，Activity 加载的 Fragment 就不能启动；如果 Activity 销毁，这个 Activity 加载的所有 Fragment 也会销毁。

说了这么多，使用 Fragment 有哪些好处呢？

- 处理在不同屏幕显示的 UI 问题，例如手机和平板电脑的适配问题。
- Activity 模块化，很多业务逻辑可以放在对应的 Fragment 中处理，而 Activity 只需要显示隐藏 Fragment 就行。
- Fragment 可以被不同 Activity 使用。当然，一个 Activity 也能加载多个 Fragment。

4.2 Fragment 生命周期

Fragment 的生命周期如图 4-1 所示。对上一章节中 Activity 组件掌握熟练的读者，肯定会发现 Fragment 生命周期方法跟 Activity 有很多相似之处。

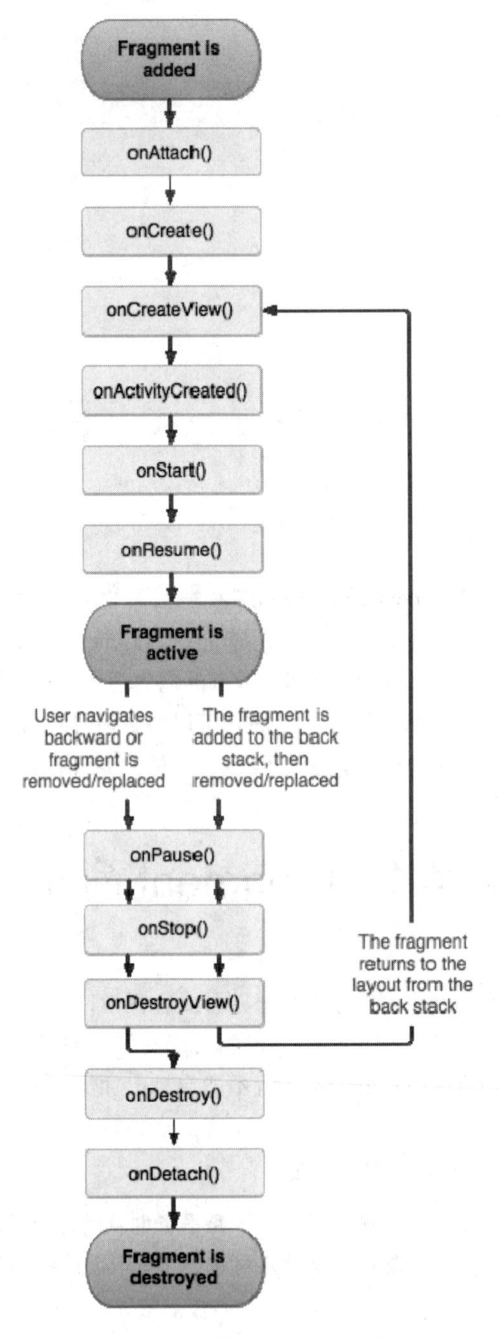

图 4-1　Fragment 的生命周期

再来一张 Fragment 与 Activity 对比的图片，如图 4-2 所示，左边是 Activity 的生命周期调用方法，右边是 Fragment 生命周期调用方法。

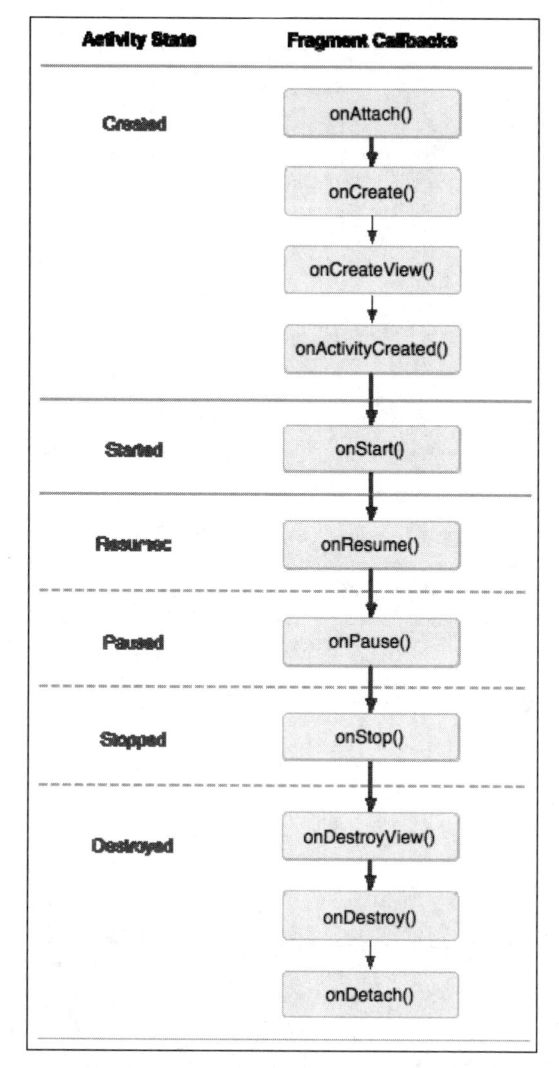

图 4-2　Fragment 与 Activity 生命周期方法对比

从图 4-2 中看到，Fragment 相比 Activity 多了以下几个方法：

- onAttach：执行该方法时，Fragment 与 Activity 已经完成绑定。
- onCreateView：返回 Fragment 显示的 View（这个方法必须要重写）。
- onActivityCreated：与 Fragment 绑定的 Activity 的 onCreate 方法已经执行完成。
- onDestroyView：销毁与 Fragment 有关的视图。
- onDetach：解除与 Activity 的绑定。

下面通过代码来证实图 4-2 中的内容，从而加深理解。

首先新建 FragmentOne 类，继承自 Fragment，重写生命周期相关函数，打印 Log 日志。

```
public class FragmentOne extends Fragment{
    @Override
    public void onAttach(Context context) {
        super.onAttach(context);
```

```
        Log.i("FragmentOne","onAttach");
    }

    @Override
    public void onCreate(@Nullable Bundle savedInstanceState) {
        super.onCreate(savedInstanceState);
        Log.i("FragmentOne","onCreate");
    }

    @Override
    public View onCreateView(LayoutInflater inflater,ViewGroup
container,Bundle savedInstanceState) {
        Log.i("FragmentOne","onCreateView");
        return inflater.inflate(R.layout.fragment_one,null);
    }

    @Override
    public void onActivityCreated(@Nullable Bundle savedInstanceState) {
        super.onActivityCreated(savedInstanceState);
        Log.i("FragmentOne","onActivityCreated");
    }

    @Override
    public void onStart() {
        super.onStart();
        Log.i("FragmentOne","onStart");
    }

    @Override
    public void onResume() {
        super.onResume();
        Log.i("FragmentOne","onResume");
    }

    @Override
    public void onPause() {
        super.onPause();
        Log.i("FragmentOne","onPause");
    }

    @Override
    public void onStop() {
        super.onStop();
        Log.i("FragmentOne","onStop");
    }

    @Override
    public void onDestroyView() {
        super.onDestroyView();
        Log.i("FragmentOne","onDestroyView");
    }

    @Override
    public void onDestroy() {
        super.onDestroy();
```

```
        Log.i("Fragment1","onDestroy");
    }

    @Override
    public void onDetach() {
        super.onDetach();
        Log.i("FragmentOne","onDetach");
    }
}
```

可以看到在 onCreateView 方法中返回了一个 View，这个 View 是通过 fragment_one 布局文件进行加载的。这个布局文件比较简单，其中仅放了一个 TextView。

```
<?xml version="1.0" encoding="utf-8"?>
<TextView xmlns:android="http://schemas.android.com/apk/res/android"
    android:layout_width="match_parent"
    android:layout_height="wrap_content"
    android:background="@color/colorAccent"
    android:gravity="center_horizontal"
    android:padding="10dp"
    android:text="静态加载 Fragment" />
```

Fragment 写好了，如何才能与 Activity 关联起来呢？这里用静态加载的方法，比较简单，在 activity_main.xml 文件中通过 fragment 控件的 name 属性指定 Fragment。必须给它设置一个 id，不然会出现错误："android.view.InflateException: Binary XML file line #5: Error inflating class fragment"。

```
<?xml version="1.0" encoding="utf-8"?>
<RelativeLayout xmlns:android="http://schemas.android.com/apk/res/android"
    android:layout_width="match_parent"
    android:layout_height="match_parent">
    <fragment
        android:id="@+id/fragment_one"
        android:layout_width="match_parent"
        android:layout_height="wrap_content"
        android:name="com.ansen.fragment.FragmentOne"/>
</RelativeLayout>
```

运行代码，打印 log 如下，回调方法过程是 onAttach →onCreate →onCreateView→onActivityCreated→onStart→onResume，这是 Fragment 创建时需要调用的方法。

```
    01-22 10:39:21.069 4987-4987/com.ansen.fragment I/FragmentOne: onAttach
    01-22 10:39:21.069 4987-4987/com.ansen.fragment I/FragmentOne: onCreate
    01-22 10:39:21.069 4987-4987/com.ansen.fragment I/FragmentOne: onCreateView
    01-22 10:39:21.081 4987-4987/com.ansen.fragment I/FragmentOne:
onActivityCreated
    01-22 10:39:21.083 4987-4987/com.ansen.fragment I/FragmentOne: onStart
    01-22 10:39:21.095 4987-4987/com.ansen.fragment I/FragmentOne: onResume
```

在当前页面按住系统 Home 键，这时手机会回到首页。打印 log 如下，回调方法过程是 onPause →onStop。

```
    01-22 10:40:25.341 4987-4987/com.ansen.fragment I/FragmentOne: onPause
    01-22 10:40:26.108 4987-4987/com.ansen.fragment I/FragmentOne: onStop
```

从手机首页找到 App，点击之后进入 App 首页，就会重新显示 FragmentOne，打印 log 如下，回调方法过程是 onStart→onResume。

```
01-22 10:41:06.193 4987-4987/com.ansen.fragment I/FragmentOne: onStart
01-22 10:41:06.194 4987-4987/com.ansen.fragment I/FragmentOne: onResume
```

接下来按住系统返回键回到手机首页，这时当前 Activity 销毁，所以 FragmentOne 也会跟着销毁。应用程序退出。打印 log 如下，回调方法过程是 onPause→onStop→onDestroyView→onDetach。

```
01-22 10:42:22.282 4987-4987/com.ansen.fragment I/FragmentOne: onPause
01-22 10:42:22.719 4987-4987/com.ansen.fragment I/FragmentOne: onStop
01-22 10:42:22.719 4987-4987/com.ansen.fragment I/FragmentOne: onDestroyView
01-22 10:42:22.719 4987-4987/com.ansen.fragment I/FragmentOne: onDetach
```

4.3 FragmentManager 与 FragmentTransaction 的使用

前面介绍了静态加载 Fragment，但是在实际工作中大部分情况都是动态加载 Fragment，例如根据用户点击不同的按钮显示不同的 Fragment。本节介绍动态操作 Fragment 的两个类 FragmentManager 与 FragmentTransaction。

4.3.1 FragmentManager（Fragment 管理类）的使用

FragmentManager 用来管理 Activity 中的 Fragment。在 Activity 中操作 Fragment 通过 Activity.getFragmentManager()获取 FragmentManager 实例。

使用 support 扩展包的时候需要使用 getSupportFragmentManager()方法获取相应的 FragmentManager，需要特别注意的是 support 扩展包中的 Fragment 和 SDK 中自带的 Fragment 是两个不同的类，两个 FragmentManager 也是不一样的类，不可混合使用。初学者很容易弄混，如果在使用中遇到提示参数类型不匹配，这时就需要检查 import 的 Fragment 和 FragmentManager 是否在同一包名下，如不一致换成同一个包路径下的即可。这两个 Fragment 类虽然包名不一致但是用法都是一样的。

FragmentManager 有以下一些常用的方法。

- beginTransaction：开启 Fragment 事务，返回 FragmentTransaction 对象。
- findFragmentById：根据 id 查找 Fragment。
- popBackStack：将 Fragment 从后台堆栈中弹出，类似按下系统返回键的操作，这个方法是异步的，底层队列实现。
- addOnBackStackChangedListener：监听后台堆栈变化。
- removeOnBackStackChangedListener：与上面那个方法对应，删除监听。

4.3.2 FragmentTransaction（Fragment 事务）的使用

FragmentTransaction 直接操作 Fragment，对 Fragment 进行增加、移除与替换，对 Fragment 操作完毕之后一定要调用 FragmentTransaction.commit 方法提交事务。

学过 SQL 语句的人都知道，为了保证一件事情的完整性，例如银行转账操作：A 用户转给 B 用户 100 元，这时需要执行两条 SQL 语句，A 用户账户减去 100，B 用户账户余额增加 100，为了保证账本正确，于是就有了事务，要么两条 SQL 语句同时执行成功，要么同时执行失败。如果有一条成功会进行回滚。

这里也是一样的，我们添加了一个 Fragment，同时又删除了一个 Fragment，只有最后调用 commit 方法时才会生效。

FragmentTransaction 常用方法有以下几种。

- add()：向 Activity 中添加一个 Fragment。
- remove()：从 Activity 中删除一个 Fragment。
- replace()：新的 Fragment 替换当前的 Fragment。
- hide()：隐藏 Fragment。
- show()：显示 Fragment。
- commit()：提交事务。

4.4 Activity 动态操作 Fragment

前面学习了 Fragment 的生命周期，以及在 XML 布局文件中静态加载 Fragment，同时介绍了动态操作 Fragment 的两个类 FragmentManager 与 FragmentTransaction。

理论知识有了，本节就通过代码来实现如何动态操作 Fragment，将在上一节写的案例代码上进行修改。

新建一个 Fragmnet，通过构造方法传入一个 int 类型的值，在 onCreateView 方法中根据传入的 int 类型参数给 TextView 设置不同的文本以及不同的背景颜色。

```
public class FragmentContainer extends Fragment {
    private int fragmentIndex;

    public FragmentContainer(){}

    @SuppressLint("ValidFragment")
    public FragmentContainer(int fragmentIndex){
        this.fragmentIndex=fragmentIndex;
    }

    @Override
    public View onCreateView(LayoutInflater inflater,ViewGroup container,
     Bundle savedInstanceState) {
```

```
        View rootView=inflater.inflate(R.layout.fragment_container,null);
        TextView tvContent = (TextView) rootView.findViewById(R.id.tv_content);
        if(fragmentIndex==1){
            tvContent.setText("第一个 Fragment");
            tvContent.setBackgroundResource(android.R.color.holo_red_light);
        }else if(fragmentIndex==2){
            tvContent.setText("第二个 Fragment");
            tvContent.setBackgroundResource(android.R.color.holo_orange_light);
        }else if(fragmentIndex==3){
            tvContent.setText("第三个 Fragment");
            tvContent.setBackgroundResource(android.R.color.holo_blue_light);
        }
        return rootView;
    }
}
```

FragmentContainer 类对应的布局文件是 fragment_container.xml，也就是 onCreateView 方法第一行 inflate 的布局，里面就包含了一个 TextView，内容如下：

```
<?xml version="1.0" encoding="utf-8"?>
<TextView xmlns:android="http://schemas.android.com/apk/res/android"
    android:layout_width="match_parent"
    android:layout_height="match_parent"
    android:id="@+id/tv_content"
    android:gravity="center"
    android:text="Fragment"
    android:background="@android:color/holo_green_dark"/>
```

再来查看首页 activity_main.xml 布局文件有哪些改动，其中增加了 5 个按钮，点击之后对应上面列出来 FragmentTransaction 的 5 个方法。再增加 FrameLayout 布局，用来显示 Fragment。我们可以看到 xml 文件中新增的 FrameLayout 是没有指定 android:name 属性的。

```
<?xml version="1.0" encoding="utf-8"?>
<LinearLayout xmlns:android="http://schemas.android.com/apk/res/android"
    android:layout_width="match_parent"
    android:layout_height="match_parent"
    android:orientation="vertical">

    <fragment
        android:id="@+id/fragment_one"
        android:name="com.ansen.fragment.FragmentOne"
        android:layout_width="match_parent"
        android:layout_height="wrap_content" />

    <LinearLayout
        android:layout_width="match_parent"
        android:layout_height="wrap_content"
        android:orientation="horizontal">

        <Button
            android:id="@+id/btn_add"
            android:layout_width="0dp"
            android:layout_height="wrap_content"
            android:layout_weight="1"
            android:text="Add" />
```

```
        <Button
            android:id="@+id/btn_remove"
            android:layout_width="0dp"
            android:layout_height="wrap_content"
            android:layout_weight="1"
            android:text="Remove" />

        <Button
            android:id="@+id/btn_replace"
            android:layout_width="0dp"
            android:layout_height="wrap_content"
            android:layout_weight="1"
            android:text="Replace" />

        <Button
            android:id="@+id/btn_hide"
            android:layout_width="0dp"
            android:layout_height="wrap_content"
            android:layout_weight="1"
            android:text="Hide" />

        <Button
            android:id="@+id/btn_show"
            android:layout_width="0dp"
            android:layout_height="wrap_content"
            android:layout_weight="1"
            android:text="Show" />
    </LinearLayout>

    <FrameLayout
        android:id="@+id/fl_container"
        android:layout_width="match_parent"
        android:layout_height="match_parent" />
</LinearLayout>
```

MainActivity.java也要进行修改，在onCreate方法中初始化三个Fragment，传入一个int类型参数就可以初始化不同背景文字的Fragment，给各个按钮设置点击事件，可以看到点击回调的onClick方法中第一行开启一个Fragment事务，最后一行调用commit进行事务提交。与做Java Web开发每次操作数据库一样，需要保证一致性。

```
    public class MainActivity extends AppCompatActivity implements
View.OnClickListener{
        private FragmentContainer fragmentOne;
        private FragmentContainer fragmentTwo;
        private FragmentContainer fragmentThree;

        @Override
        protected void onCreate(Bundle savedInstanceState) {
            super.onCreate(savedInstanceState);
            setContentView(R.layout.activity_main);

            fragmentOne=new FragmentContainer(1);
            fragmentTwo=new FragmentContainer(2);
```

```
        fragmentThree=new FragmentContainer(3);

        findViewById(R.id.btn_add).setOnClickListener(this);
        findViewById(R.id.btn_remove).setOnClickListener(this);
        findViewById(R.id.btn_replace).setOnClickListener(this);
        findViewById(R.id.btn_hide).setOnClickListener(this);
        findViewById(R.id.btn_show).setOnClickListener(this);
    }

    @Override
    public void onClick(View v) {
        //开启一个 Fragment 事务
        FragmentTransaction transaction = getSupportFragmentManager().
beginTransaction();
        if(v.getId()==R.id.btn_add){//添加 2 个 Fragment
            transaction.add(R.id.fl_container,fragmentOne);
            transaction.add(R.id.fl_container,fragmentTwo);
        }else if(v.getId()==R.id.btn_remove){//删除第 2 个
            transaction.remove(fragmentTwo);
        }else if(v.getId()==R.id.btn_replace){//替换
            transaction.replace(R.id.fl_container,fragmentThree);
        }else if(v.getId()==R.id.btn_hide){//隐藏
            transaction.hide(fragmentThree);
        }else if(v.getId()==R.id.btn_show){//显示
            transaction.show(fragmentThree);
        }
        transaction.commit();
    }
}
```

分析一下 5 个按钮依次点击之后的效果：

（1）点击添加按钮，先添加 Fragment1，再添加 Fragment2，Activity 中添加了两个 Fragment。但是同一个 View 只能显示一个 Fragment，后面添加的会显示在最上面，所以这时显示 Fragment2。

（2）点击删除按钮，删除 Fragment2，只剩下 Fragment1，所以这时显示 Fragment1。

（3）点击替换按钮，将 Fragment3 替换成 Fragment1，所以这时显示 Fragment3。

（4）点击隐藏按钮，这时只有 Fragment3，将隐藏 Fragment3。FrameLayout 控件显示为空白。

（5）点击显示按钮，重新显示 Fragment3。

4.5 Fragment 与 Activity 交互数据

Fragment 与 Activity 交互数据一般通过以下两种方法：

- 通过 Fragmentd 构造方法传递参数（适合动态加载 Fragment）。
- 通过依赖注入方式。

其实原理都差不多，Activity 拥有 Fragment 的对象，传递参数是很容易的。在 Fragment 类中写入一个 set 方法即可。不过，如果 Fragment 将数据处理完成之后又要调用 Activity，就需要通过回调。

在之前的代码上修改 FragmentOne 这个类。这里仅贴出改动的代码，增加了两个实例变量，增加了一个 setOnClickListener 方法，用于 Activity 设置点击事件，在 onCreateView 方法中将 inflater 返回的 View 赋值给 rootView 变量保存起来。在 onCreateView 中为 rootView 事件设置点击事件。

```
    private View rootView;
    private View.OnClickListener onClickListener;

    public void setOnClickListener(View.OnClickListener onClickListener) {
        this.onClickListener = onClickListener;
    }

    @Override
    public void onCreate(Bundle savedInstanceState) {
        super.onCreate(savedInstanceState);
        Log.i("FragmentOne","onCreate");
    }

    @Override
    public View onCreateView(LayoutInflater inflater,ViewGroup
container,Bundle savedInstanceState) {
        Log.i("FragmentOne","onCreateView");
        rootView=inflater.inflate(R.layout.fragment_one,null);
        return rootView;
    }

    @Override
    public void onActivityCreated(Bundle savedInstanceState) {
        super.onActivityCreated(savedInstanceState);
        Log.i("FragmentOne","onActivityCreated");
        if(onClickListener!=null){
            rootView.setOnClickListener(onClickListener);
        }
    }
```

MainActivity 修改比较简单，在 onCreate 方法中增加了如下代码，获取静态的 Fragment，然后注入点击事件，这样最终点击 Fragment 时就会调用 onClick 方法。

```
        staticFragment= (FragmentOne) getSupportFragmentManager().
                        findFragmentById (R.id.fragment_one);
        staticFragment.setOnClickListener(new View.OnClickListener() {
            @Override
            public void onClick(View v) {
                Log.i("MainActivity","动态加载的 Fragment 被点击了");
            }
        });
```

4.6 Fragment案例——实现底部导航栏

底部导航栏在 App 开发中是经常碰到的。国内的 App 大部分都有底部导航栏（例如 QQ、微信、支付宝），以便用户随时切换界面、查看不同的内容。

Android 原生操作系统（从 nexus 系列到 pxiel 系列）会预装很多 Google 自带的 App。Google 推荐用侧滑菜单，但是 YouTube 跟相册这两款 App 却都使用了底部导航栏。底部导航栏的实现方式有很多种，本节将用 TextView+Frgment 方式来实现底部导航栏。这种方式的优点是修改方便、实现简单。

4.6.1 分析需求

效果图如图 4-3 和图 4-4 所示。图 4-3 是 App 刚启动时的效果，从中可以看到底部有四个 tab（首页、动态、消息、我），点击某个 Tab 之后会显示 Tab 对应的 Fragment。图 4-4 就是点击底部 Tab“我”之后的效果。四个 Tab 中间还有一个图片，点击这个图片之后直接跳转到 Activity。实现的具体细节如下：

- 底部 Tab 在屏幕中的宽度是一致的，应该想到 LinearLayout 的 weight 属性，利用线性布局的权重。
- 中间的图片突出了一点点，考虑在最外层用 RelativeLayout 或者 FrameLayout，因为 FrameLayout 灵活性没有 RelativeLayout 强，所以优先选择 RelativeLayout。
- 点击 tab 之后底部导航栏文字颜色图片都有变化。用淡绿代表选中，灰色代表不选中。
- Tab 点击之后内容区域变化，需要切换 Fragment 显示。

图 4-3　App 启动显示首页

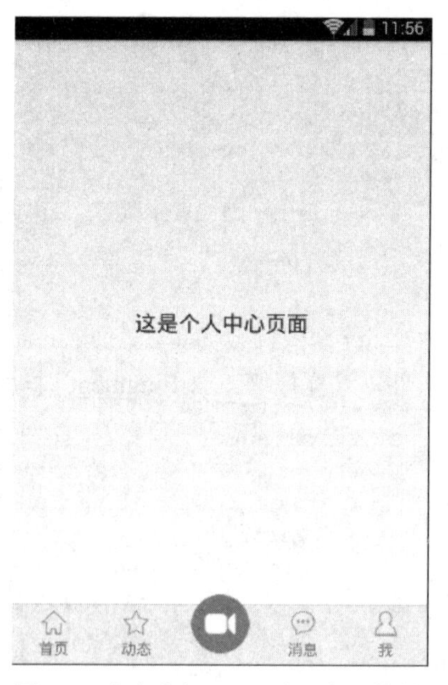

图 4-4　点击底部 Tab“我”之后的效果

4.6.2 代码实现

先看布局文件 activity_main.xml：

```
<RelativeLayout xmlns:android="http://schemas.android.com/apk/res/android"
    android:layout_width="match_parent"
```

```
    android:layout_height="match_parent" >

    <FrameLayout
        android:id="@+id/main_container"
        android:layout_width="match_parent"
        android:layout_height="match_parent"
        android:layout_above="@+id/view_line"/>

    <View
        android:id="@+id/view_line"
        android:layout_height="1dp"
        android:layout_width="match_parent"
        android:background="#DCDBDB"
        android:layout_above="@+id/rl_bottom"/>

    <LinearLayout
        android:id="@+id/rl_bottom"
        android:layout_width="match_parent"
        android:layout_height="wrap_content"
        android:layout_alignParentBottom="true"
        android:paddingTop="5dp"
        android:paddingBottom="5dp"
        android:background="#F2F2F2"
        android:orientation="horizontal" >

        <TextView
            android:id="@+id/tv_main"
            android:layout_width="0dp"
            android:layout_height="wrap_content"
            android:layout_gravity="center"
            android:layout_weight="1"
            android:drawableTop="@drawable/tab_item_main_img_selector"
            android:drawablePadding="@dimen/main_tab_item_image_and_text"
            android:focusable="true"
            android:gravity="center"
            android:text="@string/main"
            android:textColor="@drawable/tabitem_txt_sel" />

        <TextView
            android:id="@+id/tv_dynamic"
            android:layout_width="0dp"
            android:layout_height="wrap_content"
            android:layout_gravity="center"
            android:layout_weight="1"
            android:drawableTop="@drawable/tab_item_dynamic_img_selector"
            android:drawablePadding="@dimen/main_tab_item_image_and_text"
            android:focusable="true"
            android:gravity="center"
            android:text="@string/dynamic"
            android:textColor="@drawable/tabitem_txt_sel" />

        <View
            android:layout_width="0dp"
            android:layout_height="match_parent"
            android:layout_weight="1" />
```

```
        <TextView
            android:id="@+id/tv_message"
            android:layout_width="0dp"
            android:layout_height="wrap_content"
            android:layout_gravity="center"
            android:layout_weight="1"
            android:drawableTop="@drawable/tab_item_message_img_selector"
            android:drawablePadding="@dimen/main_tab_item_image_and_text"
            android:focusable="true"
            android:gravity="center"
            android:text="@string/message"
            android:textColor="@drawable/tabitem_txt_sel" />

        <TextView
            android:id="@+id/tv_person"
            android:layout_width="0dp"
            android:layout_height="wrap_content"
            android:layout_gravity="center"
            android:layout_weight="1"
            android:drawableTop="@drawable/tab_item_person_img_selector"
            android:drawablePadding="@dimen/main_tab_item_image_and_text"
            android:focusable="true"
            android:gravity="center"
            android:text="@string/person"
            android:textColor="@drawable/tabitem_txt_sel"/>
    </LinearLayout>

    <ImageView
        android:id="@+id/iv_make"
        android:layout_width="wrap_content"
        android:layout_height="wrap_content"
        android:layout_alignParentBottom="true"
        android:layout_centerHorizontal="true"
        android:paddingBottom="10dp"
        android:src="@drawable/icon_tab_make_select"/>
</RelativeLayout>
```

最外层用 RelativeLayout。

- 第一个子 View 是 FrameLayout，tab 切换的时候用它来显示 Fragment。
- 第二个 View 就是一个分割线。
- 第三个 View 是一个 LinearLayout，设置了 android:layout_alignParentBottom="true"，在父 View 的底部显示，里面有 4 个 TextView 跟一个 View，权重都是 1，但是最中间那个没有内容，是一个空的 View，只是用它在来占一个位置。
- 第 4 个 View 是一个 ImageView，也设置了 android:layout_alignParentBottom 属性，并且通过 android:layout_centerHorizontal 属性设置水平居中，在 RelativeLayout 里面如果两个 View 在同一个位置上，后面的 View 就会覆盖之前的 View，所以这个 ImageView 盖住了 LinearLayout 最中间的那个 View。

从布局文件中我们看到给每个 TextView 设置了 android:drawableTop 属性，这个属性就是在

TextView 的文本上面放一张图片。android:drawableTop 属性对应的是一个 drawable 文件，这么做的目的就是把选中跟未选中状态的图标都表示出来。最外层是一个 selector，里面有两个 item（item 可以有多个），第一个 item 是选中状态的情况，item 有一个属性叫 android:state_selected="true"，就是 View 选中状态时显示 android:drawable 对应的图片。第二个 item 是默认情况，就是非选中的情况显示。在 Java 代码中调用 TextView.setSelected（true）对应第一个 item。

```
    <?xml version="1.0" encoding="utf-8"?>
    <selector xmlns:android="http://schemas.android.com/apk/res/android">
        <!-- 选中状态显示 -->
        <item android:drawable="@drawable/icon_tab_main_select"
android:state_selected="true"/>

        <!--非选中状态显示-->
        <item android:drawable="@drawable/icon_tab_main_normal"/>
    </selector>
```

不只是图片，还有 TextView 设置文字颜色也是通过 selector 来控制的。我们可以看到底部四个 TextView 的 android:textColor 属性都引用了 drawable/tabitem_txt_sel 文件。跟图片的选择器有点类似，只不过把 android:drawable 换成了 android:color。

```
    <?xml version="1.0" encoding="utf-8"?>
    <selector xmlns:android="http://schemas.android.com/apk/res/android">
        <!-- 选中状态 -->
        <item android:color="@color/main_tab_item_text_select"
android:state_selected="true"/>

        <!--非选中状态-->
        <item android:color="@color/main_tab_item_text_normal"/>
    </selector>
```

接下来看 MainActivity 代码，继承自 FragmentActivity，这是 android-support-v4.jar 包里面的一个类，可以兼容 3.0 以下版本使用 Fragment。

```
    public class MainActivity extends FragmentActivity {
        //要切换显示的四个 Fragment
        private HomeFragment homeFragment;
        private DynamicFragment dynamicFragment;
        private MessageFragment messageFragment;
        private PersonFragment personFragment;

        private int currentId = R.id.tv_main;// 当前选中 id,默认是主页

        private TextView tvMain, tvDynamic, tvMessage, tvPerson;//底部四个 TextView

        @Override
        protected void onCreate(Bundle savedInstanceState) {
            super.onCreate(savedInstanceState);
            setContentView(R.layout.activity_main);

            tvMain = (TextView) findViewById(R.id.tv_main);
            tvMain.setSelected(true);//首页默认选中
            tvDynamic = (TextView) findViewById(R.id.tv_dynamic);
            tvMessage = (TextView) findViewById(R.id.tv_message);
```

```
        tvPerson = (TextView) findViewById(R.id.tv_person);

        //默认加载首页
        homeFragment = new HomeFragment();
        getSupportFragmentManager().beginTransaction().add(R.id.
          main_container, homeFragment).commit();

        tvMain.setOnClickListener(tabClickListener);
        tvDynamic.setOnClickListener(tabClickListener);
        tvMessage.setOnClickListener(tabClickListener);
        tvPerson.setOnClickListener(tabClickListener);
        findViewById(R.id.iv_make).setOnClickListener(onClickListener);
    }

    private OnClickListener onClickListener=new OnClickListener() {
        @Override
        public void onClick(View v) {
            switch (v.getId()) {
            case R.id.iv_make:
                Toast.makeText(MainActivity.this,"点击了制作按钮",
                        Toast.LENGTH_SHORT).show();
                break;
            }
        }
    };

    private OnClickListener tabClickListener = new OnClickListener() {
        @Override
        public void onClick(View v) {
            if (v.getId() != currentId) {//如果当前选中跟上次选中的一样,不需要处理
                changeSelect(v.getId());//改变图标跟文字颜色的选中
                changeFragment(v.getId());//fragment 的切换
                currentId = v.getId();//设置选中 id
            }
        }
    };

    /**
     * 改变 fragment 的显示
     * @param resId
     */
    private void changeFragment(int resId) {
        FragmentTransaction transaction = getSupportFragmentManager().
         beginTransaction();//开启一个 Fragment 事务

        hideFragments(transaction);//隐藏所有 fragment
        if(resId==R.id.tv_main){//主页
            if(homeFragment==null){//如果为空先添加进来，不为空则直接显示
                homeFragment = new HomeFragment();
                transaction.add(R.id.main_container,homeFragment);
            }else {
                transaction.show(homeFragment);
            }
        }else if(resId==R.id.tv_dynamic){//动态
            if(dynamicFragment==null){
```

```
            dynamicFragment = new DynamicFragment();
            transaction.add(R.id.main_container,dynamicFragment);
        }else {
            transaction.show(dynamicFragment);
        }
    }else if(resId==R.id.tv_message){//消息中心
        if(messageFragment==null){
            messageFragment = new MessageFragment();
            transaction.add(R.id.main_container,messageFragment);
        }else {
            transaction.show(messageFragment);
        }
    }else if(resId==R.id.tv_person){//我
        if(personFragment==null){
            personFragment = new PersonFragment();
            transaction.add(R.id.main_container,personFragment);
        }else {
            transaction.show(personFragment);
        }
    }
    transaction.commit();//一定要记得提交事务
}

/**
 * 显示之前隐藏所有fragment
 * @param transaction
 */
private void hideFragments(FragmentTransaction transaction){
    if (homeFragment != null)//不为空才隐藏,如果不判断第一次会有空指针异常
        transaction.hide(homeFragment);
    if (dynamicFragment != null)
        transaction.hide(dynamicFragment);
    if (messageFragment != null)
        transaction.hide(messageFragment);
    if (personFragment != null)
        transaction.hide(personFragment);
}

/**
 * 改变TextView选中颜色
 * @param resId
 */
private void changeSelect(int resId){
    tvMain.setSelected(false);
    tvDynamic.setSelected(false);
    tvMessage.setSelected(false);
    tvPerson.setSelected(false);

    switch (resId) {
    case R.id.tv_main:
        tvMain.setSelected(true);
        break;
    case R.id.tv_dynamic:
        tvDynamic.setSelected(true);
        break;
```

```
            case R.id.tv_message:
                tvMessage.setSelected(true);
                break;
            case R.id.tv_person:
                tvPerson.setSelected(true);
                break;
        }
    }
}
```

首先在 onCreate 方法中设置布局文件，查找底部的四个 TextView，给首页的 TextView 设置为选中状态，并且默认加载首页的 Fragment，最后给底部的四个 Tab 设置点击事件。还有最中间的那个 ImageView。

tabClickListener 处理 Tab 点击事件，先判断这次点击的 Tab 跟上次点击的是否一致，如果跟当前的是一样的就不需要处理，否则的话需要改变图标跟文字选中状态，还有 Fragment 的切换，并且把当前点击 View 的 id 设置成当前选中的 id。

changeSelect 方法改变 TextView 选中颜色，首先全部设置成未选中，然后判断当前选中的哪个。

changeFragment 改变 Fragment 的显示，首先调用 hideFragments 方法隐藏所有的 Fragment，然后根据当前选中的 Tab 来决定显示哪个 Fragment。显示的时候需要先判断这个 Fragment 有没有显示过，如果没有显示过需要新建一个新的 Fragment，然后调用 FragmentTransaction 的 add 方法添加进去。如果之前有添加过的，直接调用 show 方法就行。

hideFragments 方法就是隐藏所有的 Fragment。先判断 Fragment 是否为 null，不为 null 就调用 transaction.hide 方法隐藏 Fragment。

底部的四个 Tab（TextView）对应四个 Fragment，分别是 HomeFragment、DynamicFragment、MessageFragment、PersonFragment。因为是 Demo，所以都只显示了一个 TextView。我们看看首页的 Fragment，其他三个就不贴代码了。

HomeFragment 显示的是布局文件 fragment_home.xml，里面只有一个 TextView。

```
<?xml version="1.0" encoding="utf-8"?>
<RelativeLayout xmlns:android="http://schemas.android.com/apk/res/android"
    android:layout_width="match_parent"
    android:layout_height="match_parent">
    <TextView
        android:layout_width="wrap_content"
        android:layout_height="wrap_content"
        android:layout_centerInParent="true"
        android:text="这是首页"
        android:textSize="20sp"/>
</RelativeLayout>
HomeFragment.java
public class HomeFragment extends Fragment{
    @Override
    public View onCreateView(LayoutInflater inflater, ViewGroup container,
      Bundle savedInstanceState) {
        View rootView = inflater.inflate(R.layout.fragment_home, null);
        return rootView;
    }
}
```

4.7 本章小结

本章重点介绍了Fragment的使用方法。将Fragment单独拿出来讲解，足以说明其重要性。本章主要学习了Fragment的生命周期，Activity如何通过FragmentManager与FragmentTransaction这两个类来管理Fragment，以及Activity管理的多个Fragment之间如何交互数据。最后用TextView+Fragment实现底部导航栏，让大家更熟练地使用Fragment。底部导航栏也是现在主流App的一种展示方式。

第 5 章

Android 多线程开发

本章学习了在 Android 中如何创建多线程、多线程中更新 UI，用 Handler 方式实现两个线程之间的通信，同时从源码的角度学习了 Handler 的实现原理（Handler、Looper 与 MessageQueue 三者的关系），还使用了系统给我们封装的 AsyncTask 异步任务处理类，最后学习了线程池的使用，有多个异步任务时，合理地使用线程池会减少系统资源的使用，增加程序的流畅性。

5.1 多线程的创建

每个应用启动时，Android 会启动一个对应的主线程来处理 UI 相关的事情，例如用户的按键事件、用户接触屏幕的事件以及屏幕绘图事件，并且把相关的事件分发到对应的组件进行处理，所以主线程通常又称为 UI 线程。

如果比较耗时的任务（例如下载文件）也在主线程完成，就会影响用户体验，会阻塞主线程，这种情况下就需要创建一个子线程，把耗时任务放在子线程中执行。

在 Android 中创建一个线程与在 Java 中创建线程是一样的。首先在 MainActivity 的 onCreate 中增加如下代码：

```
new Thread(new Runnable(){
        @Override
        public void run(){
            for(int i=1;i<=100;i++){
                Log.i("MainActivity","当前值是:"+i);
                try {
                    Thread.sleep(200);
                } catch (InterruptedException e) {
                    e.printStackTrace();
                }
            }
```

```
            }
    }).start();
```

这里开启了一个线程，在线程中写了一个 for 循环 100 次，然后打印当前的值，并且每次循环时设置了延迟 200 毫秒。

运行软件，可以看到打印的日志如下：

```
    02-19 08:07:14.952 20477-20497/com.ansen.multithread I/MainActivity: 当前值是:1
    02-19 08:07:15.160 20477-20497/com.ansen.multithread I/MainActivity: 当前值是:2
    02-19 08:07:15.360 20477-20497/com.ansen.multithread I/MainActivity: 当前值是:3
    --------
    02-19 08:07:34.824 20477-20497/com.ansen.multithread I/MainActivity: 当前值是:100
```

5.2 子线程中更新 UI 的四种方法

在 Android 开发中，子线程是不能直接更新 UI 界面的，如果在子线程中操作 UI，程序就会崩溃，并且抛出以下异常：

```
    android.view.ViewRootImpl$CalledFromWrongThreadException: Only the original
thread that created a view hierarchy can touch its views.
    at android.view.ViewRootImpl.checkThread(ViewRootImpl.java:6024)
    at android.view.ViewRootImpl.requestLayout(ViewRootImpl.java:820)
```

一种常用的处理方法就是在子线程中执行耗时任务，执行完成之后发送消息给主线程，通知主线程更新界面。例如，下载文件，开启一个子线程去下载，下载完成之后发送一个消息通知主线程更新界面，提示用户文件已下载完成。

在 Android 中子线程更新 UI 的方法有以下四种：

- 用 Activity 对象的 runOnUiThread 方法。
- View.post（Runnable r）。
- AsyncTask 系统 SDK 提供的处理耗时任务的类。
- Handler 解决多线程间的通信。

上面介绍了子线程中更新 UI 的四种方法，其实前面三种方式的底层原理也是通过 Handler 实现的。AsyncTask 与 Handler 两种方法放到本章后面单独讲解。这里先学习前面两种方法。

5.2.1 用 Activity 对象的 runOnUiThread 方法

新建一个项目，修改 activity_main.xml 文件，内容如下：

```
    <?xml version="1.0" encoding="utf-8"?>
    <LinearLayout xmlns:android="http://schemas.android.com/apk/res/android"
        android:layout_width="match_parent"
        android:layout_height="match_parent"
        android:orientation="vertical"
        android:padding="10dp">
```

```
    <TextView
        android:id="@+id/tv_content"
        android:layout_width="wrap_content"
        android:layout_height="wrap_content"
        android:text="Hello World!" />
</LinearLayout>
```

布局文件简单，最外层是 LinearLayout，包含一个 TextView。

接着修改 MainActivity.java 文件，在 onCreate 方法中通过 id 查找 TextView，再开启一个子线程，在子线程的 run 方法中调用 Activity 的 runOnUiThread 方法，runOnUiThread 方法需要传入一个 Runnable 对象，用内部类的方式实现 Runnable 重写 run 方法，然后在 Runnable 的 run 方法中更新 UI 界面。

```
@Override
protected void onCreate(Bundle savedInstanceState) {
    super.onCreate(savedInstanceState);
    setContentView(R.layout.activity_main);
    tvContent=findViewById(R.id.tv_content);
    new Thread(){
        @Override
        public void run() {
            //用 activity 的 runOnUiThread 方法更新 ui 底层也是 handler 实现
            runOnUiThread(new Runnable() {
                @Override
                public void run() {
                    tvContent.setText("runOnUiThread 更新 ui");
                }
            });
        }
    }.start();
}
```

当代码执行到 Runnable 的 run 方法时，其实当前线程就已经是主线程了，我们继续修改代码，在三个地方打印线程 id，修改后代码如下：

```
@Override
protected void onCreate(Bundle savedInstanceState) {
    super.onCreate(savedInstanceState);
    setContentView(R.layout.activity_main);
    tvContent=findViewById(R.id.tv_content);
    Log.i("MainActivity","主线程 id:"+android.os.Process.myTid());
    new Thread(){
        @Override
        public void run() {
            Log.i("MainActivity","子线程 id:"+android.os.Process.myTid());
            //用 activity 的 runOnUiThread 方法更新 ui 底层也是 handler 实现
            runOnUiThread(new Runnable() {
                @Override
                public void run() {
                    Log.i("MainActivity","主线程
id:"+android.os.Process.myTid());
                    tvContent.setText("runOnUiThread 更新 ui");
                }
```

```
            });
        }
    }.start();
}
```

打印 log 如下，从 log 中可以看到 onCreate 中的线程 id 跟 Runnable 中 run 方法打印的线程 id 是一致的。

```
05-01 11:39:58.100 2026-2026/? I/MainActivity: 主线程 id:2026
05-01 11:39:58.100 2026-2040/? I/MainActivity: 子线程 id:2040
05-01 11:39:58.108 2026-2026/? I/MainActivity: 主线程 id:2026
```

5.2.2　View.post 的使用

同样开启一个子线程，在子线程中调用要更新的控件的 Post 方法。传入一个 Runnable 对象，在 Runnable 对象的 run 方法中更新 UI。

```
new Thread(){
    @Override
    public void run() {
        tvContent.post(new Runnable() {
            @Override
            public void run() {
                tvContent.setText("View Post 方式");
            }
        });
    }
}.start();
```

5.3　Handler 的使用

Handler 可以发送和处理消息对象或 Runnable 对象，这些消息对象和 Runnable 对象与一个线程相关联。每个 Handler 的实例都关联了一个线程和线程的消息队列。当创建了一个 Handler 对象时，一个线程或消息队列同时也被创建，该 Handler 对象将发送和处理这些消息或 Runnable 对象。

Handler 类有两种主要用途：

- 执行 Runnable 对象，还可以设置延迟。
- 两个线程之间发送消息，主要用来给主线程发送消息更新 UI。

5.3.1　为什么要用 Handler

解决多线程并发问题，假设如果在一个 Activity 中有多个线程去更新 UI，并且都没有加锁机制，那么界面显示肯定会不正常。因此，Android 官方就封装了一套更新 UI 的机制，也可以使用 Handler 来实现多个线程之间的消息发送。

5.3.2 使用 Handler

Handler 常用的方法如下：

- post（Runnable）。
- postAtTime（Runnable，long）。
- postDelayed（Runnable，long）。
- sendEmptyMessage（int）。
- sendMessage（Message）。
- sendMessageAtTime（Message，long）。
- sendMessageDelayed（Message，long）。

这些方法主要分为两类：一类是传入一个 Runnable 对象，另一类是传入一个 Message 对象。

1. 用代码学习 Post 一个 Runnable 对象

先创建 Handler 对象，直接新建即可：

```
private Handler Handler=new Handler();
```

实现 Runnable 接口，使用匿名实现方式，重写 run 方法，打印一个字符串。

```
private Runnable runnable=new Runnable() {
        @Override
        public void run() {
            Log.i("MainActivity","Handler Runnable");
        }
};
```

然后调用 Handler 的 Post 方法。这里需要注意的是，Post 一个 Runnable 对象，底层用的是回调，不会开启一个新的线程。所有 Runnable 的 run 方法还是在主线程中，是可以更新 UI 的。

```
Handler.post(runnable);//执行
Handler.postDelayed(runnable,2000);//延迟 2 秒后执行
```

运行程序，控制台打印的日志如下：

```
05-18 19:17:14.901 17750-17750/com.ansen.Handler I/MainActivity: Handler Runnable
05-18 19:17:16.901 17750-17750/com.ansen.Handler I/MainActivity: Handler Runnable
```

从上面的日志中可以看到两条日志的时间相差两秒。这是因为用 postDelayed 方法时，第二个参数设置了两秒的延迟。

2. 使用 sendMessage 方法发送消息

sendMessage 方法可以理解为用来发送消息，这种方法在 Android 中使用频率比较高。因为在 Android 多线程中是不能更新 UI 的，所以必须通过 Handler 把消息发送给 UI 线程，才能更新 UI。当然，也可以用 Handler 实现两个子线程发送消息。

新建一个项目，给 activity_main.xml 文件中的 TextView 控件设置一个 id。

```
<?xml version="1.0" encoding="utf-8"?>
```

```
<RelativeLayout xmlns:android="http://schemas.android.com/apk/res/android"
    android:layout_width="match_parent"
    android:layout_height="match_parent">

    <TextView
        android:id="@+id/textview"
        android:layout_width="wrap_content"
        android:layout_height="wrap_content"
        android:text="Hello World!" />
</RelativeLayout>
```

继续修改MainActivity.java代码：

```
public class MainActivity extends AppCompatActivity {
    private TextView textview;
    public static final int UPDATE_UI=1;
    private Handler handler=new Handler(){
        @Override
        public void handleMessage(Message msg) {
            if(msg.what==UPDATE_UI){
                textview.setText("当前值是:"+msg.obj);
            }
        }
    };

    @Override
    protected void onCreate(Bundle savedInstanceState) {
        super.onCreate(savedInstanceState);
        setContentView(R.layout.activity_main);

        textview= (TextView) findViewById(R.id.textview);
        new Thread(new Runnable(){
            @Override
            public void run(){
                for(int i=1;i<=100;i++){
                    Log.i("MainActivity","当前值是:"+i);
                    Message message=handler.obtainMessage();
                    message.what=UPDATE_UI;
                    message.obj=i;
                    handler.sendMessage(message);
                    try {
                        Thread.sleep(200);
                    } catch (InterruptedException e) {
                        e.printStackTrace();
                    }
                }
            }
        }).start();
    }
}
```

首先用内部类方式新建一个Handler对象，重写handleMessage方法，在handleMessage方法中先判断传入过来的Message对象中what的值是不是我们定义的常量UPDATE_UI的值，如果是就更新ui，把obj对象的值加上字符串赋值给TextView。

接下来看 onCreate 方法，根据 id 查找 TextView 控件，然后开启一个子线程，在子线程的 run 方法中用了一个 for 循环，循环 100 次，在循环体中，首先调用 handler.obtainMessage()获取一个 Message 对象，该方法是从消息池中返回一个 Message 对象，只有消息池中没有时才会创建 Message 对象。Message 对象的 what 属性是必须要赋值的，是一个 int 类型。这里我们赋值自己定义的常量，当 Handler 接收到这个 Message 时也是凭这个值来区分消息从哪里来的，Message 对象还有一个 obj 属性，obj 属性用来传递参数，这是一个 Object 类型，这里我们传入 i 的值，然后调用 handler.sendMessage（message）方法。当 sendMessage 方法调用之后 Handler 的 handleMessage 方法就会回调（更新 ui），最后调用 Thread.sleep（200）延迟 200 毫秒。

运行代码，每 200 毫秒更新一次界面，一共更新 100 次，for 循环结束，子线程退出。

5.3.3 Handler、Looper 与 MessageQueue 三者的关系

前面已对 Handler 进行介绍，也讲解了如何使用 Handler，但是并不知道它的实现原理。本节从源码的角度来分析是如何实现的。

首先需要知道 Handler、Looper 与 MessageQueue 三者之间的关系：

- Handler 封装了消息的发送，也负责接收消息。内部会与 Looper 关联。
- Looper 封装了线程中的消息循环，内部包含了 MessageQueue，负责从 MessageQueue 取出消息，然后交给 Handler 处理。
- MessageQueue 就是一个消息队列，负责存储消息，收到消息就存储起来，Looper 会循环地从 MessageQueue 读取消息。

1. 源码分析

当新建一个 Handler 对象时，查看它的构造方法中做了什么。默认的无参构造调用了自己两个有参数的构造方法。

```
public Handler() {
    this(null, false);
}
```

继续跟踪两个有参数的构造方法，第一个参数是一个 Callback 对象，用来拦截消息，后面讲 dispatchMessage 方法时就会明白，第二个参数用于说明是否异步。

```
public Handler(Callback callback, boolean async) {
    if (FIND_POTENTIAL_LEAKS) {
        final Class<? extends Handler> klass = getClass();
        if ((klass.isAnonymousClass() || klass.isMemberClass() ||
klass.isLocalClass()) && (klass.getModifiers() & Modifier.STATIC) == 0) {
            Log.w(TAG, "The following Handler class should be static or leaks
might occur: " + klass.getCanonicalName());
        }
    }

    mLooper = Looper.myLooper();
    if (mLooper == null) {
        throw new RuntimeException(
```

```
            "Can't create Handler inside thread that has not called Looper.prepare()");
    }
    mQueue = mLooper.mQueue;
    mCallback = callback;
    mAsynchronous = async;
}
```

在构造方法中会调用 Looper.myLooper 方法获取一个 Looper 对象，然后从 Looper 对象获取到 MessageQueue 对象。

2. 了解 Looper myLooper()

Looper.myLooper()是一个静态方法，能够以“类名.方法名”的形式直接调用。

```
public static @Nullable Looper myLooper() {
    return sThreadLocal.get();
}
```

这个方法中就有一行代码，从 sThreadLocal 中获取一个 Looper 对象，sThreadLocal 是一个 ThreadLocal 对象，可以在一个线程中存储变量。底层是 ThreadLocalMap，既然是 Map 类型，肯定先发送一个 Looper 对象，然后才能从 sThreadLocal 对象中获取一个 Looper 对象。

3. 了解 ActivityThread main()

说到这里，需要介绍一个 ActivityThread 新类。它是 Android App 进程的初始类，main 函数是 App 进程的入口。下面看一下 main 函数。

```
public static final void main(String[] args) {
        ------
        Looper.prepareMainLooper();
        if (sMainThreadHandler == null) {
            sMainThreadHandler = new Handler();
        }

        ActivityThread thread = new ActivityThread();
        thread.attach(false);

        if (false) {
            Looper.myLooper().setMessageLogging(new
                    LogPrinter(Log.DEBUG, "ActivityThread"));
        }

        Looper.loop();
        -----
}
```

在第 2 行代码调用了 Looper.prepareMainLooper()方法，第 13 行调用了 Looper.loop()方法。

4. 了解 Looper prepareMainLooper()

继续跟进 Looper.prepareMainLooper()方法，在该方法中第一行代码调用了内部的 prepare 方法。prepareMainLooper 有点类似单例模式中的 getInstance 方法，只不过 getInstance 会返回一个对象，而 prepareMainLooper 会新建一个 Looper 对象并存储在 sThreadLocal 中。

```
public static void prepareMainLooper() {
```

```
        prepare(false);
        synchronized (Looper.class) {
            if (sMainLooper != null) {
                throw new IllegalStateException("The main Looper has already been
prepared.");
            }
            sMainLooper = myLooper();
        }
    }
```

5. 了解 Looper prepare()

继续跟进 prepare 方法，查看第 5 行代码，新建了一个 Looper 对象，调用 sThreadLocal.set 方法将 Looper 对象保存起来。看到这里，肯定明白了为什么新建 Handler 对象时调用 Looper.myLooper()方法能够从 sThreadLocal 对象中获取 Looper 对象。

```
    private static void prepare(boolean quitAllowed) {
        if (sThreadLocal.get() != null) {
            throw new RuntimeException("Only one Looper may be created per thread");
        }
        sThreadLocal.set(new Looper(quitAllowed));
    }
```

6. Looper 构造方法

本章开头讲到 Looper 内部包含了 MessageQueue，其实就是在新建 Looper 对象的同时就新建了一个 MessageQueue 对象。

```
    private Looper(boolean quitAllowed) {
        mQueue = new MessageQueue(quitAllowed);
        mThread = Thread.currentThread();
    }
```

7. 了解 Looper loop()

ActivityThread 类 main 方法中调用了 Looper 的两个方法，前面已经解释了 prepareMainLooper()方法，现在查看第二个方法 loop()。

```
    public static void loop() {
        final Looper me = myLooper();//获取 Looper 对象
        if (me == null) {
            throw new RuntimeException("No Looper; Looper.prepare() wasn't called
on this thread.");
        }
        final MessageQueue queue = me.mQueue;//从 Looper 对象获取 MessageQueue 对象

        // Make sure the identity of this thread is that of the local process,
        // and keep track of what that identity token actually is.
        Binder.clearCallingIdentity();
        final long ident = Binder.clearCallingIdentity();

        for (;;) {//死循环，一直从 MessageQueue 中遍历消息
            Message msg = queue.next(); // might block
            if (msg == null) {
                return;
            }
```

```
        // This must be in a local variable, in case a UI event sets the logger
        final Printer logging = me.mLogging;
        if (logging != null) {
            logging.println(">>>>> Dispatching to " + msg.target + " " +
                    msg.callback + ": " + msg.what);
        }

        final long traceTag = me.mTraceTag;
        if (traceTag != 0 && Trace.isTagEnabled(traceTag)) {
            Trace.traceBegin(traceTag, msg.target.getTraceName(msg));
        }
        try {
            //调用 Handler 的 dispatchMessage 方法，将消息交给 Handler 处理
            msg.target.dispatchMessage(msg);
        } finally {
            if (traceTag != 0) {
                Trace.traceEnd(traceTag);
            }
        }

        if (logging != null) {
            logging.println("<<<<< Finished to " + msg.target + " " +
msg.callback);
        }

        // Make sure that during the course of dispatching the
        // identity of the thread wasn't corrupted.
        final long newIdent = Binder.clearCallingIdentity();
        if (ident != newIdent) {
            Log.wtf(TAG, "Thread identity changed from 0x"
                    + Long.toHexString(ident) + " to 0x"
                    + Long.toHexString(newIdent) + " while dispatching to "
                    + msg.target.getClass().getName() + " "
                    + msg.callback + " what=" + msg.what);
        }

        msg.recycleUnchecked();
    }
}
```

这个方法的代码比较多，给代码加上了一些注释。其实就是一个死循环，一直会从 MessageQueue 中获取消息。如果获取到消息，就会执行 msg.target.dispatchMessage(msg)这行代码，msg.target 就是 Handler，也就是调用 Handler 的 dispatchMessage 方法，然后将从 MessageQueue 中获取的消息传入。

8. Handler dispatchMessage()方法

前面的源码分析中，如果获取到了 Message，就调用 Handler 的 dispatchMessage 方法，dispatchMessage 方法对消息进行最后处理，如果是 post 类型，就调用 handlerCallback 方法处理，否则是 sendMessage 发送的消息。先看有没有拦截消息，如果没有，最终就调用 handlerMessage 方法进行处理。

```
public void dispatchMessage(Message msg) {
    //如果 callback 不为空，说明发送消息时是发送一个 Runnable 对象
    if (msg.callback != null) {
        handlerCallback(msg);
    } else {
        if (mCallback != null) {  //这是用来拦截消息的
            if (mCallback.handlerMessage(msg)) {
                return;
            }
        }
        handlerMessage(msg);  //最终调用重写的 HandlerMessage 方法
    }
}
```

9. 了解 Handler handlerCallback()

看到这里就应该知道为什么发送一个 Runnable 对象时 run 方法执行的代码在主线程了，因为底层根本就没有开启线程，只是调用了 run 方法而已。

```
private static void HandlerCallback(Message message) {
    message.callback.run();
}
```

前面介绍了创建 Handler 对象、创建 Looper 以及创建 MessageQueue 的整个流程。现在分析一下，当调用 post 以及 sendMessage 方法时如何将消息添加到 MessageQueue。Handler post()调用 getPostMessage 方法，将 Runnable 传递进去。

```
public final boolean post(Runnable r)
{
   return  sendMessageDelayed(getPostMessage(r), 0);
}
```

10. 了解 Handler getPostMessage()

首先调用 Message.obtain()方法，取出一个 Message 对象，然后将 Runnable 对象赋值为 Message 对象的 callback 属性。看到这里，也应该明白 dispatchMessage 方法为什么要先判断 callback 是否为空了。

```
private static Message getPostMessage(Runnable r) {
    Message m = Message.obtain();
    m.callback = r;
    return m;
}
```

11. 了解 Handler enqueueMessage()

在 post 方法中调用 sendMessageDelayed 方法，其实最终调用的是 enqueueMessage 方法，所以这里直接看 enqueueMessage 方法源码。第一行代码就把 Handler 自己赋值给 Message 对象的 target 属性，然后调用 MessageQueue 的 enqueueMessage 方法将当前的 Messgae 添加进去。

```
    private boolean enqueueMessage(MessageQueue queue, Message msg, long
uptimeMillis) {
        msg.target = this;
        if (mAsynchronous) {
            msg.setAsynchronous(true);
```

```
    }
    return queue.enqueueMessage(msg, uptimeMillis);
}
```

12. 小结

通过一步步的源码跟踪分析，相信读者对 Handler 的实现原理有了更深入的理解。一句话做一个总结：Handler 负责发送消息，Looper 负责接收 Handler 发送的消息，并直接将消息回传给 Handler 自己。MessageQueue 就是一个存储消息的容器。

5.4　使用 AsyncTask 创建后台线程

前面在多线程中更新 UI，用 Handler 类来发送消息，然后更新 UI。这种方式对于整个过程灵活控制，但是也存在缺点，代码比较臃肿。当多个任务同时执行时，不易对线程进行精确控制。

为了简化操作，Android 1.5 提供了工具类 android.os.AsyncTask，在代码上要比 Handler 轻量级。AsyncTask 底层是一个线程池，执行多任务时消耗资源较少。

AsyncTask 定义了三种泛型类型：Params、Progress 和 Result。

- Params：启动任务需要的参数，例如下载链接地址。
- Progress：执行的进度。
- Result：执行结果。

AsyncTask 需要重写以下四个方法：onPreExecute、doInBackground、onProgressUpdate 和 onPostExecute。

- onPreExecute：执行任务之前调用。
- doInBackground：执行任务的方法，该方法是多线程调用，所以耗时操作都写在这个方法中，通过调用 publishProgress 方法更新进度。
- onProgressUpdate：更新任务进度。在调用 publishProgress()时，这个方法才会被调用，该方法可以直接把数据更新到 UI 控件上。
- onPostExecute：任务执行完毕调用。

上面说了这么多的理论，想必听得也是一头雾水，直接贴代码来查看怎么使用，我们模拟从网上下载文件。

```
private class DownloadFilesTask extends AsyncTask<String,Integer,Long> {
    @Override
    protected void onPreExecute() {
        Log.i("DownloadFilesTask","执行任务之前");
    }

    protected Long doInBackground(String... url) {
        int count = url[0].length();//第一个字符串
        long totalSize = 0;
        for (int i = 0; i < count; i++) {
```

```
                totalSize += i;
                publishProgress(i);//执行此方法,会调用 onProgressUpdate 方法更新下载进度
                // 如果取消就结束任务
                if (isCancelled()) break;
            }
            return totalSize;
        }

        protected void onProgressUpdate(Integer... progress){
            Log.i("DownloadFilesTask","当前下载进度:"+progress[0].intValue());
        }

        protected void onPostExecute(Long result) {
            Log.i("DownloadFilesTask","下载完成:"+result);
        }
    }
```

在 MainActivity 中写了一个内部类 DownloadFilesTask 继承 AsyncTask。定义泛型，重写方法。接下来在 onCreate 中调用这个类。

```
new DownloadFilesTask().execute("www.downloadfile.com");
```

运行代码，查看打印的日志就知道代码的执行顺序。

```
    02-26 07:32:05.626 11871-11871/com.ansen.asynctask I/DownloadFilesTask: 执行
任务之前
    02-26 07:32:05.687 11871-11871/com.ansen.asynctask I/DownloadFilesTask: 当前
下载进度:0
    02-26 07:32:05.687 11871-11871/com.ansen.asynctask I/DownloadFilesTask: 当前
下载进度:1
    .......
    02-26 07:32:05.687 11871-11871/com.ansen.asynctask I/DownloadFilesTask: 当前
下载进度:18
    02-26 07:32:05.687 11871-11871/com.ansen.asynctask I/DownloadFilesTask: 当前
下载进度:19
    02-26 07:32:05.687 11871-11871/com.ansen.asynctask I/DownloadFilesTask: 下载
完成:190
```

5.5 线程池的使用

线程池是一种多线程处理形式，处理过程中将任务添加到队列，然后在创建线程后自动启动这些任务。

什么情况会用到线程池呢？假如你现在做一个音乐类 App，用户需要下载歌曲。下载歌曲很耗时，需要启动一个新线程进行下载。我们之前可能会采用下面的代码：

```
new Thread(new Runnable(){
    @Override
    public void run(){
        //下载歌曲
    }
}).start();
```

如果要下载 1000 首歌曲，是否要同时开启 1000 个线程？这样会产生什么问题呢？

- 每下载一首歌曲，就要新建一个线程，导致频繁地创建线程与销毁线程，使程序卡住或者程序崩溃。
- 这样创建的线程没法统一管理。
- 不方便统计，例如已下载完成歌曲的数量。

如果使用线程池就能完美地解决以上问题。线程池的优点如下：

- 重用已创建线程，不会频繁创建线程与销毁线程。
- 对线程统一管理、分配、调优和监控。
- 控制线程数量，合理使用系统资源。这样不会造成程序卡住或者程序崩溃。

Android 中常用的线程操作都是通过 ThreadPoolExecutor 类来实现的，此类最长的构造方法有 7 个参数：

```
public ThreadPoolExecutor(int corePoolSize,
                          int maximumPoolSize,
                          long keepAliveTime,
                          TimeUnit unit,
                          BlockingQueue<Runnable> workQueue,
                          ThreadFactory threadFactory,
                          RejectedExecutionHandler handler)
```

- corePoolSize：核心线程数。
- maximumPoolSize：线程池最大线程数。
- keepAliveTime：线程空闲时保持时间。如果线程池中线程数量超过核心线程数量，并且多出的线程空闲时间超过 keepAliveTime，就会结束多出的线程，从而及时销毁多出来的空闲线程，减少资源消耗。如果线程池任务增多，就重新创建新线程。这个参数也可以通过 setKeepAliveTime(long, TimeUnit)方法动态设置。默认情况下，这个 keepAliveTime 参数针对的是非核心线程。如果调用 allowCoreThreadTimeOut(boolean)方法，传入 true，keepAliveTime 超时策略就会运用在核心线程上。
- unit：第三个参数的单位。这是一个枚举，常用的有 TimeUnit.SECONDS（秒），具体还有以下值：
    - TimeUnit.DAYS（天）。
    - TimeUnit.HOURS（小时）。
    - TimeUnit.MICROSECONDS（微秒）。
    - TimeUnit.MILLISECONDS（毫秒）。
    - TimeUnit.MINUTES（分钟）。
    - TimeUnit.NANOSECONDS（纳秒）。
    - TimeUnit.SECONDS（秒）。
- workQueue：线程池中的任务队列。这个队列保存线程池提交的任务，它的使用与线程池中的线程数量有关，具体规则如下：
    - 如果当前的线程池运行的线程数少于 corePoolSize，则 execute 方法执行任务时会开启一个核心线程进行处理。

- ➢ 如果线程池中的线程数量达到核心线程数，并且 workQueue 未满，当 execute 方法执行任务时不会开启新线程，而是将任务加入 workQueue 队列中等待处理。
- ➢ 当 execute 方法执行任务时，线程池中的线程数已达到核心线程数，并且 workQueue 队列已满。这时就会判断线程池中的线程数是否大于 maximumPoolSize，如果没有大于 maximumPoolSize，就会开启非核心线程处理任务；如果大于 maximumPoolSize，就拒绝执行该任务。

- BlockingQueue：阻塞队列，有以下三个常用的 BlockingQueue 实现类。
  - ➢ SynchronousQueue：一种无缓冲的等待队列，它的特别之处在于内部没有容器，当它生产产品（put）时，如果当前没有人想要消费产品（当前没有线程执行 take），此生产线程必然阻塞，等待一个消费线程调用 take 操作，take 操作将会唤醒该生产线程，同时消费线程会获取生产线程的产品（数据传递），这样的一个过程称为一次配对过程（当然，也可以先 take 后 put，原理是一样的）。
  - ➢ LinkedBlockingQueue：无界队列。调用 execute 方法执行任务，线程池中的核心线程都在运行时，使用无界队列（例如创建时没有指定大小）会将新的任务加入 inkedBlockingQueue 中等待执行。当然，这个队列创建时，构造方法中也可以指定大小。
  - ➢ ArrayBlockingQueue：必须要指定大小的队列，构造方法要传入一个 int 类型参数，设置队列的大小。存储在 ArrayBlockingQueue 中的元素按照 FIFO（先进先出）的方式来进行存取。
- threadFactory：创建新线程的工厂类，可以通过 Executors.defaultThreadFactory()获取系统给封装的线程工厂。我们也可以自己手动实现一个，继承 ThreadFactory 接口，实现 newThread 方法。
- handler：拒绝任务。调用 execute(Runnable)方法提交新任务时，如果线程池关闭或者线程池线程数量等于 maximumPoolSize 并且队列满了，execute 内部就会调用 RejectedExecutionHandler 接口实现类的 rejectedExecution(Runnable, ThreadPoolExecutor)方法。

针对拒绝任务，SQK 提供了以下几种策略：

- ThreadPoolExecutor.AbortPolicy：添加任务被拒绝，这是默认策略，如果不传递 handler 参数，默认就是这个值。
- ThreadPoolExecutor.CallerRunsPolicy：提供一个反馈机制，告诉调用者可以减慢提交新任务的速度。
- ThreadPoolExecutor.DiscardPolicy：任务无法执行被丢弃。
- ThreadPoolExecutor.DiscardOldestPolicy：如果线程池没有被关闭，丢弃队列最前面的任务，然后重新尝试执行任务（可能会再次失败，导致重复执行）。

我们也可以自定义拒绝策略，编写一个类实现 RejectedExecutionHandler 接口，重写 rejectedExecution 方法。

```
    public static class MyRejectedExecutionHandler implements
RejectedExecutionHandler {
        public void rejectedExecution(Runnable r, ThreadPoolExecutor e) {
            throw new RejectedExecutionException("任务被拒绝");
        }
    }
```

学习了这么多理论知识，下面使用一段代码进行实践：

```
    MyRejectedExecutionHandler handler=new MyRejectedExecutionHandler();

    ThreadPoolExecutor threadPoolExecutor = new ThreadPoolExecutor(
          2, 5, 30, TimeUnit.SECONDS,new LinkedBlockingQueue<Runnable>
          (128), sThreadFactory,handler);

    for(int i=0;i<10;i++){
       final int iValue=i;
       Runnable runnable=new Runnable(){
          @Override
          public void run() {
             SystemClock.sleep(1000);
             Log.i("ansen","当前线程 id:"+android.os.Process.myTid()+"
iValue:"+iValue);
          }
       };
       threadPoolExecutor.execute(runnable);
    }
```

我们主要查看 new ThreadPoolExecutor 对象时传入的参数：核心线程为 2，最大线程数为 5，线程空闲时保持时间 30 秒；LinkedBlockingQueue 为无界队列，其长度为 128；sThreadFactory 参数与 handler 参数没有用默认值，都自己重写了。

接下来是 for 循环 10 次，在循环中调用 execute 方法执行 Runnable。

sThreadFactory 参数使用内部类方式实现了 ThreadFactory 接口：

```
    private static final ThreadFactory sThreadFactory = new ThreadFactory() {
       //可以在并发情况下达到原子更新，避免使用 synchronized，而且性能非常高
       private final AtomicInteger mCount = new AtomicInteger(1);

       public Thread newThread(Runnable r) {
          return new Thread(r,"ThreadPoolExecutor new Thread #"+mCount.
getAndIncrement());
       }
    };
```

handler 参数是自定义拒绝策略类，这个类的实现前面已经见过，就不贴出相关代码了。

运行代码，打印结果如下：

```
    03-09 14:58:24.038 15397-15526/... I/ansen: 当前线程 id:15526 iValue:0
    03-09 14:58:24.038 15397-15527/... I/ansen: 当前线程 id:15527 iValue:1
    03-09 14:58:25.040 15397-15526/... I/ansen: 当前线程 id:15526 iValue:2
    03-09 14:58:25.040 15397-15527/... I/ansen: 当前线程 id:15527 iValue:3
    03-09 14:58:26.042 15397-15526/... I/ansen: 当前线程 id:15526 iValue:4
    03-09 14:58:26.043 15397-15527/... I/ansen: 当前线程 id:15527 iValue:5
    03-09 14:58:27.044 15397-15526/... I/ansen: 当前线程 id:15526 iValue:6
    03-09 14:58:27.045 15397-15527/... I/ansen: 当前线程 id:15527 iValue:7
    03-09 14:58:28.046 15397-15526/... I/ansen: 当前线程 id:15526 iValue:8
    03-09 14:58:28.046 15397-15527/... I/ansen: 当前线程 id:15527 iValue:9
```

从结果中可以看到一共就开启了两个线程，线程 id 是 15526 与 15527，新建 ThreadPoolExecutor 对象时 corePoolSize 也是传入的 2，LinkedBlockingQueue 长度为 128，execute 只循环执行了 10 次。

这也印证了我们之前的结论，如果 execute 提交的任务数量小于 BlockingQueue 长度，那么线程池中的线程数量只会等于核心线程数，其余加入 BlockingQueue 队列，依次等待执行。

修改代码，将 LinkedBlockingQueue 的长度从之前的 128 修改为 6。

```
ThreadPoolExecutor threadPoolExecutor = new ThreadPoolExecutor(
2, 5, 30, TimeUnit.SECONDS,
new LinkedBlockingQueue<Runnable>(6), sThreadFactory,handler);
```

修改后，运行打印结果如下：

```
03-09 15:07:28.445 15836-15890/ I/ansen: 当前线程 id:15890 iValue:0
03-09 15:07:28.445 15836-15891/ I/ansen: 当前线程 id:15891 iValue:1
03-09 15:07:28.446 15836-15892/ I/ansen: 当前线程 id:15892 iValue:8
03-09 15:07:28.446 15836-15893/ I/ansen: 当前线程 id:15893 iValue:9
03-09 15:07:29.447 15836-15890/ I/ansen: 当前线程 id:15890 iValue:3
03-09 15:07:29.447 15836-15891/ I/ansen: 当前线程 id:15891 iValue:2
03-09 15:07:29.447 15836-15892/ I/ansen: 当前线程 id:15892 iValue:4
03-09 15:07:29.449 15836-15893/ I/ansen: 当前线程 id:15893 iValue:5
03-09 15:07:30.449 15836-15891/ I/ansen: 当前线程 id:15891 iValue:7
03-09 15:07:30.449 15836-15890/ I/ansen: 当前线程 id:15890 iValue:6
```

当我们调用 execute 方法调用前两次时，会开启两个核心线程进行处理，循环到第 6 次时还能加入 LinkedBlockingQueue 队列，循环到第 7 次时发现队列满了，这时就会开启非核心线程去处理任务。

我们继续修改 ThreadPoolExecutor 参数，将 maximumPoolSize 最大线程数改为 3。

```
ThreadPoolExecutor threadPoolExecutor = new ThreadPoolExecutor(
2, 3, 30, TimeUnit.SECONDS,
new LinkedBlockingQueue<Runnable>(6), sThreadFactory,handler);
```

运行结果如下：

```
03-09 15:42:18.944 19397-19397/... E/AndroidRuntime: FATAL EXCEPTION: main
Process: ..., PID: 19397
java.util.concurrent.RejectedExecutionException: 任务被拒绝
at com.ansen.threadpoolexecutor.MainActivity$MyRejectedExecutionHandler.
   rejectedExecution(MainActivity.java:78)
/////
at com.android.internal.os.RuntimeInit$MethodAndArgsCaller.run(RuntimeInit.
   java:438)
at com.android.internal.os.ZygoteInit.main(ZygoteInit.java:807)
03-09 15:42:19.943 19397-19589/... I/ansen: 当前线程 id:19589 iValue:1
03-09 15:42:19.943 19397-19588/... I/ansen: 当前线程 id:19588 iValue:0
03-09 15:42:19.944 19397-19590/... I/ansen: 当前线程 id:19590 iValue:8
03-09 15:42:20.943 19397-19589/... I/ansen: 当前线程 id:19589 iValue:2
03-09 15:42:20.944 19397-19588/... I/ansen: 当前线程 id:19588 iValue:3
03-09 15:42:20.944 19397-19590/... I/ansen: 当前线程 id:19590 iValue:4
03-09 15:42:21.944 19397-19589/... I/ansen: 当前线程 id:19589 iValue:5
03-09 15:42:21.945 19397-19588/... I/ansen: 当前线程 id:19588 iValue:6
03-09 15:42:21.946 19397-19590/... I/ansen: 当前线程 id:19590 iValue:7
```

从 ThreadPoolExecutor 构造方法中可以看到最大线程数是 3，LinkedBlockingQueue 队列的长度是 6，9 个任务是处理的极限了。当 execute 方法启动第 10 个任务时会抛出异常，拒绝添加任务。

自己配置 ThreadPoolExecutor 类，面对这么多的参数，当用到线程池时也不知道核心线程数给多少合适，用 LinkedBlockingQueue 队列好还是 ArrayBlockingQueue 好。还记得我们前面学过 AsyncTask 创建后台线程吧，其实它内部就用到了线程池，我们跟踪到源码里面看看是如何实现的。

```
private static final int CPU_COUNT = Runtime.getRuntime().availableProcessors();
private static final int CORE_POOL_SIZE = Math.max(2, Math.min(CPU_COUNT - 1, 4));
private static final int MAXIMUM_POOL_SIZE = CPU_COUNT * 2 + 1;
private static final int KEEP_ALIVE_SECONDS = 30;
private static final ThreadFactory sThreadFactory = new ThreadFactory() {
    private final AtomicInteger mCount = new AtomicInteger(1);

    public Thread newThread(Runnable r) {
        return new Thread(r, "AsyncTask #" + mCount.getAndIncrement());
    }
};
private static final BlockingQueue<Runnable> sPoolWorkQueue =
        new LinkedBlockingQueue<Runnable>(128);

ThreadPoolExecutor threadPoolExecutor = new ThreadPoolExecutor(
CORE_POOL_SIZE, MAXIMUM_POOL_SIZE, KEEP_ALIVE_SECONDS, TimeUnit.SECONDS,
sPoolWorkQueue, sThreadFactory);
threadPoolExecutor.allowCoreThreadTimeOut(true);
```

核心线程数跟 CPU 核数有关，结果在 2~4。最大线程数就是 CPU 核数*2+1，线程空闲时保持 30 秒，用的是 LinkedBlockingQueue 队列，设置长度是 128，并且调用了 allowCoreThreadTimeOut(true) 方法，也就是说核心线程闲置时间超过 keepAliveTime 之后照样回收。

除了我们根据 AsyncTask 类来配置参数之外，其实系统已经给我们封装了几个常用的线程池。

1. Executors.newCachedThreadPool() 队列大小不固定线程池（线程超时自动回收）

源码如下：

```
public static ExecutorService newCachedThreadPool() {
    return new ThreadPoolExecutor(0, Integer.MAX_VALUE,
    60L, TimeUnit.SECONDS,
    new SynchronousQueue<Runnable>());
}
```

从源码中可以看到没有核心线程，最大线程数是 int 类型的最大值，几乎是无穷大，线程空闲超时时间是 30 秒，并且用 SynchronousQueue 作为队列。

这个队列有什么优点呢？当我们调用 execute 方法执行一个新任务时，线程池就会开启一个新线程去处理。有多少个任务就会启动多少个线程。最大线程数是无穷大，所以可以提交大量任务，需要注意的是没有核心线程，并且设置了线程空闲时间为 30 秒，所以线程池会自动回收空闲线程。当所有任务执行完毕并且没有提交新任务时，线程池中一个线程也没有。

2. Executors.newFixedThreadPool(int)创建固定线程数的线程池

源码如下：

```
public static ExecutorService newFixedThreadPool(int nThreads) {
    return new ThreadPoolExecutor(nThreads, nThreads,
                                  0L, TimeUnit.MILLISECONDS,
                                  new LinkedBlockingQueue<Runnable>());
}
```

从源码中我们可以看到核心线程数跟最大线程数是一样的值，就是我们传入的线程数。线程空闲时保持时间是 0 毫秒，并且 LinkedBlockingQueue 队列没指定长度。

这个线程池有什么特点呢？如果线程池中所有的线程处于忙碌状态，新添加的任务就会保存在队列中，队列没有指定长度基本可以无限添加。线程空闲时也不会超时销毁，所有线程池中线程数量永远固定并且一直存在。除非调用 shutdown()方法才会销毁线程。这个线程池不会频繁地创建和销毁线程，线程数永远保持在一个固定的数量。

3. Executors.newSingleThreadExecutor()单线程线程池

源码如下：

```
public static ExecutorService newSingleThreadExecutor() {
    return new FinalizableDelegatedExecutorService
        (new ThreadPoolExecutor(1, 1,
        0L, TimeUnit.MILLISECONDS,
        new LinkedBlockingQueue<Runnable>()));
}
```

源码跟 newFixedThreadPool 线程池有点相似，唯一的区别就是核心线程数跟最大线程数都是 1，也就是说这个线程池永远只有一个线程，保证所有任务按照指定顺序执行。

4. ScheduledThreadPoolExecutor()定时定期执行任务功能的线程池

ScheduledThreadPoolExecutor 是一个具有定时定期执行任务功能的线程池，是 ThreadPoolExecutor 的子类，跟前面介绍的其他线程池一样，也是基于 ThreadPoolExecutor 类实现的线程池。源码如下：

```
private static final long DEFAULT_KEEPALIVE_MILLIS = 10L;
public ScheduledThreadPoolExecutor(int corePoolSize) {
    super(corePoolSize, Integer.MAX_VALUE,
          DEFAULT_KEEPALIVE_MILLIS, MILLISECONDS,
          new DelayedWorkQueue());
}
```

构建 ScheduledThreadPool 线程池必须要指定核心线程数，最大线程数是无穷大，非核心线程空闲时间是 10 毫秒，队列用的是 DelayedWorkQueue。DelayedWorkQueue 是无界的 BlockingQueue，用于放置实现了 Delayed 接口的对象，其中的对象只能在其到期时才能从队列中取走。这种队列是有序的，即队头对象的延迟到期时间最长。

写个简单的 demo 来实现一下：

```
ScheduledThreadPoolExecutor executorService=new ScheduledThreadPoolExecutor(1);
Runnable runnable=new Runnable() {
    @Override
    public void run() {
        Log.i("ansen", "ScheduledThreadPoolExecutor 任务执行完成");
    }
};
Log.i("ansen", "ScheduledThreadPoolExecutor 任务开始完成");
executorService.schedule(runnable,5,TimeUnit.SECONDS);
```

创建 ScheduledThreadPoolExecutor 对象时指定核心线程数，这里我们传入 1。需要注意的是我们调用 schedule 方法执行任务，有三个参数，整体意思就是 5 秒后执行任务。

- command Runnable 对象，要执行的任务。
- delay，任务延迟多久执行。
- unit，第二个参数的单位。

运行代码，打印 log 如下：

```
03-10 17:25:40.125 ... I/ansen: ScheduledThreadPoolExecutor 任务开始完成
03-10 17:25:45.128 ... I/ansen: ScheduledThreadPoolExecutor 任务执行完成
```

从两条 log 中可以看到时间差不多相差 5 秒。

5. 线程池其他常用方法

- shutDown：已经执行的任务继续执行，队列中的任务停止执行，并且不再接收新任务。
- shutdownNow：尝试停止所有正在执行的任务，暂停队列中的任务，返回正在执行的任务列表。
- execute：执行任务。
- submit：执行任务并且返回 Future 接口，通过这个 Future 的 get 方法取得返回值。通过它的 isDone 方法可以判断是否执行成功！
- setKeepAliveTime：线程空闲时销毁时间。
- allowCoreThreadTimeOut(boolean value)：设置为 true，线程空闲销毁时间针对核心线程；设置为 false，线程空闲销毁时间针对非核心线程。

第 6 章

Android 网络编程与数据存储

随着互联网飞速的发展，很多行业都离不开网络，应用市场上大部分 App 也都需要服务器支持。本章就学习在 Android 中如何发送网络请求，以及网络请求开源库 OKHttp 的使用、对 OKHttp 进行简单封装。同时，学习数据存储的三种方式以及不同之处，以便我们在不同的业务场景下选择不同的存储方式。

6.1 基于 Android 平台的 HTTP 通信

HTTP（Hypertext Transfer Protocol，超文本传输协议）是互联网上应用最为广泛的一种网络协议，也是手机联网常用的协议之一，所有的 WWW 文件必须遵守这个标准。HTTP 协议是建立在 TCP 协议之上的一种协议。简单来说，HTTP 协议就是客户端和服务器端之间数据传输的格式规范。

HTTP 协议具有以下特点：

- 支持 B/S 以及 C/S 模式。
- 简单快速：客户向服务器请求服务时，只需传送请求方法和路径。常用的请求方法有 Get、Head 和 Post。
- 灵活：HTTP 允许传输任意类型的数据对象，正在传输的类型由 Content-Type 加以标记。
- 无状态：HTTP 协议是无状态协议。无状态是指协议对于事务处理没有记忆能力。如果后续处理需要前面的信息，就必须重传，这样可能会导致每次连接传送的数据量增大。

HTTP 协议包括两个具体的请求方式 Get 以及 Post，主要有以下区别：

- Get 通常是从服务器上获取数据。Post 通常是向服务器传送数据。
- Get 将参数拼接在 URL 后面。Post 是将参数作为 HTTP 请求的内容发送到指定的 URL 中。
- Get 传送数据量比较小，最多只能是 2048 字节（不同的浏览器略有区别）。Post 传送的数据量比较大，一般被默认为不受限制。

- Get 安全性非常低。Post 安全性较高。

一般企业开发中，从服务器获取数据时使用 Get 请求，而增加、删除、修改时使用 Post 请求。

6.1.1　使用 Get 方式向服务器提交数据

使用 Get 请求方式访问服务器的登录接口。在 Android 中，提供了标准 Java 接口 HttpURLConnection，通过 url.openConnection()方法就能够获取。

因为要访问网络，所以要在 AndroidManifest.xml 文件中增加访问网络的权限：

```
<uses-permission android:name="android.permission.INTERNET"/>
```

新建一个项目，在 MainActivity 中增加一个方法 getUserInfo(String userid)。首先新建一个 URL 对象，在构造方法中传入一个请求路径，请求路径根据 userid 拼接。接下来，调用 url.openConnection()方法，打开一个链接，强转成 HttpURLConnection 对象。这个对象有很多方法，例如 setConnectTimeout 设置连接超时时间，setRequestMethod 设置请求方式，这里设置的是 Get 请求。最后判断返回的状态码是不是 200。200 代表请求成功，从连接中获取输入流，通过 dealResponseResult 方法读取出来并且转换成字符串。

```
    private String getUserInfo(String userid){
            //get 方式提交就是 url 拼接的方式
            String path="http://139.196.35.30:8080/OkHttpTest/getUserInfo.do?userid=
                        "+userid;
            try {
               URL url = new URL(path);
               HttpURLConnection connection = (HttpURLConnection)
url.openConnection();
               connection.setConnectTimeout(5000);//设置连接超时时间
               connection.setRequestMethod("GET");//设置以 Get 方式提交数据
               if(connection.getResponseCode() ==200){//请求成功
                  InputStream is = connection.getInputStream();
                  return dealResponseResult(is);
               }
            }catch (Exception e) {
               e.printStackTrace();
            }
            return null;
        }
```

上面用到的 dealResponseResult 方法代码如下。从输入流中循环读取 byte 数组，存入输出流，最后将输出流转换成字符串返回。

```
    /**
     * 处理服务器的响应结果（将输入流转化成字符串）
     * @param inputStream 服务器的响应输入流
     * @return
     */
    private String dealResponseResult(InputStream inputStream) {
       String resultData = null;      //存储处理结果
```

```
        ByteArrayOutputStream byteArrayOutputStream = new ByteArrayOutputStream();
        byte[] data = new byte[1024];
        int len = 0;
        try {
            while((len = inputStream.read(data)) != -1) {
                byteArrayOutputStream.write(data, 0, len);
            }
        } catch (IOException e) {
            e.printStackTrace();
        }
        resultData = new String(byteArrayOutputStream.toByteArray());
        return resultData;
    }
```

6.1.2 使用 Post 方式向服务器提交数据

在 MainActivity 中继续增加 login（String username,String password）方法，这是一个 Post 请求。Post 请求与 Get 请求的代码差不多，只是传递参数的方式不一样。setRequestMethod 方法传入的是 Post，并且需要通过 setRequestProperty 方法设置请求属性，然后从 connection 中获取输出流，通过 write 方法将参数写进去，后面请求成功并且解析成字符串与 Get 请求一致。

```
    private String login(String username,String password){
        String path = "http://139.196.35.30:8080/OkHttpTest/login.do";
        try {
            URL url = new URL(path);
            HttpURLConnection connection = (HttpURLConnection)
url.openConnection();
            connection.setConnectTimeout(5000);//设置连接超时时间
            connection.setRequestMethod("POST");//设置以 Post 方式提交数据
            String data = "username="+username+"&password="+password;//请求数据

            //至少要设置的两个请求头
            connection.setRequestProperty("Content-Type","application/
             x-www-form- urlencoded");
            connection.setRequestProperty("Content-Length",data.length()+"");

            //post 方式提交的实际上是以流的方式提交给服务器
            connection.setDoOutput(true);
            OutputStream outputStream = connection.getOutputStream();
            outputStream.write(data.getBytes());

            if(connection.getResponseCode() ==200){//状态码==200 请求成功
                InputStream is = connection.getInputStream();
                return dealResponseResult(is);
            }
        }catch (Exception e) {
            e.printStackTrace();
        }
        return null;
    }
```

两种访问方式封装了两个方法。接下来在 MainActivity 中通过点击按钮来调用这两个方法。在

Android 中不能在主线程中访问网络（防止 UI 堵塞），所以需要开启一个线程来访问网络。

```
@Override
public void onClick(View view) {
    //在 Android 中不能在主线程中访问网络，所以需要开启一个线程
    new TestGetOrPostThread(view).start();
}

public class TestGetOrPostThread extends Thread{
        private View view;
        public TestGetOrPostThread(View view){
            this.view=view;
        }

        @Override
        public void run() {
            switch (view.getId()){
                case R.id.btn_get://get 请求
                    String getResult=getUserInfo("123");
                    Log.i("MainActivity","Get 获取用户信息:"+getResult);
                    break;
                case R.id.btn_post://post 请求
                    String postResult=login("ansen","123");
                    Log.i("MainActivity","Post 登录结果:"+postResult);
                    break;
            }
        }
}
```

点击“Get 请求”按钮打印日志如下，结果 JSON 只是截取了一部分：

```
I/MainActivity: Get 获取用户信息:{"errorReason":"","password":"123",
"username": "ansen"}
```

点击“Post 请求”按钮，打印日志如下：

```
I/MainActivity: Post 登录结果:{"errorReason":"登录成功","password":"123",
"username": "ansen"}
```

这两个请求的接口是用 Java Web 编写的服务器，部署在云服务器上，所以大家测试时，只要网络环境是正常的，就能够访问这两个接口并且返回 JSON 数据。

6.1.3 使用 GSON 解析 JSON 格式的数据

普通的 JSON 解法是通过 JsonObject 和 JsonArray 这两个对象配合完成的，有一个缺点，即当服务器返回的数据比较复杂时解析起来很麻烦，如果层次很多就需要一层层遍历。接下来讲解如何用 GSON 包解析 JSON 数据。

1. JSON 介绍

JSON（JavaScript Object Notation, JS 对象标记）是一种轻量级的数据交换格式。它是基于 ECMAScript 规范的一个子集，采用完全独立于编程语言的文本格式来存储和表示数据。简洁和清晰的层次结构使得 JSON 成为理想的数据交换语言，易于阅读和编写，同时也易于机器解析和生成，

并有效地提升网络传输效率。现在大部分 App 都使用 JSON 给前端返回数据。

2. GSON 包

GSON 是 Google 提供的用来在 Java 对象和 JSON 数据之间进行映射的 Java 类库，可以将一个 JSON 字符串转换成一个 Java 对象，或者反过来转换。GSON 包解析 JSON 数据主要有以下几个优点：

- 快速、高效。
- 代码量少、简洁。
- 面向对象（直接把 JSON 转换成对象）。

3. 几种常见的 JSON 数据如何解析成 Java 对象

使用 Android Studio 开发时，可以通过在线引用库的方式。在 project/app/build.gradle 文件中的 dependencies 下加入一句代码：

```
compile 'com.google.code.gson:gson:2.8.0'
```

Gradle 会默认从 Jcenter Maven 仓库获取 aar 文件。

（1）解析对象

现在有一个 JSON 字符串 "{'name':'Ansen', 'age':20}"，包含 name 与 age 两个属性，可以写一个实体类 User 进行对应，并重写 toString 方法。输出对象时会输出所有属性，而不是一个 hash 值。

```
public class User {
    private String name;//姓名
    private int age;//年龄

    public String getName() {
        return name;
    }

    public void setName(String name) {
        this.name = name;
    }

    public int getAge() {
        return age;
    }

    public void setAge(int age) {
        this.age = age;
    }

    @Override
    public String toString() {
        return "User{" +
                "name='" + name + '\'' +
                ", age=" + age +
                '}';
    }
}
```

使用 GSON 包将 JSON 字符串转换成 User 对象。这里使用的 GSON 对象是 MainActivity 的实例变量。调用 fromJson 方法就能转换。

```
    String jsonStr="{'name':'Ansen', 'age':20}";
    User user=gson.fromJson(jsonStr, User.class);
    Log.i("MainActivity","parseObject user:"+user.toString());
```

（2）解析成数组

上面将 JSON 字符串转换成了对象，在实际开发中，服务器给用户一个数组格式的 JSON 数据也很常见，与转换对象基本是一样的。

```
    String jsonStr="[{'name':'Uini', 'age':30},{'name':'Lina', 'age':10}]";
    List<User> users=gson.fromJson(jsonStr,new TypeToken<List<User>>()
{}.getType());
    for(int i=0;i<users.size();i++){
       Log.i("MainActivity","parseArrayList user:"+users.get(i));
    }
```

（3）解析成 Map

```
    String jsonStr="{'1':{'name':'haha', 'age':11},'2':{'name':'nihao', 'age':22}}";
    Map<String, User> users = gson.fromJson(jsonStr, new
TypeToken<Map<String,User>>() {}.getType());
    for(String key:users.keySet()){
       Log.i("MainActivity","parseMap key:"+key+" user:"+users.get(key));
    }
```

（4）对象解析成 JSON 字符串

我们已经知道了将 JSON 字符串转换成对象的方法，那么如何将对象转换成 JSON 字符串呢？其实也很简单，GSON 包都封装好了。新建一个 User 对象，然后调用 GSON 类的 toJson 方法就能将 User 对象转换成 JSON 字符串。

```
    User user=new User();
    user.setAge(111);
    user.setName("nime");
    String jsonStr=gson.toJson(user);
    Log.i("MainActivity","jsonStr:"+jsonStr);
```

运行以上代码，打印的日志如下：

```
    05-15 09:56:20.430 5630-5630/? I/MainActivity: parseObject
user:User{name='Ansen', age=20}
    05-15 09:56:20.440 5630-5630/? I/MainActivity: parseArrayList
user:User{name='Uini', age=30}
    05-15 09:56:20.440 5630-5630/? I/MainActivity: parseArrayList
user:User{name='Lina', age=10}
    05-15 09:56:20.440 5630-5630/? I/MainActivity: parseMap key:1
user:User{name='haha', age=11}
    05-15 09:56:20.440 5630-5630/? I/MainActivity: parseMap key:2
user:User{name='nihao', age=22}
    05-15 09:56:20.440 5630-5630/? I/MainActivity: jsonStr:{"name":"nime","age":111}
```

6.1.4　OkHttp 开源项目的使用

Android 系统提供了两种 HTTP 通信类：HttpURLConnection 和 HttpClient。HttpURLConnection

相对来说比 HttpClient 难用，Google 自从 2.3 版本之后一直推荐使用 HttpURLConnection，并且在 6.0 版本的 SDK 中直接删除了 HttpClient 类。

不过，上面两个类库和 OkHttp 相比起来就弱爆了，因为 OkHttp 不仅具有高效的请求效率，并且节省宽带，还提供了很多开箱即用的网络疑难杂症解决方案。

- 支持 HTTP/2。HTTP/2 通过使用多路复用技术在一个单独的 TCP 连接上支持并发，通过在一个连接上一次性发送多个请求来发送或接收数据。
- 如果 HTTP/2 不可用，连接池减少请求延迟。
- 支持 GZIP，可以压缩下载体积。
- 响应缓存可以避免重复请求网络。
- 从很多常用的连接问题中自动恢复，如果服务器配置了多个 IP 地址，当第一个 IP 连接失败时，OkHttp 会自动尝试下一个 IP。
- OkHttp 还处理了代理服务器问题和 SSL 握手失败问题。

1. OkHttp 的基本使用

这里将讲解 OkHttp 的基本使用方法，大家有其他需求时可以自行扩展。以下的所有请求都是异步请求服务器，在实际开发中，基本都是异步。

（1）依赖

Android Studio 可以在线依赖，在 app/build.gradle 文件中加上下面这句代码即可。

```
compile 'com.squareup.okhttp3:okhttp:3.8.0'
```

（2）Get 请求

首先创建一个全局的 OkHttpClient 对象，所有的 Http 请求都共用这个对象即可。

```
private OkHttpClient client = new OkHttpClient();
```

一般从服务器获取信息的接口都是 Get 请求，这里调用获取用户的信息接口。

```
private void getUserInfo(){
    //创建一个请求
    Request.Builder builder = new Request.Builder().url
                ("http://139.196.35.30:8080/OkHttpTest/getUserInfo.do");
    execute(builder);
}

//执行请求
private void execute(Request.Builder builder){
    Call call = client.newCall(builder.build());
    call.enqueue(callback);//加入调度队列
}

//请求回调
private Callback callback=new Callback(){
        @Override
        public void onFailure(Call call, IOException e) {
            Log.i("MainActivity","onFailure");
            e.printStackTrace();
        }
```

```
        @Override
        public void onResponse(Call call, Response response) throws IOException{
            //从 response 获取服务器返回的数据，转换成字符串处理
            String str = new String(response.body().bytes(),"utf-8");
            Log.i("MainActivity","onResponse:"+str);

            //通过 Handler 更新 UI
            Message message=handler.obtainMessage();
            message.obj=str;
            message.sendToTarget();
        }
};
```

- 创建一个 request 对象，通过 request 设置请求 url，通过这个类还可以设置更多的请求信息。
- 通过 Request 去构造一个 Call 对象。
- 调用 enqueue 执行异步请求，有一个参数设置回调。请求成功或者失败会调用 Callback 接口的 onResponse 与 onFailure 方法，因为这是异步请求，在回调方法中不能直接更新 UI，所以需要通过 Handler 去更新 UI。

Handler 的代码很简单，就是将请求的结果显示在 TextView 上：

```
private Handler handler=new Handler(){
    @Override
    public void handleMessage(Message msg) {
        String result= (String) msg.obj;
        tvResult.setText(result);
    }
};
```

（3）Post 请求

通过调用登录接口发送一个 Post 请求。与 Get 不一样的地方就是传递参数不一样，Post 请求需要将参数封装到 RequestBody 对象，调用 Request 对象的 Post 方法将 RequestBody 传入进去。最后调用 execute 方法执行请求，这个方法在前面讲解 Get 请求时介绍过。

```
private void login(){
    //将请求参数封装到 RequestBody 中
    FormBody.Builder formBuilder = new FormBody.Builder();
    formBuilder.add("username","ansen");//请求参数一
    formBuilder.add("password","123");//请求参数二
    RequestBody requestBody = formBuilder.build();

    Request.Builder builder = new Request.Builder().url("http:
            //139.196.35.30:8080/ OkHttpTest/login.do").post(requestBody);
    execute(builder);
}
```

（4）文件上传

上传文件需要用到 MultipartBody 对象，通过调用 addFormDataPart 方法添加表单参数，通过 setType 方法设置内容类型，这里设置 form 表单类型，调用自己的 getUploadFileBytes 方法获取文件 byte 数组。通过 addFormDataPart 方法添加文件，后面的流程与之前的 Post 请求一样。

```
    private void uploadFile(){
        MultipartBody.Builder builder = new MultipartBody.Builder();
        builder.addFormDataPart("username", "ansen");//表单参数
        builder.addFormDataPart("password", "123456");//表单参数

        builder.setType(MultipartBody.FORM);
        MediaType mediaType = MediaType.parse("application/octet-stream");

        byte[] bytes=getUploadFileBytes();//获取文件内容存入 byte 数组
        //上传文件 参数 1:name 参数 2:文件名称 参数 3:文件 byte 数组
        builder.addFormDataPart("upload_file", "ansen.txt",RequestBody.create
(mediaType,bytes));
        RequestBody requestBody = builder.build();
        Request.Builder requestBuider = new Request.Builder();
        requestBuider.url("http://139.196.35.30:8080/OkHttpTest/uploadFile.do");
        requestBuider.post(requestBody);
        execute(requestBuider);
    }
```

如何证明文件上传到服务器呢？只需打开浏览器，输入下面这个地址，就能看到文件内容了。如果是本地服务器记得把 139.196.35.30 改成 localhost。

```
    http://139.196.35.30:8080/OkHttpTest/upload/ansen.txt
```

通过 HTTP 协议请求服务器数据，常用的就这几种请求，如果有特殊需求，可以自己扩展（如下载文件、从服务器下载图片等）。

（5）服务器接口

这三个接口的服务器代码是作者自己用 Java Web 编写的，开发工具用的是 IntelliJ IDEA，服务器是 Tomcat，已经部署在云上，139.196.35.30 是云服务器的外网 IP，方便大家测试。服务器代码已经放在 GitHub 上，扩展接口或者查看源码都很方便。

2. OKHttp 封装

前面介绍了 OKHttp 的基本使用方法，并在 Activity 中写了大量访问网络的代码。这种代码写起来很无聊，并且对技术没有什么提升。在实际开发中，可以将这些代码封装起来，制作成一个库，便于 Activity 调用。

封装之前，需要考虑以下问题：

- 封装基本的公共方法给外部调用，例如 Get 请求、Post 请求、PostFile 等。
- 官方建议 OkHttpClient 实例仅新建一次，所以网络请求库可以做成单例模式。
- 如果同一时间访问同一个 API 多次，我们是不是应该取消之前的请求？
- 如果用户连接 HTTP 代理了，就不让访问，防止用户通过抓包工具查看我们的接口数据。
- 每个接口都要带上的参数如何封装？例如 App 版本号、设备号和登录之后的用户 token，这些参数可能每次请求都要带上。
- 返回的 JSON 字符串转换成实体对象。
- 访问服务器是异步请求，如何在主线程中调用回调接口？

（1）代码实现

首先需要在线引用以下三个依赖库：

```
compile 'com.squareup.okhttp3:okhttp:3.2.0' //okhttp
compile 'com.google.code.gson:gson:2.7' //解析 jsons 数据
compile 'io.github.lizhangqu:coreprogress:1.0.2' //上传下载回调监听
```

新建一个 HttpConfig 类，用来配置一些请求参数（存储请求服务器的公共参数、设置连接时间等）。

```
public class HttpConfig {
    private boolean debug=false;//true:debug 模式
    private String userAgent="";//用户代理，它是一个特殊字符串头，使得服务器能够识别
客户使用的操作系统及版本、CPU 类型、浏览器及版本、浏览器渲染引擎、浏览器语言、浏览器插件等。
    private boolean agent=true;
              //有代理的情况能不能访问。true:有代理能访问。false:有代理不能访问
    private String tagName="Http";

    private int connectTimeout=10;//连接超时时间，单位:秒
    private int writeTimeout=10;//写入超时时间，单位:秒
    private int readTimeout=30;//读取超时时间，单位:秒

    //通用字段
    private List<NameValuePair> commonField=new ArrayList<>();

    public boolean isDebug() {
        return debug;
    }

    public void setDebug(boolean debug) {
        this.debug = debug;
    }

    public String getUserAgent() {
        return userAgent;
    }

    public void setUserAgent(String userAgent) {
        this.userAgent = userAgent;
    }

    public boolean isAgent() {
        return agent;
    }

    public void setAgent(boolean agent) {
        this.agent = agent;
    }

    public String getTagName() {
        return tagName;
    }

    public void setTagName(String tagName) {
```

```
        this.tagName = tagName;
    }

    public List<NameValuePair> getCommonField() {
        return commonField;
    }

    public int getConnectTimeout() {
        return connectTimeout;
    }

    public void setConnectTimeout(int connectTimeout) {
        this.connectTimeout = connectTimeout;
    }

    public int getWriteTimeout() {
        return writeTimeout;
    }

    public void setWriteTimeout(int writeTimeout) {
        this.writeTimeout = writeTimeout;
    }

    public int getReadTimeout() {
        return readTimeout;
    }

    public void setReadTimeout(int readTimeout) {
        this.readTimeout = readTimeout;
    }

    /**
     * 更新参数
     * @param key
     * @param value
     */
    public void updateCommonField(String key,String value){
        boolean result = true;
        for(int i=0;i<commonField.size();i++){
            NameValuePair nameValuePair = commonField.get(i);
            if(nameValuePair.getName().equals(key)){
                commonField.set(i,new NameValuePair(key,value));
                result = false;
                break;
            }
        }
        if(result){
            commonField.add(new NameValuePair(key,value));
        }
    }

    /**
     * 删除公共参数
     * @param key
     */
```

```
    public void removeCommonField(String key){
        for(int i=commonField.size()-1;i>=0;i--){
            if(commonField.get(i).equals("key")){
                commonField.remove(i);
            }
        }
    }

    /**
     * 添加请求参数
     * @param key
     * @param value
     */
    public void addCommonField(String key,String value){
        commonField.add(new NameValuePair(key,value));
    }
}
```

我们给服务器提交表单数据都是通过 name 与 value 的形式提交的，封装了 NameValuePair 类。如果是上传文件，将 isFile 设置为 true 即可。

```
public class NameValuePair {
    private String name;//请求名称
    private String value;//请求值
    private boolean isFile=false;//是否是文件

    public NameValuePair(String name, String value){
        this.name=name;
        this.value = value;
    }

    public NameValuePair(String name, String value,boolean isFile){
        this.name=name;
        this.value = value;
        this.isFile=isFile;
    }

    public String getName() {
        return name;
    }

    public void setName(String name) {
        this.name = name;
    }


    public String getValue() {
        return value;
    }

    public void setValue(String value) {
        this.value = value;
    }

    public boolean isFile() {
```

```
        return isFile;
    }

    public void setFile(boolean file) {
        isFile = file;
    }
}
```

请求服务器成功或者失败都会回调 Callback 接口，我们封装 HttpResponseHandler 抽象类来实现这个接口，对服务器返回的数据以及状态码进行简单的过滤，最终调用自己的 onFailure 与 onSuccess 方法。

```
public abstract class HttpResponseHandler implements Callback {
    public HttpResponseHandler(){
    }

    public void onFailure(Call call, IOException e){
        onFailure(-1,e.getMessage().getBytes());
    }

    public void onResponse(Call call, Response response) throws IOException {
        int code =response.code();
        byte[] body = response.body().bytes();
        if(code>299){
            onFailure(response.code(),body);
        }else{
            Headers headers = response.headers();
            Header[] hs = new Header[headers.size()];

            for (int i=0;i<headers.size();i++){
                hs[i] = new Header(headers.name(i),headers.value(i));
            }
            onSuccess(code,hs,body);
        }
    }

    public void onFailure(int status,byte[] data){
    }

//  public void onProgress(int bytesWritten, int totalSize) {
//  }

    public abstract void onSuccess(int statusCode, Header[] headers, byte[]
responseBody);
}
```

（2）Get 请求封装

封装后的 HTTPCaller 代码如下：（现在只有 Get 请求的代码，如果将 Post 请求与 postFile 方法贴进来，代码有点多。）

```
public class HTTPCaller {
    private static HTTPCaller _instance = null;
    private OkHttpClient client;//okhttp 对象
    private Map<String,Call> requestHandleMap = null;//以 URL 为 KEY 存储的请求
```

```
        private CacheControl cacheControl = null;//缓存控制器

        private Gson gson = null;

        private HttpConfig httpConfig=new HttpConfig();//配置信息

        private HTTPCaller() {}

        public static HTTPCaller getInstance(){
            if (_instance == null) {
                _instance = new HTTPCaller();
            }
            return _instance;
        }

        /**
         * 设置配置信息 这个方法必须调用一次
         * @param httpConfig
         */
        public void setHttpConfig(HttpConfig httpConfig) {
            this.httpConfig = httpConfig;

            client = new OkHttpClient.Builder()
                    .connectTimeout(httpConfig.getConnectTimeout(),
TimeUnit.SECONDS)
                    .writeTimeout(httpConfig.getWriteTimeout(), TimeUnit.SECONDS)
                    .readTimeout(httpConfig.getReadTimeout(), TimeUnit.SECONDS)
                    .build();

            gson = new Gson();
            requestHandleMap = Collections.synchronizedMap(new WeakHashMap<String,
Call>());
            cacheControl =new CacheControl.Builder().noStore().noCache().build();
            //不使用缓存
        }

        public <T> void get(Class<T> clazz,final String url,Header[] header,final
RequestDataCallback<T> callback) {
            this.get(clazz,url,header,callback,true);
        }

        /**
         * get 请求
         * @param clazz json 对应类的类型
         * @param url 请求 url
         * @param header 请求头
         * @param callback 回调接口
         * @param autoCancel 是否自动取消。true:同一时间请求一个接口多次，只保留最后一个
         * @param <T>
         */
        public <T> void get(final Class<T> clazz,final String url,Header[]
header,final RequestDataCallback<T> callback, boolean autoCancel){
            if (checkAgent()) {
                return;
            }
```

```
            add(url,getBuilder(url, header, new
MyHttpResponseHandler(clazz,url,callback)),autoCancel);
        }

        private Call getBuilder(String url, Header[] header, HttpResponseHandler
responseCallback) {
            url=Util.getMosaicParameter(url,httpConfig.getCommonField());
                //拼接公共参数
            Request.Builder builder = new Request.Builder();
            builder.url(url);
            builder.get();
            return execute(builder, header, responseCallback);
        }

        private Call execute(Request.Builder builder, Header[] header, Callback
responseCallback) {
            boolean hasUa = false;
            if (header == null) {
                builder.header("Connection","close");
                builder.header("Accept", "*/*");
            } else {
                for (Header h : header) {
                    builder.header(h.getName(), h.getValue());
                    if (!hasUa && h.getName().equals("User-Agent")) {
                        hasUa = true;
                    }
                }
            }
            if (!hasUa&&!TextUtils.isEmpty(httpConfig.getUserAgent())){
                builder.header("User-Agent",httpConfig.getUserAgent());
            }
            Request request = builder.cacheControl(cacheControl).build();
            Call call = client.newCall(request);
            call.enqueue(responseCallback);
            return call;
        }

        public class MyHttpResponseHandler<T> extends HttpResponseHandler {
            private Class<T> clazz;
            private String url;
            private RequestDataCallback<T> callback;

            public MyHttpResponseHandler(Class<T> clazz,String
url,RequestDataCallback<T> callback){
                this.clazz=clazz;
                this.url=url;
                this.callback=callback;
            }

            @Override
            public void onFailure(int status, byte[] data) {
                clear(url);
                try {
                    printLog(url + " " + status + " " + new String(data, "utf-8"));
                } catch (UnsupportedEncodingException e) {
```

```
                e.printStackTrace();
            }
            sendCallback(callback);
        }

        @Override
        public void onSuccess(int status,final Header[] headers, byte[]
responseBody) {
            try {
                clear(url);
                String str = new String(responseBody,"utf-8");
                printLog(url + " " + status + " " + str);
                T t = gson.fromJson(str, clazz);
                sendCallback(status,t,responseBody,callback);
            } catch (Exception e){
                if (httpConfig.isDebug()) {
                    e.printStackTrace();
                    printLog("自动解析错误:" + e.toString());
                }
                sendCallback(callback);
            }
        }
    }

    private void autoCancel(String function){
        Call call = requestHandleMap.remove(function);
        if (call != null) {
            call.cancel();
        }
    }

    private void add(String url,Call call) {
        add(url,call,true);
    }

    /**
     * 保存请求信息
     * @param url 请求 url
     * @param call http 请求 call
     * @param autoCancel 自动取消
     */
    private void add(String url,Call call,boolean autoCancel) {
        if (!TextUtils.isEmpty(url)){
            if (url.contains("?")) {//get 请求需要去除后面的参数
                url=url.substring(0,url.indexOf("?"));
            }
            if(autoCancel){
                autoCancel(url);//如果同一时间对 API 进行多次请求，自动取消之前的
            }
            requestHandleMap.put(url,call);
        }
    }

    private void clear(String url){
        if (url.contains("?")) {//get 请求需要去除后面的参数
```

```
            url=url.substring(0,url.indexOf("?"));
        }
        requestHandleMap.remove(url);
    }

    private void printLog(String content){
        if(httpConfig.isDebug()){
            Log.i(httpConfig.getTagName(),content);
        }
    }

    /**
     * 检查代理
     * @return
     */
    private boolean checkAgent() {
        if (httpConfig.isAgent()){
            return false;
        } else {
            String proHost = android.net.Proxy.getDefaultHost();
            int proPort = android.net.Proxy.getDefaultPort();
            if (proHost==null || proPort<0){
                return false;
            }else {
                Log.i(httpConfig.getTagName(),"有代理,不能访问");
                return true;
            }
        }
    }

    //更新字段值
    public void updateCommonField(String key,String value){
        httpConfig.updateCommonField(key,value);
    }

    public void removeCommonField(String key){
        httpConfig.removeCommonField(key);
    }

    public void addCommonField(String key,String value){
        httpConfig.addCommonField(key,value);
    }

    private <T> void sendCallback(RequestDataCallback<T> callback){
        sendCallback(-1,null,null,callback);
    }

    private <T> void sendCallback(int status,T data,byte[]
body,RequestDataCallback<T> callback){
        CallbackMessage<T> msgData = new CallbackMessage<T>();
        msgData.body = body;
        msgData.status = status;
        msgData.data = data;
        msgData.callback = callback;
```

```
        Message msg = handler.obtainMessage();
        msg.obj = msgData;
        handler.sendMessage(msg);
    }

    private Handler handler=new Handler(){
        @Override
        public void handleMessage(Message msg) {
            CallbackMessage data = (CallbackMessage)msg.obj;
            data.callback();
        }
    };

    private class CallbackMessage<T>{
        public RequestDataCallback<T> callback;
        public T data;
        public byte[] body;
        public int status;

        public void callback(){
            if(callback!=null){
                if(data==null){
                    callback.dataCallback(null);
                }else{
                    callback.dataCallback(status,data,body);
                }
            }
        }
    }
}
```

- getInstance：用来获取对象类的对象，因为是单例模式，所以无论这个方法调用多少次都只会创建一个对象。
- setHttpConfig：创建 HTTPCaller 对象时必须调用该方法来初始化一些对象，通过参数 httpConfig 获取连接时间来初始化 OkHttpClient 对象，用 WeakHashMap 来存储每一次的请求，还需要初始化 GSON 对象。
- get：有两个同名方法，即方法重载，如果想同一时间多次访问同一接口，就采用下面的方法，最后逐个参数传 false；如果没有要求，就用上面的方法。我们看到在 get 方法中首先调用 getBuilder 发起了 get 请求，返回一个 Call 对象。这个方法的第三个参数传入一个 MyHttpResponseHandler 对象，实现了 Callback 接口，将请求服务器成功或者失败的结果回调给这个类。最外层调用 add 方法，请求 url 作为关键字、Call 作为 value 保存到 map 中。
- getBuilder：这个方法的第一行调用 Util 的 getMosaicParameter 方法拼接公共参数，然后根据 url 生成 Request.Builder 对象，继续调用 execute。
- execute：首先设置请求头信息，一般没有特殊要求都不需要设置请求，通过 builder 对象的 cacheControl 方法设置缓存控制，通过 build 方法生成 Request，通过 OkHttpClient 对象的 newCall 方法生成一个 call 对象，最后调用 enqueue 方法进行异步请求，传入一个回调接口。
- MyHttpResponseHandler 类继承自己写的 HttpResponseHandler 抽象类，重写两个方法 onFailure 与 onSuccess。

  - onFailure（请求失败后调用）：先从 map 中删除这个请求，打印错误日志，发送回调。
  - onSuccess（请求成功后调用）：先从 map 中删除这个请求，把 json 字符串转换成对象，发送回调。
- autoCancel：从 map 中删除这个请求，调用 Call.call 取消 http 请求。
- add：有两个同名的方法，方法重载。我们来分析后面这个。首先判断这个 url 是否为空，若为空则什么也干不了，然后需要截取前面的信息，不然参数不一致将无法判断是不是同一时间多次请求同一接口了。接下来判断要不要自动取消，最后存入 map。经过前面的分析都知道发起 get 请求时会调用 add 方法将 call 请求保存到 map 中，请求成功回调中会将 call 从 map 中删除。假如手机网络状态不稳定，同一时间请求了 10 次同一个接口，都没有回调回来，这也是为什么需要在添加时会自动取消之前的请求。
- clear：从 map 中删除请求。
- printLog：打印日志。
- sendCallback：有两个方法（方法名一致），这里用到了方法重载。在 sendCallback 方法中将服务器返回内容（byte 数组）、请求状态、用 GSON 包解析 JSON 后的对象这三个值封装到 CallbackMessage 对象中，然后调用 handler. sendMessage 方法将 CallbackMessage 对象发给自己，其实就是把服务器返回结果发送到主线程中（因为我们用的是异步请求，通过前面章节我们知道在多线程中无法更新 UI），Handler 的 handleMessage 方法收到消息之后，调用 CallbackMessage 对象的 callback 方法。

（3）Post 请求封装

通过以上对 HTTPCaller 类各个方法的解释，相信已经知道 Get 请求服务器的整个调用流程。现在对 HTTPCaller 类增加 Post 请求的方法以及上传文件的方法。对 postFile 方法增加了 ProgressUIListener 参数，通过这个接口监听上传文件进度的回调。

```
    public <T> void post(final Class<T> clazz, final String url, Header[] header,
List<NameValuePair> params, final RequestDataCallback<T> callback) {
            this.post(clazz,url, header, params, callback,true);
    }

    /**
     *
     * @param clazz json对应类的类型
     * @param url  请求url
     * @param header 请求头
     * @param params 参数
     * @param callback 回调
     * @param autoCancel 是否自动取消。true:同一时间请求一个接口多次，只保留最后一个
     * @param <T>
     */
    public <T> void post(final Class<T> clazz,final String url, Header[] header,
final List<NameValuePair> params, final RequestDataCallback<T> callback, boolean
autoCancel) {
        if (checkAgent()) {
            return;
        }
        add(url,postBuilder(url, header, params, new
HTTPCaller.MyHttpResponseHandler(clazz,url,callback)),autoCancel);
```

```
    }

    private Call postBuilder(String url, Header[] header, List<NameValuePair> form,
HttpResponseHandler responseCallback) {
        try {
            if (form == null) {
                form = new ArrayList<>(2);
            }
            form.addAll(httpConfig.getCommonField());//添加公共字段
            FormBody.Builder formBuilder = new FormBody.Builder();
            for (NameValuePair item : form) {
                formBuilder.add(item.getName(), item.getValue());
            }
            RequestBody requestBody = formBuilder.build();
            Request.Builder builder = new Request.Builder();
            builder.url(url);
            builder.post(requestBody);
            return execute(builder, header, responseCallback);
        } catch (Exception e) {
            if (responseCallback != null)
                responseCallback.onFailure(-1, e.getMessage().getBytes());
        }
        return null;
    }

    /**
     * 上传文件
     * @param clazz json对应类的类型
     * @param url 请求url
     * @param header 请求头
     * @param form 请求参数
     * @param callback 回调
     * @param <T>
     */
    public <T> void postFile(final Class<T> clazz, final String url, Header[] header,
List<NameValuePair> form,final RequestDataCallback<T> callback) {
        postFile(url, header, form, new HTTPCaller.MyHttpResponseHandler
        (clazz,url, callback),null);
    }

    /**
     * 上传文件
     * @param clazz json对应类的类型
     * @param url 请求url
     * @param header 请求头
     * @param form 请求参数
     * @param callback 回调
     * @param progressUIListener  上传文件进度
     * @param <T>
     */
    public <T> void postFile(final Class<T> clazz, final String url, Header[] header,
List<NameValuePair> form,final RequestDataCallback<T>
callback,ProgressUIListener progressUIListener) {
        add(url, postFile(url, header, form, new
HTTPCaller.MyHttpResponseHandler(clazz, url,callback),progressUIListener));
```

```
    }

    /**
     * 上传文件
     * @param clazz json对应类的类型
     * @param url 请求url
     * @param header 请求头
     * @param name 名字
     * @param fileName 文件名
     * @param fileContent 文件内容
     * @param callback 回调
     * @param <T>
     */
    public <T> void postFile(final Class<T> clazz,final String url,Header[]
header,String name,String fileName,byte[] fileContent,final
RequestDataCallback<T> callback){
        postFile(clazz,url,header,name,fileName,fileContent,callback,null);
    }


    /**
     * 上传文件
     * @param clazz json对应类的类型
     * @param url 请求url
     * @param header 请求头
     * @param name 名字
     * @param fileName 文件名
     * @param fileContent 文件内容
     * @param callback 回调
     * @param progressUIListener 回调上传进度
     * @param <T>
     */
    public <T> void postFile(Class<T> clazz,final String url,Header[]
header,String name,String fileName,byte[] fileContent,final
RequestDataCallback<T> callback,ProgressUIListener progressUIListener) {
        add(url,postFile(url, header,name,fileName,fileContent,new
HTTPCaller.MyHttpResponseHandler(clazz,url,callback),progressUIListener));
    }

    private Call postFile(String url, Header[] header,List<NameValuePair>
form,HttpResponseHandler responseCallback, ProgressUIListener
progressUIListener){
        try {
            MultipartBody.Builder builder = new MultipartBody.Builder();
            builder.setType(MultipartBody.FORM);
            MediaType mediaType = MediaType.parse("application/octet-stream");

            form.addAll(httpConfig.getCommonField());//添加公共字段

            for(int i=form.size()-1;i>=0;i--){
                NameValuePair item = form.get(i);
                if(item.isFile()){//上传文件
                    File myFile = new File(item.getValue());
                    if (myFile.exists()){
                        String fileName = Util.getFileName(item.getValue());
```

```
                    builder.addFormDataPart(item.getName(),
fileName,RequestBody. create(mediaType, myFile));
                }
            }else{
                builder.addFormDataPart(item.getName(), item.getValue());
            }
        }

        RequestBody requestBody;
        if(progressUIListener==null){//不需要回调进度
            requestBody=builder.build();
        }else{//需要回调进度
            requestBody = ProgressHelper.withProgress(builder.build(),
progressUIListener);
        }
        Request.Builder requestBuider = new Request.Builder();
        requestBuider.url(url);
        requestBuider.post(requestBody);
        return execute(requestBuider, header, responseCallback);
    } catch (Exception e) {
        e.printStackTrace();
        Log.e(httpConfig.getTagName(),e.toString());
        if (responseCallback != null)
            responseCallback.onFailure(-1, e.getMessage().getBytes());
    }
    return null;
}

private Call postFile(String url, Header[] header,String name,String
filename,byte[] fileContent, HttpResponseHandler
responseCallback,ProgressUIListener progressUIListener) {
    try {
        MultipartBody.Builder builder = new MultipartBody.Builder();
        builder.setType(MultipartBody.FORM);
        MediaType mediaType = MediaType.parse("application/octet-stream");
        builder.addFormDataPart(name,filename,RequestBody.create(mediaType,
fileContent));

        List<NameValuePair> form = new ArrayList<>(2);
        form.addAll(httpConfig.getCommonField());//添加公共字段
        for (NameValuePair item : form) {
            builder.addFormDataPart(item.getName(),item.getValue());
        }

        RequestBody requestBody;
        if(progressUIListener==null){//不需要回调进度
            requestBody=builder.build();
        }else{//需要回调进度
            requestBody = ProgressHelper.withProgress(builder.build(),
                    progressUIListener);
        }
        Request.Builder requestBuider = new Request.Builder();
        requestBuider.url(url);
        requestBuider.post(requestBody);
        return execute(requestBuider, header,responseCallback);
```

```
        } catch (Exception e) {
            if (httpConfig.isDebug()) {
                e.printStackTrace();
                Log.e(httpConfig.getTagName(), e.toString());
            }
            if (responseCallback != null)
                responseCallback.onFailure(-1, e.getMessage().getBytes());
        }
        return null;
    }
```

（4）增加下载文件的方法

get 请求与 post 请求都有了，上传文件就是 post 请求，继续增加 downloadFile 方法用来下载文件，通用的 MyHttpResponseHandler 不能处理回调，又增加了 DownloadFileResponseHandler 类处理下载回调。

```
    public void downloadFile(String url, String saveFilePath, Header[] header,
ProgressUIListener progressUIListener) {
            downloadFile(url,saveFilePath, header, progressUIListener,true);
        }

    public void downloadFile(String url,String saveFilePath, Header[] header,
ProgressUIListener progressUIListener,boolean autoCancel) {
        if (checkAgent()) {
            return;
        }
        add(url,downloadFileSendRequest(url,saveFilePath, header,
progressUIListener), autoCancel);
    }

    private Call downloadFileSendRequest(String url, final String saveFilePath,
Header[] header, final ProgressUIListener progressUIListener){
        Request.Builder builder = new Request.Builder();
        builder.url(url);
        builder.get();
        return execute(builder, header, new HTTPCaller.DownloadFileResponseHandler
         (url, saveFilePath,progressUIListener));
    }

    public class DownloadFileResponseHandler implements Callback {
        private String saveFilePath;
        private ProgressUIListener progressUIListener;
        private String url;

        public DownloadFileResponseHandler(String url,String saveFilePath,
ProgressUIListener progressUIListener){
            this.url=url;
            this.saveFilePath=saveFilePath;
            this.progressUIListener=progressUIListener;
        }

        @Override
        public void onFailure(Call call, IOException e) {
            clear(url);
```

```
        try {
            printLog(url + " " + -1 + " " + new String(e.getMessage().getBytes(),
"utf-8"));
        } catch (UnsupportedEncodingException encodingException) {
            encodingException.printStackTrace();
        }
    }

    @Override
    public void onResponse(Call call, Response response) throws IOException {
        printLog(url + " code:" + response.code());
        clear(url);

        ResponseBody responseBody = ProgressHelper.withProgress
          (response.body(), progressUIListener);
        BufferedSource source = responseBody.source();

        File outFile = new File(saveFilePath);
        outFile.delete();
        outFile.createNewFile();

        BufferedSink sink = Okio.buffer(Okio.sink(outFile));
        source.readAll(sink);
        sink.flush();
        source.close();
    }
}
```

RequestDataCallback 接口封装了三个方法，自己想要什么数据就去重写对应的方法。

```
public abstract class RequestDataCallback<T> {
    //返回 json 对象
    public void dataCallback(T obj) {
    }

    //返回 http 状态和 json 对象
    public void dataCallback(int status, T obj) {
        dataCallback(obj);
    }

    //返回 http 状态、json 对象和 http 原始数据
    public void dataCallback(int status, T obj, byte[] body) {
        dataCallback(status, obj);
    }
}
```

Util 公共类有三个静态方法。

```
public class Util {
    /**
     * 获取文件名称
     * @param filename
     * @return
     */
    public static String getFileName(String filename){
        int start=filename.lastIndexOf("/");
```

```
//        int end=filename.lastIndexOf(".");
        if(start!=-1){
            return filename.substring(start+1,filename.length());
        }else{
            return null;
        }
    }

    /**
     * 拼接公共参数
     * @param url
     * @param commonField
     * @return
     */
    public static String getMosaicParameter(String url, List<NameValuePair>
commonField){
        if (TextUtils.isEmpty(url))
            return "";
        if (url.contains("?")) {
            url = url + "&";
        } else {
            url = url + "?";
        }
        url += getCommonFieldString(commonField);
        return url;
    }

    private static String getCommonFieldString(List<NameValuePair>
commonField){
        StringBuffer sb = new StringBuffer();
        try{
            int i=0;
            for (NameValuePair item:commonField) {
                if(i>0){
                    sb.append("&");
                }
                sb.append(item.getName());
                sb.append('=');
                sb.append(URLEncoder.encode(item.getValue(),"utf-8"));
                i++;
            }
        }catch (Exception e){
        }
        return  sb.toString();
    }
}
```

3. OkHttp 封装后的使用方法

OkHttp 代码都封装好了，我们在 Activity 中如何调用呢？

（1）依赖

如果是 Android Studio 开发工具，我们封装的这个库就支持在线依赖（已经将项目添加到 jcenter）：

```
compile 'com.ansen.http:okhttpencapsulation:1.0.1'
```

如果是eclipse，那么先将ide切换到Android Studio。不嫌麻烦的话，也可以将源码module的源码复制出来，反正也就几个类。

(2) 初始化HTTPCaller类

初始化的工作可以放入Application，新建MyApplication类继承Application。初始化时通过HttpConfig设置一些参数，也可以添加公共参数。

```
public class MyApplication extends Application{
    @Override
    public void onCreate() {
        super.onCreate();

        HttpConfig httpConfig=new HttpConfig();
        httpConfig.setAgent(true);//有代理的情况能不能访问
        httpConfig.setDebug(true);//是否为debug模式，如果是debug模式，就打印日志
        httpConfig.setTagName("ansen");//打印日志的tagname

        //可以添加一些公共字段，每个接口都会带上
        httpConfig.addCommonField("pf","android");
        httpConfig.addCommonField("version_code","1");

        //初始化HTTPCaller类
        HTTPCaller.getInstance().setHttpConfig(httpConfig);
    }
}
```

因为自定义Application，需要给AndroidManifest.xml文件application标签中的android:name属性赋值，指定自己重写的MyApplication。

(3) 发送get请求

发送get请求只需以下一行代码。

```
    HTTPCaller.getInstance().get(User.class,
"http://139.196.35.30:8080/OkHttpTest/getUserInfo.do?per=123", null,
requestDataCallback);
```

(4) 请求回调

http请求回调接口，无论成功或者失败都会回调。因为是测试，所以都用在这个接口进行回调。在真实开发中，根据不同的请求使用不同的回调。

```
    private RequestDataCallback requestDataCallback = new
RequestDataCallback<User>() {
        @Override
        public void dataCallback(User user) {
            if(user==null){
                Log.i("ansen", "请求失败");
            }else{
                Log.i("ansen", "获取用户信息:" + user.toString());
            }

        }
    };
```

(5) 发送 post 请求

post 请求参数不是跟在 url 后面，所以需要把请求参数放到集合中。因为登录接口也是返回的用户信息，所以可以与 get 请求使用同一回调。

```
List<NameValuePair> postParam = new ArrayList<>();
postParam.add(new NameValuePair("username","ansen"));
postParam.add(new NameValuePair("password","123"));
HTTPCaller.getInstance().post(User.class, "http://139.196.35.30:8080/
  OkHttpTest/login.do", null, postParam, requestDataCallback);
```

(6) 上传文件

① 上传文件不带回调进度：

```
updaloadFile(null);
```

② 上传文件回调上传进度：

```
updaloadFile(new ProgressUIListener(){
    @Override
    public void onUIProgressChanged(long numBytes, long totalBytes, float
percent, float speed) {
        Log.i("ansen","numBytes:"+numBytes+" totalBytes:"+totalBytes+"
percent:"+percent+" speed:"+speed);
    }
});
```

上传文件与其他表单参数不一样的地方就是新建 NameValuePair 对象时需要传入三个参数，最后一个参数需要设置成 true。

```
private void updaloadFile(ProgressUIListener progressUIListener){
    List<NameValuePair> postParam = new ArrayList<>();
    postParam.add(new NameValuePair("username", "ansen"));
    postParam.add(new NameValuePair("password", "123"));
    String filePath=copyFile();//复制一份文件到 sdcard 上，并且获取文件路径
    postParam.add(new NameValuePair("upload_file",filePath,true));
    if(progressUIListener==null){//上传文件没有回调进度条
        HTTPCaller.getInstance().postFile(User.class, "http://139.196.35.30:
      8080/OkHttpTest/uploadFile.do", null, postParam, requestDataCallback);
    }else{//上传文件并且回调上传进度
        HTTPCaller.getInstance().postFile(User.class, "http://139.196.35.30:
        8080/OkHttpTest/uploadFile.do", null, postParam,
        requestDataCallback,progressUIListener);
    }
}
```

(7) 上传文件（传入 byte 数组）

```
byte[] bytes=getUploadFileBytes();//获取文件内容存入 byte 数组
            HTTPCaller.getInstance().postFile(User.class,
"http://139.196.35.30:8080/OkHttpTest/uploadFile.do", null,
"upload_file","test.txt",bytes,requestDataCallback);
```

(8) 上传文件（传入 byte 数组）&&回调上传进度

```
byte[] bytes=getUploadFileBytes();//获取文件内容存入 byte 数组
HTTPCaller.getInstance().postFile(User.class,
```

```
"http://139.196.35.30:8080/OkHttpTest/uploadFile.do", null, "upload_file",
"test.txt", bytes, requestDataCallback, new ProgressUIListener() {
        @Override
        public void onUIProgressChanged(long numBytes, long totalBytes, float
percent, float speed) {
            Log.i("ansen","upload file content numBytes:"+numBytes+"
totalBytes:"+totalBytes+" percent:"+percent+" speed:"+speed);
        }
    });
```

（9）下载文件&&回调下载进度

```
    String saveFilePath=Environment.getExternalStorageDirectory() +
"/test/test222.txt";
    HTTPCaller.getInstance().downloadFile("http://139.196.35.30:8080/OkHttpTes
t/upload/test.txt",saveFilePath,null,new ProgressUIListener(){
        @Override
        public void onUIProgressChanged(long numBytes, long totalBytes, float
percent, float speed) {
            Log.i("ansen","dowload file content numBytes:"+numBytes+"
totalBytes:"+totalBytes+" percent:"+percent+" speed:"+speed);
        }
    });
```

（10）修改公共参数

```
    HTTPCaller.getInstance().updateCommonField("version_code","2");//更新公共字段版
本号的值
```

6.2　数据存储

当我们开发一个产品级 App 时，有时需要在手机本地保存一些数据，例如用户登录后的用户信息，这样的好处就是当手机没有网络时打开 App 也能看到用户信息。

在 Android 中，本地数据存储主要有 SharedPreferences、SQLite 数据库、文件存储三种方式。

6.2.1　SharedPreferences

SharedPreferences 以键值对的形式存储在 xml 文件中，存储文件路径位于“/data/data/应用程序包/shared_prefs”目录下。正常情况下，其他应用程序没有操作权限，所以相对比较安全。当然用户卸载 App 或者系统设置里清除应用数据都会将 SharedPreferences 文件删除。

SharedPreferences 可以存一些简单的数据，例如用户登录状态、登录后用户信息等。

1. 获取 SharedPreferences 的两种方式

- 调用 Context 对象的 getSharedPreferences()方法：调用 Context 对象的 getSharedPreferences()方法获得的 SharedPreferences 对象可以被同一应用程序下的其他组件共享。
- 调用 Activity 对象的 getPreferences()方法：调用 Activity 对象的 getPreferences()方法获得的 SharedPreferences 对象只能在该 Activity 中使用。

2. 调用 Context 对象的 getSharedPreferences()方法的代码

在工作中可能会存储用户登录成功后的信息和登录状态，这时可以使用 SharedPreferences 进行存储。

首先通过 Context 的 getSharedPreferences 方法获取 SharedPreferences 对象，需要传入两个参数：第一个是文件名，第二个是操作模式。操作模式有以下几种：

- Context.MODE_PRIVATE：默认操作模式，代表该文件是私有数据，只能被应用本身访问。在该模式下，写入的内容会覆盖原文件的内容。
- Context.MODE_APPEND：该模式会检查文件是否存在，若存在就向文件中追加内容，否则创建新文件。
- Context.MODE_WORLD_READABLE：表示当前文件可以被其他应用读取。
- Context.MODE_WORLD_WRITEABLE：表示当前文件可以被其他应用写入。

如果是存储数据，还需要获取 SharedPreferences 的内部类 Editor，通过 Editor 来 put 数据，支持常用的 String、int、boolean、float、long、Set 集合。最后记得调用 commit 方法进行提交。以下代码是存储一个用户名：

```
    SharedPreferences
sp=getApplicationContext().getSharedPreferences("filename",
Context.MODE_PRIVATE);
    SharedPreferences.Editor edit=sp.edit();
    edit.putString("username","ansen");
    edit.commit();
```

上一步存储了 username，那么肯定会想办法提取出来，不然存储就变得没有意义。首先是获取 SharedPreferences 对象，然后调用 getString 方法获取，第一个参数是 key，第二个参数是默认值，也就是如果 username 没有找到就返回第二个参数。

```
    SharedPreferences
sp=getApplicationContext().getSharedPreferences("filename",
Context.MODE_PRIVATE);
    String username=sp.getString("username","");
```

6.2.2 SQLite 数据库

SQLite 是一款轻型的数据库，设计目标是嵌入式的，而且目前已经在很多嵌入式产品中使用了它。它占用资源非常低，在嵌入式设备中，可能只需要几百千字节的内存就够了。它支持 Windows/Linux/UNIX 等主流的操作系统，同时能够与很多程序语言相结合，比如 C#、PHP、Java 等，还提供 ODBC 接口。比起 Mysql、PostgreSQL 这两款开源的世界著名数据库管理系统，它的处理速度更快。SQLite 第一个 Alpha 版本诞生于 2000 年 5 月，至 2015 年已经有 15 个年头，SQLite 又迎来了一个版本 SQLite3。

1. SQLite 优点

SQLite 具有以下优点：

（1）轻量级：SQLite 和 C/S 模式的数据库软件不同，它是进程内的数据库引擎，因此不存在数据库的客户端和服务器。使用 SQLite 一般只需要带上它的一个动态库，就可以享受它的全部功能，而且那个动态库的尺寸也挺小，以版本 3.6.11 为例，Windows 下为 487KB、Linux 下为 347KB。

（2）不需要“安装”：SQLite 的核心引擎本身不依赖第三方的软件，使用它也不需要“安装”，有点类似绿色软件。

（3）单一文件：数据库中所有的信息（比如表、视图等）都包含在一个文件内。这个文件可以自由复制到其他目录或机器上。

（4）跨平台/可移植性：除了主流操作系统 Windows、Linux 之后，SQLite 还支持其他一些不常用的操作系统。

（5）弱类型的字段：同一列中的数据可以是不同类型。

（6）开源：源码完全开源，可以用于任何用途，包括出售。

SQLite 可以存相对多一点的数据，例如 QQ，可以把用户之前的聊天记录保存在本地 SQLite 数据库中。

2. 在 Android 中使用 SQlite 数据库

Android 系统自带了 SQLite 数据库，所以 Google 封装了操作 SQLite 数据库的 API，包含在 android.database.sqlite 包下。

首先重写一个类 SQLiteHelper，需要继承 SQLiteOpenHelper 类；重写两个方法 onCreate 与 onUpgrade，这两个方法的作用都有注释，相信一看就懂。

```
    public class SQLiteHelper extends SQLiteOpenHelper{
        private static final int VERSION = 1;//数据库版本号
        private static String db_name="test";//数据库名称

        public SQLiteHelper(Context context) {
            super(context, db_name, null, VERSION);
        }

        //当第一次建库的时候，调用该方法
        @Override
        public void onCreate(SQLiteDatabase db) {
            //创建数据库的时候把学生表创建好
            String sql = "create table user(id int,name varchar(20),age int)";
            db.execSQL(sql);

            Log.i("SQLiteHelper", "onCreate.....");
        }

        //当更新数据库版本号的时候就会执行该方法
        @Override
        public void onUpgrade(SQLiteDatabase sqLiteDatabase, int oldVersion, int
newVersion) {
            Log.i("SQLiteHelper", "update Database.....");
        }
    }
```

创建数据库，只需要实例化 SQLiteHelper 类即可（第一次会调用 oncreate 方法，这时创建好学生表），再通过 getReadableDatabase 方法获取 SQLiteDatabase 对象，对数据库的增加、删除、修改、查找都需要这个实例来实现。

```
    public void createDb(View view){
        SQLiteHelper sqLiteHelper=new SQLiteHelper(this);
        sqLiteDatabase=sqLiteHelper.getReadableDatabase();
    }
```

向学生表增加一条记录的操作如下：

```
        public void addRecord(View view){
            ContentValues values = new ContentValues();
            //参数1:表的字段  参数2:字段对应的值
            values.put("id",1);
            values.put("name","test1");
            values.put("age",11);
            //参数1:表名  参数3:插入的数据
            sqLiteDatabase.insert("user",null,values);

            Log.i("MainActivity","往学生表添加一条数据");
        }
```

更新记录，将用户表 name="test1"的记录年龄修改成 88：

```
    public void updateRecord(View view){
        Log.i("MainActivity","updateRecord");

        ContentValues contentValues = new ContentValues();
        contentValues.put("age", "88");
        //参数1:表名 参数2:修改的值 参数3:查询条件 参数4:查询条件需要的参数
        sqLiteDatabase.update("user", contentValues, "name=?", new
String[]{"test1"});
    }
```

查找用户表所有数据，并且打印出来：

```
    public void findStudent(View view){
            Cursor cursor=sqLiteDatabase.query("user",new
String[]{"id","name","age"},null,null,null,null,null);
            while (cursor.moveToNext()){//判断下一条有没有数据
                int id = cursor.getInt(0);
                String name = cursor.getString(1);
                int age = cursor.getInt(2);
                Log.i("MainActivity","id:"+id+" name:"+name+" age:"+age);
            }
            cursor.close();
    }
```

删除一条记录的操作如下：

```
    public void delete(View view){
        sqLiteDatabase.delete("user","name=?",new String[]{"test1"});
        Log.i("MainActivity","删除一条数据");
    }
```

这里使用系统 API 给大家讲解了数据库的增加、删除、修改与查找，也是在工作中常用的方式，但是有些复杂的查询还需要手动去写 SQL 语句。SQLiteDatabase 本身支持执行 SQL，调用 execSQL 方法即可，有需要时可上网去学习。

6.2.3 文件存储

在 Android 中，文件存储一般有两种选择，SD 卡目录或者应用程序下的存储目录。两者之间的比较如下：

- SD 卡目录：应用程序卸载了，之前创建的文件还在，通常手机 SD 卡的内存很大，有 16GB、32GB、64GB 等。
- 应用程序下的存储目录：存储内存很小，应用程序删除了，之前创建的文件也会删除。

1. SD 卡目录

首先判断有没有 SD 卡，然后获取 SD 卡目录。

```
if(Environment.getExternalStorageState().equals(Environment.MEDIA_MOUNTED)) {
 //判断有没有 sdcard
    File sdCard = Environment.getExternalStorageDirectory();
    Log.i("MainActivity",sdCard.getPath());
}
```

打印的日志如下：

```
I/MainActivity: /storage/emulated/0
```

2. 应用程序下的存储目录

```
File getFilesDir =getFilesDir();
Log.i("MainActivity",getFilesDir.getPath());
```

打印的日志如下：

```
I/MainActivity: /data/data/com.ansen.sdcard/files
```

为了正常读取 SD 卡，需要在 AndroidManifest.xml 中声明读写权限：

```
<uses-permission android:name="android.permission.WRITE_EXTERNAL_STORAGE"/>
<uses-permission android:name="android.permission.READ_EXTERNAL_STORAGE"/>
```

6.3 本章小结

本章主要学习了 Android 平台的 HTTP 通信与数据存储。

HTTP 协议请求数据主要用到了 HttpURLConnection 类，还有两种请求方式：Get 请求与 Post 请求。现在主流的应用程序前后端交互数据基本都用 JSON，我们学习了 Google 开源库 GSON 的使用，最后学习了 Android 网络开源框架 OkHttp 的使用，并对 OkHttp 进一步封装。

数据存储主要学习了三种主要的存储方式：SharedPreferences 键值对存储、SQLite 数据库存储以及文件存储（主要存储在 SD 卡上）。SharedPreferences 和 SQLite 存储的内容与应用程序的生命周期一致，程序卸载后数据就没有了。存储在 SD 卡上的文件即使程序卸载了也会存在。

第 7 章

Android 高级应用

本章内容比较丰富，并且知识点相对前面的章节来说稍微难一点，是一些必须要掌握的内容但是又不常用。首先介绍 Notification（通知）的使用，使用 Notification 让我们的 App 在后台也能通知用户，让应用程序（App）更人性化。同时介绍多媒体开发，调用官方 API，简单播放音视频只需几行代码就能完成；后面还将介绍 WebView，在原生 App 中嵌入网页；通过 NDK 工具，使用 jni 技术，使 Android 和 C++能够进行交互。最后是 SourceTree（Git 可视化）工具，从此不再需要记住很长的 Git 命令。

7.1 Notification（通知）使用

通知是显示在应用外的 UI 界面，当你的 App 向系统发送一个通知时，先以图标的形式显示在通知区域中，如图 7-1 所示。用户可以下拉通知栏查看通知的详细信息，如图 7-2 所示。通知栏是由系统控制的，用户可以随便查看。

图 7-1　通知区域中的通知

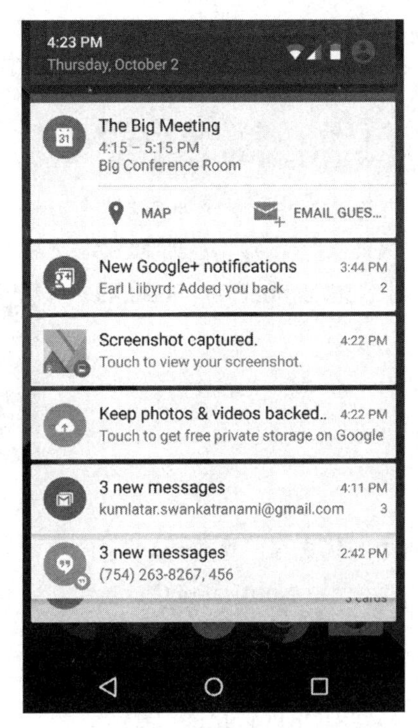

图 7-2　下拉通知栏显示通知列表

需要显示通知的场景有很多，例如收到短信、邮箱时通过通知栏提醒用户，音乐 App 下载歌曲，通知栏显示下载进度等。

7.1.1　创建通知

首先我们创建一个通知，通过 NotificationCompat.Builder 对象操作通知的 UI。要创建通知，将调用 NotificationCompat.Builder.build()，返回具体的 Notification 对象。最后调用 NotificationManager.notify()向系统发出通知。

设置通知内容一般会调用以下三个方法：

- setSmallIcon：设置小图标。
- setContentTitle：设置标题。
- setContentText：设置内容。

我们创建一个简单的通知，代码如下：

```
NotificationCompat.Builder mBuilder =
new NotificationCompat.Builder(this)
       .setSmallIcon(R.mipmap.ic_launcher)//小图标
       .setContentTitle("标题")
       .setContentText("内容");

Intent intent=new Intent(this,NotificationActivity.class);
PendingIntent ClickPending = PendingIntent.getActivity(this, 0, intent, 0);
```

```
mBuilder.setContentIntent(ClickPending);
mBuilder.setAutoCancel(true);//点击这条通知自动从通知栏中取消
NotificationManager mNotificationManager =
(NotificationManager) getSystemService(Context.NOTIFICATION_SERVICE);
mNotificationManager.notify(id, mBuilder.build());
```

最后一行代码调用 notify 方法向系统发送一个通知，这个方法有两个参数：第一个参数是通知的唯一 ID，我们可以通过这个 ID 来修改/删除当前通知；第二个参数是 Notification 对象，通过 NotificationCompat.Builder 的 build 方法获取。

运行效果如图 7-3 所示。

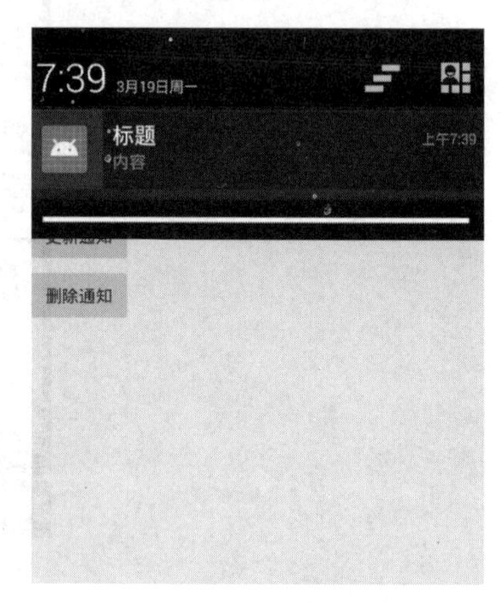

图 7-3 创建一个简单的通知

7.1.2 通知优先级

我们可以通过代码来设置通知的优先级，优先级相当于一个提示，提醒手机如何显示通知。调用 NotificationCompat.Builder.setPriority()并传入一个 NotificationCompat 优先级常量。

有 5 个优先级别，范围从 PRIORITY_MIN (-2) 到 PRIORITY_MAX (2)，如果未设置，则优先级默认为 PRIORITY_DEFAULT (0)。

7.1.3 更新通知

更新通知与创建通知调用的方法是一样的，都是调用 notify()，这个方法的第一个参数与创建时传入的 ID 一致即可，只需要更新或创建 NotificationCompat.Builder 对象。如果这个 ID 之前已经存在通知，那么系统就会更新之前的通知；如果之前通知已取消或者未创建，系统就会创建一个新通知。

写一个简单的例子，每次更新通知栏的次数进行累加显示出来。下面这段代码与创建通知栏的代码几乎一样，改变的只是 NotificationCompat.Builder 对象的内容。

```
NotificationCompat.Builder mBuilder =
new NotificationCompat.Builder(this)
    .setSmallIcon(R.mipmap.ic_launcher)//小图标
    .setContentTitle("更新通知-标题"+(++number))
    .setContentText("更新通知-内容").setNumber(number);

Intent intent=new Intent(this,NotificationActivity.class);
PendingIntent ClickPending = PendingIntent.getActivity(this, 0, intent, 0);
mBuilder.setContentIntent(ClickPending);

NotificationManager mNotificationManager =
(NotificationManager) getSystemService(Context.NOTIFICATION_SERVICE);
mNotificationManager.notify(id, mBuilder.build());
```

7.1.4 删除通知

使用代码删除通知有以下几种方法：

- 创建通知时调用 setAutoCancel(true)方法，这样当用户点击这条通知时会从通知栏自动删除。
- 根据 ID 删除这条通知，调用 NotificationManager.cancel(int id)方法。
- 删除所有通知，调用 NotificationManager.cancelAll()方法。

7.1.5 自定义通知布局

系统的通知栏样式有很多缺点，首先各个版本的样式不一致，各种手机的样式也不统一，国内手机厂商定制严重，但是又想统一样式怎么办呢？谷歌早帮我们想好了，让我们自己去自定义通知栏显示的布局文件，通过 xml 文件想怎么设计就怎么设计。

自定义布局通知栏的可用高度取决于通知视图。一般布局限制为 64dp，最高可用设置为 256dp。

首先创建 layout_custom_notification.xml 文件，其中包含三个控件：icon、标题、内容。

```
<?xml version="1.0" encoding="utf-8"?>
<RelativeLayout xmlns:android="http://schemas.android.com/apk/res/android"
   android:layout_width="match_parent"
   android:layout_height="wrap_content">

   <ImageView
      android:id="@+id/imageview"
      android:layout_width="wrap_content"
      android:layout_height="wrap_content"/>

   <TextView
      android:id="@+id/tv_title"
      android:layout_alignParentRight="true"
      android:layout_width="wrap_content"
      android:layout_height="wrap_content"/>

   <TextView
      android:id="@+id/tv_content"
      android:layout_below="@+id/tv_title"
      android:layout_marginTop="10dp"
      android:layout_alignParentRight="true"
      android:layout_width="wrap_content"
      android:layout_height="wrap_content"/>
</RelativeLayout>
```

我们来看看 Java 代码是如何实现的：

```
//RemoteViews 加载 xml
RemoteViews remoteViews = new RemoteViews(getPackageName(),
R.layout.layout_custom_notification);
//设置图片，参数 1 是 xml 中 ImageView 设置的 id，参数 2 是资源 id
remoteViews.setImageViewResource(R.id.imageview,R.mipmap.ic_launcher);
remoteViews.setTextViewText(R.id.tv_title,"这是标题");
remoteViews.setTextViewText(R.id.tv_content,"这是内容");
```

```
    Intent intent=new Intent(this,NotificationActivity.class);
    PendingIntent clickIntent = PendingIntent.getActivity(this, 0, intent, 0);

    Notification notification = new Notification();
    //必须要设置一个图标，通知区域中需要显示
    notification.icon = android.R.drawable.ic_media_play;
    notification.contentView = remoteViews;//自定义布局
    notification.contentIntent = clickIntent;//点击跳转 Intent
    NotificationManager mNotificationManager = (NotificationManager)
getSystemService(Context.NOTIFICATION_SERVICE);
    mNotificationManager.notify(id, notification);
```

与之前的不同，这里没有创建 NotificationCompat.Builder 对象，而是构造一个 RemoteViews 对象，这个类的构造方法有两个参数，包名与布局文件 id，调用 setImageViewResource 方法设置图片资源，调用 setTextViewText 方法设置文本内容。接下来构造 Notification 对象，设置通知区域显示的 icon，设置自定义布局，设置点击 PendingIntent，最后调用 notify 方法显示在系统通知栏。运行代码，效果如图 7-4 所示。

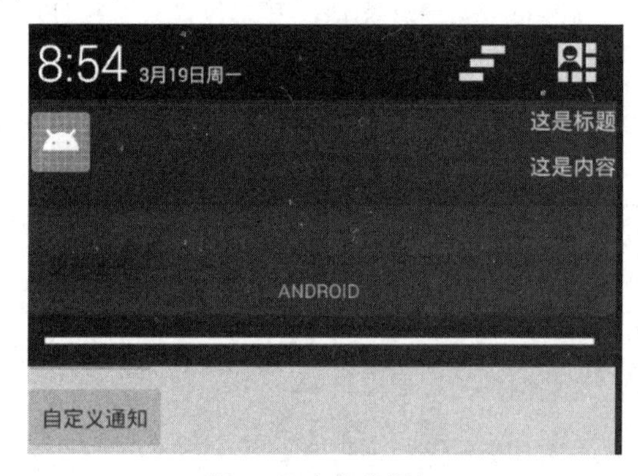

图 7-4　自定义通知

7.2　多媒体开发

多媒体（Multimedia）是多种媒体的综合，一般包括文本、声音和图像等多种媒体形式。

在移动设备上，随着性能的提高、宽带的增长、流量的提升，音频跟视频的使用场景非常多，本节就使用 Android SDK 自带的 API 实现音视频的播放。

7.2.1　播放音频

音乐播放器是手机中最常见的应用，基本每一部 Android 手机都有自带的音乐播放器，有时我们开发的 App 也有播放音乐的需求，例如播放一段提示音。因此，学习在 Android 手机上播放音乐是非常有必要的，本小节学习一些基本知识，例如音乐播放、拖动进度条、暂停播放、继续播放等。

新建项目，首先修改布局文件 activity_main.xml：

```
<?xml version="1.0" encoding="utf-8"?>
<LinearLayout xmlns:android="http://schemas.android.com/apk/res/android"
   android:layout_width="match_parent"
   android:layout_height="match_parent"
   android:padding="10dp"
   android:orientation="vertical">

   <TextView
      android:layout_width="wrap_content"
      android:layout_height="wrap_content"
      android:text="歌曲名称:成都"/>

   <RelativeLayout
      android:layout_marginTop="10dp"
      android:layout_marginBottom="10dp"
      android:layout_width="match_parent"
      android:layout_height="wrap_content">

      <TextView
         android:id="@+id/tv_start_time"
         android:layout_width="wrap_content"
         android:layout_height="wrap_content"
         android:layout_centerVertical="true"
         android:text="00:00:00"/>

      <SeekBar
         android:id="@+id/seekbar"
         android:layout_toRightOf="@+id/tv_start_time"
         android:layout_toLeftOf="@+id/tv_end_time"
         android:layout_width="match_parent"
         android:layout_height="wrap_content" />

      <TextView
         android:id="@+id/tv_end_time"
         android:layout_alignParentRight="true"
         android:layout_width="wrap_content"
         android:layout_height="wrap_content"
         android:layout_centerVertical="true"
         android:text="00:00:00"/>
   </RelativeLayout>

   <ImageView
      android:id="@+id/iv_play"
      android:layout_width="wrap_content"
      android:layout_height="wrap_content"
      android:src="@mipmap/icon_play"/>
</LinearLayout>
```

最外层是一个 LinearLayout（垂直排列），最上面的 TextView 显示歌曲名称，中间的 RelativeLayout 用来设置开始时间和结束时间，还有一个 SeekBar 用来显示播放的进度，并且能拖动进度条。底部的 ImageView 用来点击播放或暂停。

布局文件对应的 MainActivity.java 内容如下：

```
public class MainActivity extends AppCompatActivity {
```

```
private boolean isPlay=false;//播放状态 true:播放 false:未播放
private boolean isPause=false;//暂停状态 true:暂停 false:未暂停

private MediaPlayer mediaPlayer;
private ImageView ivPlay;
private SeekBar seekBar;

private Handler handler=new Handler();

private TextView tvStartTime,tvEndTime;//开始时间与结束时间

private final Runnable mTicker = new Runnable(){
    public void run() {
        long now = SystemClock.uptimeMillis();
        long next = now + (1000 - now % 1000);

        handler.postAtTime(mTicker,next);
                                //延迟一秒再次执行 runnable,与计时器效果一样

        if(mediaPlayer!=null&&isPlay&&!isPause){
            seekBar.setProgress(mediaPlayer.getCurrentPosition());//更新播放进度

            tvStartTime.setText(getTimeStr(mediaPlayer.getCurrentPosition()));
        }
    }
};

@Override
protected void onCreate(Bundle savedInstanceState) {
    super.onCreate(savedInstanceState);
    setContentView(R.layout.activity_main);

    ivPlay= (ImageView) findViewById(R.id.iv_play);
    ivPlay.setOnClickListener(onClickListener);
    seekBar= (SeekBar) findViewById(R.id.seekbar);
    seekBar.setOnSeekBarChangeListener(onSeekBarChangeListener);
            //seekbar 改变监听

    tvStartTime= (TextView) findViewById(R.id.tv_start_time);
    tvEndTime= (TextView) findViewById(R.id.tv_end_time);
}

private View.OnClickListener onClickListener=new View.OnClickListener() {
    @Override
    public void onClick(View v) {
        switch (v.getId()){
            case R.id.iv_play:
                if(isPlay){
                    if(isPause){//暂停中
                        mediaPlayer.start();//开始播放
                        isPause=false;//
                        ivPlay.setImageResource(R.mipmap.icon_stop);
                        handler.post(mTicker);//更新进度
                    }else{//未暂停
                        mediaPlayer.pause();//暂停播放
```

```
                    isPause=true;
                    ivPlay.setImageResource(R.mipmap.icon_play);
                    handler.removeCallbacks(mTicker);
                     //删除执行的 Runnable 终止计时器
                }
            }else{
                playMusic();//播放音乐
                ivPlay.setImageResource(R.mipmap.icon_stop);
                isPlay=true;
            }
            break;
        }
    }
};

private void playMusic(){
    try {
        mediaPlayer= new MediaPlayer();
        Uri uri =
Uri.parse("android.resource://com.ansen.playmusic/"+R.raw. chengdu);
        mediaPlayer.setDataSource(MainActivity.this,uri);
               //设置播放资源(可以是应用的资源文件 / url / sdcard 路径)
        mediaPlayer.setAudioStreamType(AudioManager.STREAM_MUSIC);//设置播放类型
        mediaPlayer.setOnCompletionListener(onCompletionListener);//播放完成监听
        mediaPlayer.setOnPreparedListener(new
          MediaPlayer.OnPreparedListener() {//预加载监听
            @Override
            public void onPrepared(MediaPlayer mp){//预加载完成
                seekBar.setMax(mediaPlayer.getDuration());//设置总进度
                mediaPlayer.start();//开始播放
                handler.post(mTicker);//更新进度
                tvEndTime.setText(getTimeStr(mediaPlayer.getDuration()));
            }
        });
        mediaPlayer.prepare();
    } catch (IOException e) {
        e.printStackTrace();
    }
}

private SeekBar.OnSeekBarChangeListener onSeekBarChangeListener=new
SeekBar.OnSeekBarChangeListener() {
    @Override
    public void onProgressChanged(SeekBar seekBar, int progress, boolean
fromUser) {//进度改变
    }

    @Override
    public void onStartTrackingTouch(SeekBar seekBar) {//开始拖动 seekbar
    }

    @Override
    public void onStopTrackingTouch(SeekBar seekBar) {//停止拖动 seekbar
        if(mediaPlayer!=null&&isPlay){//播放中
            mediaPlayer.seekTo(seekBar.getProgress());
```

```
            }
        }
    };

    private MediaPlayer.OnCompletionListener onCompletionListener=new
MediaPlayer.OnCompletionListener() {
        @Override
        public void onCompletion(MediaPlayer mp) {
            //播放完成，进入暂停状态，恢复初始值
            isPause=true;
            ivPlay.setImageResource(R.mipmap.icon_play);
            mediaPlayer.seekTo(0);
            seekBar.setProgress(0);
            tvStartTime.setText("00:00:00");
            handler.removeCallbacks(mTicker);//删除执行的 Runnable，终止计时器
        }
    };

    @Override
    protected void onDestroy() {
        if(mediaPlayer!=null){//如果不为空，释放资源
            mediaPlayer.stop();
            mediaPlayer.reset();
            mediaPlayer.release();
            mediaPlayer = null;
        }
        super.onDestroy();
    }

    /**
     * @param time 时间戳
     * @return 分秒字符串
     */
    private String getTimeStr(long time){
        SimpleDateFormat simpleDateFormat=new SimpleDateFormat("mm:ss");
        return simpleDateFormat.format(time);
    }
}
```

- onCreate：Activity 的入口，查找控件，给 ivPlay 设置点击监听，给 seekBar 设置改变监听。
- onClickListener：播放按钮的回调，逻辑稍微复杂点，首先判断是否播放，如果未播放则肯定是第一次播放，将调用 playMusic 方法进行播放，把播放按钮改成停止图片，状态设置为 true，也就是播放状态。如果已经播放，又分暂停中和没有暂停两种状态，暂停时就继续播放，用 handler 执行一个 Runnable，里面是死循环，类似计时器，每 1 秒钟去更新进度条。没有暂停点击的话就暂停。暂停播放就调用 handler 删除 Runnable。
- playMusic：这个方法用来播放音乐，只有第一次点击播放按钮的时候才会调用，首先初始化 MediaPlayer，设置播放资源（已经复制了一个 chengdu.mp3 文件放到了 res/raw 文件夹中），设置播放类型，设置播放完成监听，预加载完成监听，最后调用 prepare 方法去预加载。在预加载完成监听回调方法中，设置 seekbar 的总进度条，调用 mediaPlayer 的 start 方法开始播放，用 ghandler 启动计时器来设置总时间，将时间戳转成分秒显示。

- onSeekBarChangeListener：在进度条结束拖动方法onStopTrackingTouch中设置当前播放进度，告诉mediaPlayer跳转到用户想播放的位置。
- onCompletionListener：播放完成回调，状态设置成暂停状态，播放图片显示未播放图片，给mediaPlayer、seekBar、tvStartTime设置初始值，用handler终止计时器。
- onDestroy：Activity销毁回调，用来释放mediaPlayer资源。
- getTimeStr：给一个时间戳返回分秒格式的字符串。
- mTicker：类似计时器，1秒钟之后调用自己，如果音频在播放并且未暂停的情况下更新seekbar的播放进度，则更新开始时间。

我们简单地实现了音乐播放、拖动进度条、暂停、继续播放。其实播放音频的代码并不多，MediaPlayer都封装好了，主要是需要自己去实现这个逻辑。这个Demo音频播放的生命周期是跟着Activity走的，但是市面上的App都可以在后台播放，就算用户关闭了软件返回到手机主界面，音乐还在播放。这是使用service实现的，有兴趣的读者可以自己去实现。

7.2.2 播放视频的三种方式

在Android中播放视频有三种方式，可根据自己的需求去选择。

1. 自带播放器

使用自带播放器播放视频比较简单，使用隐式Intent来调用它，通过构造方法传入一个Action，调用Intent的setDataAndType方法传入一个URL与视频格式。

```
String path=Environment.getExternalStorageDirectory()+"/ansen.mp4";
Intent intent = new Intent(Intent.ACTION_VIEW);
intent.setDataAndType(Uri.parse("file://"+path), "video/mp4");
startActivity(intent);
```

2. 使用VideoView方式实现

这种方式不怎么灵活，很多东西不可控。

```
Uri uri = Uri.parse(Environment.getExternalStorageDirectory()+"/ansen.mp4");
VideoView videoView = (VideoView)this.findViewById(R.id.video_view);
videoView.setMediaController(new MediaController(this));
videoView.setVideoURI(uri);
videoView.start();
```

3.MediaPlayer+TextureView

如果想显示一段在线视频或者任意的数据流，例如视频或者OpenGL场景，可以使用Android中的TextureView实现。

（1）TextureView的兄弟SurfaceView

应用程序的视频或者OpenGL内容往往显示在一个特别的UI控件中：SurfaceView。SurfaceView的工作方式是创建一个置于应用窗口之后的新窗口。这种方式的效率非常高，因为SurfaceView窗口刷新时不需要重绘应用程序的窗口（Android普通窗口的视图绘制机制是一层一层的，任何一个子元素或者局部的刷新都会导致整个视图结构全部重绘一次，因此效率非常低，不过满足普通应用界面的需求还是绰绰有余的），当然，SurfaceView 也有一些非常不便的限制。因为 SurfaceView 的内容不在应用窗口中，所以不能使用变换（平移、缩放、旋转等），也难以放在 ListView 或者 ScrollView 中，将不能使用 UI 控件的一些特性，例如 View.setAlpha()。

（2）TextureView

为了解决前面的问题，Android 4.0 中引入了 TextureView。与 SurfaceView 相比，TextureView 并没有创建单独的 Surface 来绘制，这使得它可以像普通的 View 一样执行一些变换操作，如设置透明度等。另外，TextureView 必须在硬件加速开启的窗口中。

在项目中，经常会碰到一些问题，例如：

- 用 SurfaceView 播放视频时，从图片切换到播放视频会出现黑屏的现象。
- SurfaceView 灵活性没有 TextureView 好。

（3）代码是最好的老师

首先修改布局文件 activity_main.xml，最外层是一个 LinearLayout，里面有一个 FrameLayout 与 SeekBar，通过 android:layout_weight 属性来控制它们的高度。FrameLayout 中有 TextureView 与 ImageView 两个属性，TextureView 用来显示视频，ImageView 用来显示视频的第一帧，也就是视频的第一张图片，只有在视频"播放前"和"播放完成后"两个状态下才会显示。

```
<LinearLayout xmlns:android="http://schemas.android.com/apk/res/android"
   android:layout_width="match_parent"
   android:layout_height="match_parent"
   android:orientation="vertical">

   <FrameLayout
      android:layout_width="match_parent"
      android:layout_weight="30"
      android:layout_height="0dp">

      <TextureView
         android:id="@+id/textureview"
         android:layout_width="match_parent"
         android:layout_height="match_parent" />

      <ImageView
         android:id="@+id/video_image"
         android:layout_width="match_parent"
         android:layout_height="match_parent"
         android:scaleType="centerCrop"
         android:src="@drawable/cover"/>
   </FrameLayout>

   <SeekBar
      android:id="@+id/seekbar"
      android:layout_weight="1"
      android:layout_width="match_parent"
      android:layout_height="0dp"/>
</LinearLayout>
```

布局文件对应的 MainActivity.java 实现：

```
    public class MainActivity extends Activity{
        private final String Tag = MainActivity.class.getSimpleName();

        private MediaPlayer mMediaPlayer;
        private Surface surface;

        private ImageView videoImage;
        private SeekBar seekBar;

        private Handler handler=new Handler();

        private final Runnable mTicker = new Runnable() {
            public void run() {
                long now = SystemClock.uptimeMillis();
                long next = now + (1000 - now % 1000);

                handler.postAtTime(mTicker,next);
                 //延迟一秒再次执行 runnable，与计时器效果一样

                if(mMediaPlayer!=null&&mMediaPlayer.isPlaying()){
                    seekBar.setProgress(mMediaPlayer.getCurrentPosition());//更新播放进度
                }
            }
        };

        @Override
        protected void onCreate(Bundle savedInstanceState) {
            super.onCreate(savedInstanceState);
            setContentView(R.layout.activity_main);

            TextureView textureView=(TextureView) findViewById(R.id.textureview);
            textureView.setSurfaceTextureListener(surfaceTextureListener);
                    //设置监听函数，重写 4 个方法

            videoImage=(ImageView) findViewById(R.id.video_image);

            seekBar= (SeekBar) findViewById(R.id.seekbar);
            seekBar.setOnSeekBarChangeListener(onSeekBarChangeListener);//seekbar 改变监听
        }

        private SurfaceTextureListener surfaceTextureListener=new
SurfaceTextureListener() {
            //SurfaceTexture 可用
            @Override
            public void onSurfaceTextureAvailable(SurfaceTexture surfaceTexture,
int width,int height) {
                surface=new Surface(surfaceTexture);
                new PlayerVideoThread().start();//开启一个线程去播放视频

                handler.post(mTicker);//更新进度
            }

            @Override
            public void onSurfaceTextureSizeChanged(SurfaceTexture surface, int
width,int height) {//尺寸改变
```

```
        }

        @Override
        public boolean onSurfaceTextureDestroyed(SurfaceTexture surfaceTexture) {
        //销毁
            surface=null;
            mMediaPlayer.stop();
            mMediaPlayer.release();
            mMediaPlayer=null;
            return true;
        }

        @Override
        public void onSurfaceTextureUpdated(SurfaceTexture surfaceTexture){//更新
        }
    };

    private class PlayerVideoThread extends Thread{
        @Override
        public void run(){
             try {
                 mMediaPlayer= new MediaPlayer();
                 Uri uri = Uri.parse("android.resource://com.example.
                         textureviewvideo/"+R.raw.ansen);
                 mMediaPlayer.setDataSource(MainActivity.this,uri);
                    //设置播放资源(可以是应用的资源文件 / url / sdcard 路径)
                 mMediaPlayer.setSurface(surface);//设置渲染画板
                 mMediaPlayer.setAudioStreamType(AudioManager.STREAM_MUSIC);
                    //设置播放类型
                 mMediaPlayer.setOnCompletionListener(onCompletionListener);
                    //播放完成监听
                 mMediaPlayer.setOnPreparedListener(new OnPreparedListener() {
                   @Override
                   public void onPrepared(MediaPlayer mp){//预加载完成
                       videoImage.setVisibility(View.GONE);
                       mMediaPlayer.start();//开始播放

                       seekBar.setMax(mMediaPlayer.getDuration());//设置总进度
                   }
                 });
                 mMediaPlayer.prepare();
             } catch (Exception e) {
                 e.printStackTrace();
             }
        }
    }

    private SeekBar.OnSeekBarChangeListener onSeekBarChangeListener=new
SeekBar.OnSeekBarChangeListener() {
        @Override
        public void onProgressChanged(SeekBar seekBar, int progress, boolean
fromUser) {//进度改变
        }

        @Override
```

```
            public void onStartTrackingTouch(SeekBar seekBar) {//开始拖动 seekbar
                Log.i(Tag,"onStartTrackingTouch");
            }

            @Override
            public void onStopTrackingTouch(SeekBar seekBar) {//停止拖动 seekbar
                if(mMediaPlayer!=null&&mMediaPlayer.isPlaying()){
                    mMediaPlayer.seekTo(seekBar.getProgress());
                }
            }
        };

        private MediaPlayer.OnCompletionListener onCompletionListener=new
MediaPlayer.OnCompletionListener() {
            @Override
            public void onCompletion(MediaPlayer mediaPlayer) {//播放完成
                videoImage.setVisibility(View.VISIBLE);
                seekBar.setProgress(0);
                handler.removeCallbacks(mTicker);
            }
        };
    }
```

这个类代码有点多，通过下面的方法进行解析：

- onCreate：首先通过 findViewById 查找 TextureView，设置 SurfaceTexture 监听，查找 SeekBar，也设置了 SeekBar 改变监听。
- surfaceTextureListener：重写了四个方法。在 onSurfaceTextureAvailable 方法中开启线程去播放视频。在 onSurfaceTextureDestroyed 方法中释放 surface 与 mMediaPlayer 对象。
- PlayerVideoThread：播放视频多线程，查看 run 方法，新建一个 MediaPlayer 对象，设置播放资源、设置渲染画板、设置播放类型、设置播放完成监听函数、设置预加载，预加载完成回调方法中隐藏图片，调用 MediaPlayer 的 start 方法开始播放。将视频总长度设置为 seek 的总进度，通过 handler 去执行 Runnable 来更新 seekbar 的进度。
- onSeekBarChangeListener：seekbar 监听，在 onStopTrackingTouch 方法中判断视频有没有在播放，如果正在播放中，就将拖动的进度设置为 MediaPlayer 当前播放的进度。
- onCompletionListener：播放完成监听，显示图片，seekBar 进度设置为 0，从 handler 中删除 Runnable，终止计时器。
- mTicker：每秒通过 handler 调用一次自己达到计时器的效果，获取当前 MediaPlayer 播放进度来更新进度条的进度。

学了这三种播放视频方式应该能够解决开发中的大部分需求了，但是实际开发中会有很多新的需求，如全屏切换、暂停、继续播放等。给大家推荐一个国内很好的视频播放开源框架（下载地址如下），这个框架能够满足 90%以上的需求。

```
    https://github.com/lipangit/JiaoZiVideoPlayer
```

7.3 调用浏览器打开网页

在 Android 中可以调用自带的浏览器，或者指定一个浏览器来打开一个链接，只需要传入一个 URL（可以是链接地址）即可。

7.3.1 启动 Android 默认浏览器

在 Android 中，可以通过发送隐式 Intent 来启动系统默认的浏览器。如果手机本身安装了多个浏览器而又没有设置默认浏览器的话，系统将让用户选择使用哪个浏览器来打开链接。

```
Uri uri = Uri.parse("https://www.baidu.com");
Intent intent = new Intent(Intent.ACTION_VIEW, uri);
startActivity(intent);
```

使用以上三行代码就能够调用系统自带浏览器。

7.3.2 启动指定浏览器打开

在 Android 中，可以通过发送显式 Intent 来启动指定的浏览器。例如，手机安装了多个浏览器，包括 QQ 浏览器、Chrome 浏览器、UC 浏览器等，可以指定用某个浏览器打开这个链接。例如，打开 QQ 浏览器，需要如下代码：

```
Uri uri = Uri.parse("https://www.baidu.com");
Intent intent = new Intent(Intent.ACTION_VIEW,uri);
//intent.setClassName("com.UCMobile","com.uc.browser.InnerUCMobile");//打开UC浏览器
intent.setClassName("com.tencent.mtt","com.tencent.mtt.MainActivity");//打开QQ浏览器
startActivity(intent);
```

使用 UC 浏览器打开，只需要将打开 QQ 浏览器那行代码注释掉，再将打开 UC 浏览器那行代码取消注释即可。

7.3.3 优先使用

正常情况下，让用户自己选择使用 QQ 浏览器打开链接。除非有特殊需求才会用到 UC 浏览器。

假如你想用 UC 浏览器打开，但是新版本的 UC 浏览器不使用原来的包名了，这时就没法打开了。另外，UC 浏览器兼容可能存在问题，跳转过去只会显示 UC 首页，而不是直接打开提供的 HTTP 链接。QQ 浏览器就没有这个问题。

7.4 WebView的使用

WebView在Android平台上是一个特殊的View，是基于Webkit引擎、展现Web页面的控件。这个类可以在App中显示一个在线网页，还可以用来开发浏览器。WebView内部实现是采用渲染引擎来展示View内容的，提供网页的前进与后退、网页放大、缩小、搜索等。Android的WebView在低版本和高版本采用了不同的Webkit版本内核，4.4版本后直接使用了Chrome。

现在很多App都内置了Web网页，比如电商平台（淘宝、京东、聚划算等）。WebView比较灵活，不需要升级客户端，只需修改网页代码即可。一些经常变化的页面可以用WebView这种方式去加载网页。例如，中秋节和国庆节打开的页面不一样，如果是用WebView显示的话，只需要后台修改html页面即可，而不需要升级客户端。

WebView功能强大，可以直接使用html文件（本地sdcard/assets目录），还可以直接加载URL，使用JavaScript可以让html与原生App互调。

下面介绍WebView的使用方法。

7.4.1 WebView加载网页的四种方式

下面这段代码介绍WebView加载网页的四种方式，既可以在线加载url，也能加载本地的html内容。

```
    webView.loadUrl("http://139.196.35.30:8080/OkHttpTest/apppackage/test.html");
     //加载url

    webView.loadUrl("file:///android_asset/test.html");//加载asset文件夹下的html

    //方式3：加载手机卡上的html页面
    webView.loadUrl("content://com.ansen.webview/sdcard/test.html");

    //方式4 使用webview显示html代码
    webView.loadDataWithBaseURL(null,"<html><head><title> 欢迎您
</title></head>" + "<body><h2>使用webview显示 html代码</h2></body></html>",
"text/html" , "utf-8", null);
```

7.4.2 WebViewClient与WebChromeClient的区别

Android WebView作为承载网页的载体控件，显示网页的过程中会产生一些事件，并回调给应用程序，以便我们在网页加载过程中做应用程序想处理的事情。比如说客户端需要显示网页加载的进度、网页加载发生的错误事件等。WebView提供两个事件回调类给应用层，分别为WebViewClient、WebChromeClient，开发者可以继承这两个类，接手相应事件处理。

WebViewClient主要帮助WebView处理各种通知、请求事件，有以下几个常用方法。

- onPageFinished：页面请求完成。
- onPageStarted：页面开始加载。
- shouldOverrideUrlLoading：拦截 url。
- onReceivedError：访问错误时回调，例如访问网页时报错 404，在这个方法回调的时候可以加载错误页面。

WebChromeClient 主要辅助 WebView 处理 Javascript 的对话框、网站图标、网站 title、加载进度等，有以下几个常用方法。

- onJsAlert：WebView 不支持 JS 的 alert 弹窗，需要自己监听后通过 dialog 弹窗。
- onReceivedTitle：获取网页标题。
- onReceivedIcon：获取网页 icon。
- onProgressChanged：加载进度回调。

7.4.3 WebView 的简单使用

前面讲解了加载 html 的四种方式，以及 WebViewClient 与 WebChromeClient 的区别，接下来通过代码来实践。

新建项目，修改首页布局文件 activity_main.xml：

```
<?xml version="1.0" encoding="utf-8"?>
<FrameLayout xmlns:android="http://schemas.android.com/apk/res/android"
    android:layout_width="match_parent"
    android:layout_height="match_parent"
    android:orientation="vertical">

    <WebView
        android:id="@+id/webview"
        android:layout_width="match_parent"
        android:layout_height="match_parent"/>

    <ProgressBar
        android:id="@+id/progressbar"
        style="@android:style/Widget.ProgressBar.Horizontal"
        android:layout_width="match_parent"
        android:layout_height="3dip"
        android:max="100"
        android:progress="0"
        android:visibility="gone"/>
</FrameLayout>
```

外层为 FrameLayout，里面有 WebView 与 ProgressBar。WebView 的宽高匹配父类，ProgressBar 横向进度条，高度 3dip，按照 FrameLayout 布局规则，ProgressBar 会覆盖在 WebView 之上，默认是隐藏不显示。

因为需要加载网页 URL，所以需要在 AndroidManifest.xml 中添加访问网络权限。

```
<uses-permission android:name="android.permission.INTERNET" />
```

布局文件对应的MainActivity.java内容如下：

```
public class MainActivity extends AppCompatActivity {
    private WebView webView;
    private ProgressBar progressBar;

    @Override
    protected void onCreate(Bundle savedInstanceState) {
        super.onCreate(savedInstanceState);
        setContentView(R.layout.activity_main);

        progressBar= (ProgressBar)findViewById(R.id.progressbar);//进度条

        webView = (WebView) findViewById(R.id.webview);
        // webView.loadUrl("file:///android_asset/test.html");
        //加载asset文件夹下的html
        webView.loadUrl("http://139.196.35.30:8080/OkHttpTest/apppackage/
test.html");//加载url

        //使用webview显示html代码
        //webView.loadDataWithBaseURL(null,"<html><head><title> 欢迎您
        //</title></head>" +"<body><h2>使用webview显示html代码</h2></body></html>",
        //"text/html" , "utf-8", null);

        webView.addJavascriptInterface(this,"android");
          //添加js监听，这样html就能调用客户端
        webView.setWebChromeClient(webChromeClient);
        webView.setWebViewClient(webViewClient);

        WebSettings webSettings=webView.getSettings();
        webSettings.setJavaScriptEnabled(true);//允许使用js

        /**
         * LOAD_CACHE_ONLY: 不使用网络，只读取本地缓存数据。
         * LOAD_DEFAULT: （默认）根据cache-control决定是否从网络上取数据。
         * LOAD_NO_CACHE: 不使用缓存，只从网络获取数据。
         * LOAD_CACHE_ELSE_NETWORK，只要本地有，无论是否过期，或者no-cache，
              都使用缓存中的数据。
         */
        webSettings.setCacheMode(WebSettings.LOAD_NO_CACHE);
                //不使用缓存，只从网络获取数据。

        //支持屏幕缩放
        webSettings.setSupportZoom(true);
        webSettings.setBuiltInZoomControls(true);

        //不显示webview缩放按钮
//  webSettings.setDisplayZoomControls(false);
    }

    //WebViewClient主要帮助WebView处理各种通知、请求事件
    private WebViewClient webViewClient=new WebViewClient(){
        @Override
        public void onPageFinished(WebView view, String url) {//页面加载完成
            progressBar.setVisibility(View.GONE);
```

```
        }

        @Override
        public void onPageStarted(WebView view, String url, Bitmap favicon) {
          //页面开始加载
            progressBar.setVisibility(View.VISIBLE);
        }

        @Override
        public boolean shouldOverrideUrlLoading(WebView view, String url) {
            Log.i("ansen","拦截url:"+url);
            if(url.equals("http://www.google.com/")){
                Toast.makeText(MainActivity.this,"国内不能访问google,拦截该url",
Toast.LENGTH_LONG).show();
                return true;//表示我已经处理过了
            }
            return super.shouldOverrideUrlLoading(view, url);
        }

    };

    //WebChromeClient主要辅助WebView处理Javascript的对话框、网站图标、网站title、
    //加载进度等
    private WebChromeClient webChromeClient=new WebChromeClient(){
        //不支持js的alert弹窗，需要自己监听，然后通过dialog弹窗
        @Override
        public boolean onJsAlert(WebView webView, String url, String message,
JsResult result) {
            AlertDialog.Builder localBuilder = new AlertDialog.Builder
             (webView.getContext());
            localBuilder.setMessage(message).setPositiveButton("确定",null);
            localBuilder.setCancelable(false);
            localBuilder.create().show();

            //注意:
            //必须要有result.confirm(),
            //表示处理结果为确定状态同时唤醒WebCore线程
            //否则不能继续点击按钮
            result.confirm();
            return true;
        }

        //获取网页标题
        @Override
        public void onReceivedTitle(WebView view, String title) {
            super.onReceivedTitle(view, title);
            Log.i("ansen","网页标题:"+title);
        }

        //加载进度回调
        @Override
        public void onProgressChanged(WebView view, int newProgress) {
            progressBar.setProgress(newProgress);
        }
    };
```

```
    @Override
    public boolean onKeyDown(int keyCode, KeyEvent event) {
        Log.i("ansen","是否有上一个页面:"+webView.canGoBack());
        if (webView.canGoBack() && keyCode == KeyEvent.KEYCODE_BACK){
            //点击返回按钮的时候判断有没有上一页
            webView.goBack(); // goBack()表示返回WebView的上一页面
            return true;
        }
        return super.onKeyDown(keyCode,event);
    }

    /**
     * JS调用Android的方法
     * @param str
     * @return
     */
    @JavascriptInterface //仍然必不可少
    public void  getClient(String str){
        Log.i("ansen","html调用客户端:"+str);
    }

    @Override
    protected void onDestroy() {
        super.onDestroy();

        //释放资源
        webView.destroy();
        webView=null;
    }
}
```

- onCreate：查找控件，给WebView设置加载url，添加JS监听，监听的名称是"android"，设置webChromeClient与webViewClient回调，通过getSettings方法获取WebSettings对象，设置允许加载JS，设置缓存模式，支持缩放。
- webViewClient：重写了几个方法，onPageFinished页面加载完成隐藏进度条，onPageStarted页面开始加载显示进度条，shouldOverrideUrlLoading拦截url，如果请求url是打开Google，则不让它请求，因为Google在国内不能访问，还不如拦截掉，直接告诉用户不能访问。
- webChromeClient：WebView不支持alert弹窗，在这个方法中用AlertDialog去弹窗。onReceivedTitle获取网页标题。onProgressChanged页面加载进度，将加载进度给progressBar。
- onKeyDown：如果点击系统自带返回键并且WebView有上一个页面，则调用goBack返回上一页内容；否则我们不处理，交给父类的onKeyDown方法处理。什么时候会有上一级页面呢？就是从首页跳转到一个新页面，点击返回时会返回首页。如果本来就位于首页，点击返回时会退出App。
- getClient：html页面的JS可以通过这个方法回调原生App。这个方法有个注解@JavascriptInterface，这是必须的。这个方法有一个字符串参数，这个方法与在onCreate中调用addJavascriptInterface传入的name一起使用。例如，html中想要回调这个方法，可以写成javascript:android.getClient（"传一个字符串给客户端"）。
- onDestroy：Activity销毁时释放WebView资源。

运行代码，效果如图 7-5 所示。

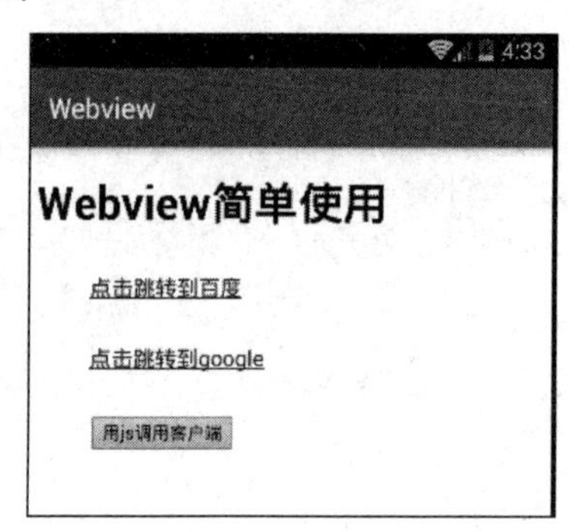

图 7-5　Webview 的简单使用效果

7.5　复制和粘贴

Android 提供的剪贴板框架，允许复制和粘贴不同类型的数据。数据可以是文本、图像、二进制流数据或其他复杂的数据类型。另外，还可以复制多条记录，其功能很强大。当然，大部分用户只需要复制文本，下面学习文本的复制。

7.5.1　复制文本

```
    String text = "abcdefg";
    ClipboardManager cmb = (ClipboardManager)
getSystemService(Context.CLIPBOARD_SERVICE);
    cmb.setPrimaryClip(ClipData.newPlainText(null,text));
```

首先获取 ClipboardManager 对象，全包名是 android.content.ClipboardManager，调用 setPrimaryClip 给剪贴板赋值。

7.5.2　粘贴文本

```
    ClipboardManager clipboardManager = (ClipboardManager)
getSystemService(Context.CLIPBOARD_SERVICE);
    String content= clipboardManager.getText().toString();
    Log.i("ansen","content:"+content);
```

首先获取 ClipboardManager 对象，然后调用 getText 获取文本信息。

7.6　定位的使用

在 Android 开发中地图跟定位是很多 App（应用程序）不可缺少的内容，这些特色功能给人们带来了很多方便。

7.6.1 定位的三种方式

在 Android 系统中定位有三种方式：GPS 定位、WiFi 定位、基站定位。

1. GPS 定位

GPS（Global Positioning System，全球定位系统）是由美国建立的一个卫星导航定位系统。利用该系统，用户可以在全球范围内实现全天候、连续、实时的三维导航定位和测速；另外，用户还能够进行高精度的时间传递和高精度的精密定位。

使用 GPS 定位，需要硬件支持，基本上每部 Android 手机都支持 GPS 定位。直接和卫星交互来获取当前经纬度。

使用 GPS 的优点如下：

- 速度快。
- 精度高。
- 可在无网络情况下使用。

使用 GPS 的缺点如下：

- 首次连接时间长。
- 只能在户外开阔地使用，设备不能被遮挡。在室内基本不能使用。
- 比较耗电。

2. WiFi 定位

WiFi 定位是根据一个固定的 WiFi MAC 地址，通过收集到的该 WiFi 热点的位置，然后访问网络上的定位服务以获得经纬度坐标。它和基站定位其实都需要使用网络，所以在 Android 也统称为 Network 方式。

使用 WiFi 定位的优点在于受环境影响较小，只要有 WiFi 的地方即可使用。

使用 WiFi 定位的缺点是需要有 WiFi、精度不准，并且需要请求第三方 API。

3. 基站定位

基站定位的方式有多种，一般手机附近的三个基站进行三角定位，由于每个基站的位置是固定的，利用电磁波在这三个基站间中转需要时间来算出手机所在的坐标。

还有一种方式是获取最近的基站信息，其中包括基站 ID、Location Area Code、Mobile Country Code、Mobile Network Code 和信号强度，将这些数据发送到 Google 的定位 Web 服务中，就能获取当前所在的位置信息。

使用基站定位的优点是：

- 受环境的影响情况较小，只要有基站的地方都能使用。一般的地方都有，除非那种鸟无人烟的地方，毕竟基站运营商也是需要考虑成本的。

使用基站定位的缺点是：

- 需要消耗流量。
- 精度没有 GPS 那么准确，大概在十几米到几十米之间。
- 需要根据基站信息请求 API 接口获取位置，Google 公开的 API 在国内没法访问。移动与联通有公开基础基站信息获取位置的 API。

7.6.2 定位的相关类

1. LocationManager

LocationManager 为位置服务管理器类，无论采用哪种方式获取位置信息，都需要获取一个 LocationManger 对象。

```
    LocationManager locationManager = (LocationManager)getSystemService(Context.
LOCATION_SERVICE);
```

LocationManager 对象有四个常量，用来区分定位方式。

- NETWORK_PROVIDER：网络定位。
- GPS_PROVIDER：通过 GPS 定位。
- PASSIVE_PROVIDER：被动地获取定位信息，通过接受其他 App 或 Service 的定位信息。
- FUSED_PROVIDER：Google 已经将这个定位方式隐藏了。

2. Location 位置对象

描述地理位置信息的类，记录了经纬度、海拔高度、获取坐标时间、速度、方位等。可以通过 LocationManager.getLastKnowLocation(provider)获取位置坐标，provider 就是定位方式，也是上文中提到的 LocationManager 的四个常量之一。

大部分情况下，通过 getLastKnowLocation 得到的 Location 对象都为 null；实时动态坐标可以在监听器 locationListener 的 onLocationChanged(Location location)方法中来获取。

3. LocationListener 位置监听接口

LocationListener 用于监听位置（包括 GPS、网络、基站等所有提供位置的）变化，监听设备开关与状态，实时动态获取位置信息。

（1）注册监听：LocationManger.requestLocationUpdates(String provider, long minTime, float minDistance, LocationListener listener)。

（2）使用完之后，需要在适当的位置移除监听（不需要获取位置信息或者程序退出时）：LocationManager .removeUpdates(LocationListener listener)。

实现 LocationListener 接口需要重写以下几个方法：

- onLocationChanged(Location location)：当位置发生变化时会自动调用该方法，参数 location 记录了最新的位置信息。
- onStatusChanged(String provider, int status, Bundle extras)：当位置提供者的状态发生改变（可用到不可用、不可用到可用）时自动调用该方法。
- onProviderEnabled(String provider)：位置信息提供者可用时自动调用，例如 GPS 开启。
- onProviderDisabled(String provider)：位置信息不可用时自动调用，例如禁用 GPS。

7.6.3　GPS 获取经纬度

前面讲述了定位的三种方式，基站定位和 WiFi 定位都需要调用第三方 API 接口获取位置信息，因此本节内容主要讲述如何使用 GPS 实现定位。

1. 手机打开 GPS 定位

首先进入系统设置界面→服务与偏好设置→位置信息→开启 GPS。设置界面开启 GPS 可能会因为手机厂商不一致，导致开启 GPS 的界面略有区别。本书截图的这款手机是 Google 和 HTC 合作的 Pixel 手机，Android 8.0 原生操作系统。

效果图如图 7-6 所示。

从图 7-6 中可以看到，GPS 已经处于打开状态。

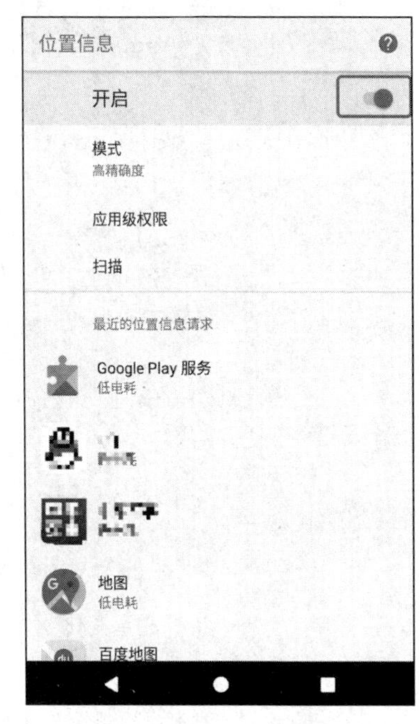

图 7-6　打开手机的 GPS 功能

2. 源码实现

新建项目，因为需要使用 GPS 定位，所以首先在 AndroidManifest.xml 中加入权限。

```
<uses-permission
android:name="android.permission.
ACCESS_FINE_LOCATION" />
```

修改首页布局文件 activity_main.xml，布局文件很简单，仅包括一个 TextView，用来显示经纬度信息。

```
<?xml version="1.0" encoding="utf-8"?>
<LinearLayout xmlns:android="http://schemas.android.com/apk/res/android"
   android:layout_width="match_parent"
   android:layout_height="match_parent"
   android:padding="10dp">

   <TextView
      android:id="@+id/tv_location"
      android:layout_width="wrap_content"
      android:layout_height="wrap_content"
      android:text="Hello World!"/>
</LinearLayout>
```

首页布局文件对应的 MainActivity.java 内容如下：

```
public class MainActivity extends AppCompatActivity {
   private LocationManager locationManager;
   private TextView tvLocation;

   @Override
   protected void onCreate(Bundle savedInstanceState) {
      super.onCreate(savedInstanceState);
      setContentView(R.layout.activity_main);
```

```
        tvLocation= (TextView) findViewById(R.id.tv_location);

        locationManager = (LocationManager)getSystemService
         (Context.LOCATION_SERVICE);
        startRequestLocation();
    }

    private void startRequestLocation(){
        boolean gps = locationManager.isProviderEnabled(LocationManager.
                    GPS_PROVIDER);
        if(!gps){
            Toast.makeText(this,"请先设置界面开启GPS定位服务",Toast.LENGTH_LONG).show();
            return ;
        }

        Location location=locationManager.getLastKnownLocation(LocationManager.
  GPS_PROVIDER); //通过GPS获取位置
        if (location != null) {
            showLocation(location);
        }

        /**
         * 监听位置变化
         * 参数1，定位方式，主要有GPS_PROVIDER和NETWORK_PROVIDER，前者是GPS,后者
是GPRS以及WIFI定位
         * 参数2，位置信息更新周期，单位是毫秒
         * 参数3，位置变化最小距离。当位置距离变化超过此值时，将更新位置信息
         * 参数4，监听
         * 备注：参数2和3，如果参数3不为0，则以参数3为准；参数3为0，则通过时间来定
时更新；两者为0，则随时刷新
         */
        locationManager.requestLocationUpdates(LocationManager.GPS_PROVIDER,
         1000, 10,locationListener);
    }

    private LocationListener locationListener=new LocationListener() {
        @Override
        public void onLocationChanged(Location location) {
            showLocation(location);
        }

        //当位置提供者的状态发生改变
        @Override
        public void onStatusChanged(String provider, int status, Bundle extras) {}

        //位置信息提供者可用时自动调用，例如GPS开启
        @Override
        public void onProviderEnabled(String provider) {
        }
        //位置信息不可用时自动调用，例如禁用GPS
        @Override
        public void onProviderDisabled(String provider) {
        }
    };
```

```
    /**
     * 显示定位结果
     * @param location
     */
    private void showLocation(Location location){
        if(location!=null){
            tvLocation.setText("经度: "+location.getLongitude()+"\n"+"纬度:
"+location.getLatitude());
        }
    }

    @Override
    protected void onDestroy() {
        super.onDestroy();
        locationManager.removeUpdates(locationListener);
    }
}
```

- onCreate：查找 TextView，获取 LocationManager 对象，然后调用 startRequestLocation 方法请求定位。
- startRequestLocation：首先判断 GPS 是否开启，若没有开启则不进行定位。通过 getLastKnownLocation 方法获取最后位置信息，第一次定位时，这个方法肯定会返回 null，最后调用 requestLocationUpdates 方法开启定位监听变化。通过第二个与第三个参数来决定回调频率。
- locationListener：在 onLocationChanged 方法中调用 showLocation 更新位置信息。
- showLocation：将当前的经纬度信息显示到 TextView 上。
- onDestroy：Activity 退出移除定位监听。

运行程序，其效果图如图 7-7 所示。

注意事项：

- 需要先设置界面开启 GPS，再打开软件。
- 因为是 GPS 定位，需要到室外，有时手机拿出窗外就能定位，主要看卫星情况。

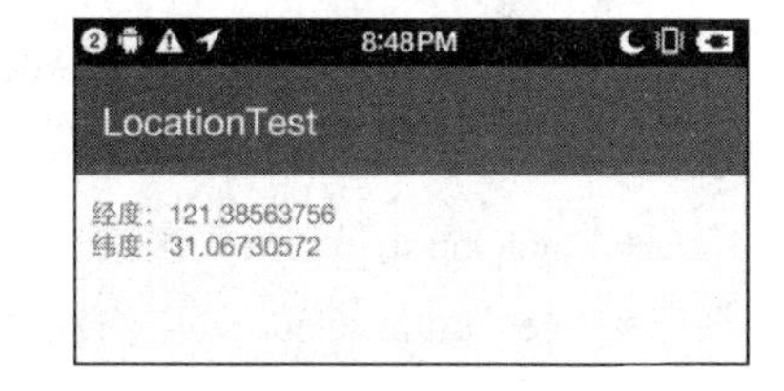

图 7-7　显示经纬度

- 第一次定位需要时间，可以多等一会。如果超过 5 分钟还没有获取到位置信息，就检查前面两条注意事项。

7.6.4　根据经纬度反向编码获取地址

当我们拿到一个经纬度时，得到的是两个 Double 类型的值，是一个空间坐标地址，这时需要转换成看得懂的值，就需要地理反编码（Reverse Geocoding）。

同理，如果知道当前所在的街道具体地址，然后想拿到空间坐标地址，就是经纬度，需要地理编码（Geocoding）。

1. 反编码相关类

（1）Geocoder

该类用于地理编码与反向编码。

Geocoder 请求的是一个后台服务，但是该服务不包括在标准 Android Framework 中。因此，如果当前设备不包含 Location Services，则 Geocoder 返回的地址或者经纬度为空。

当然，可以使用 Geocoder.isPresent()方法来判断当前设备是否包含地理位置服务。由于国内使用不了 Google Services 服务，因此一般的手机厂商都会在自己的手机内置百度地图服务或者高德地图服务来替代 Google Services 服务。

创建 Geocoder 对象有两个构造方法。

- new Geocoder(Context context, Locale locale)：参数 1 是 context，参数 2 是国家/地区代码。
- new Geocoder(Context context)：参数 1 是 context，因为没有指定国家/地区代码，初始化时会取系统设置的国家/地区代码。

这个对象有两个常用方法：

- getFromLocation(double latitude, double longitude, int maxResults)：根据经纬度获取地址，其中参数 latitude 获取纬度，longitude 获取经度，maxResults 返回地址的数目（由于同一经纬度可能对应多个地址，该参数取 1~5）。
- getFromLocationName(String locationName, int maxResults)：根据地址获取经纬度信息，其中参数 locationName 用于获取地址信息（例如，上海市闵行区 xx 路 112 号），maxResults 用于返回条数（由于同一地址可能对应多个经纬度，该参数取 1~5）。

（2）Address

一个地理位置信息，其中 Address 类中包含了各种地理位置信息，包括经纬度、国家/地区、城市、地区、街道、国家/地区编码、城市编码等。

- getCountryName：国家/地区名称。
- getCountryCode：国家/地区编码。
- getAdminArea：省份。
- getLocality：市。
- getFeatureName：道路。
- getLatitude：纬度。
- getLongitude：经度。

我们在上一节的项目代码上进行修改，首先修改 activity_main.xml 文件，添加一个 TextView，用来显示反向编码后的地址。

接下来修改 MainActivity.showLocation 方法，将获取的经纬度反向编码为地址，将国家/地区名称与城市显示到 TextView 上：

```
    private void showLocation(Location location){
        if(location!=null){
            tvLocation.setText("经度："+location.getLongitude()+"\n"+"纬度："
+location.getLatitude());
```

```
        }
        Geocoder geocoder = new Geocoder(this,Locale.CHINA);
        try {
            //参数 1:纬度 参数 2:经度 参数 3:返回地址的数目（由于同一经纬度可能对应多个地
址，该参数取 1~5）
            List<Address> addressList = geocoder.getFromLocation(location.
getLatitude(),location.getLongitude(),1);
            for (Address address : addressList) {
                tvAddress.setText(address.getCountryName()+"
"+address.getLocality());
                Log.i("ansen", address.toString());
            }
        } catch (IOException e) {
            e.printStackTrace();
        }
    }
```

修改后运行代码，效果如图 7-8 所示。

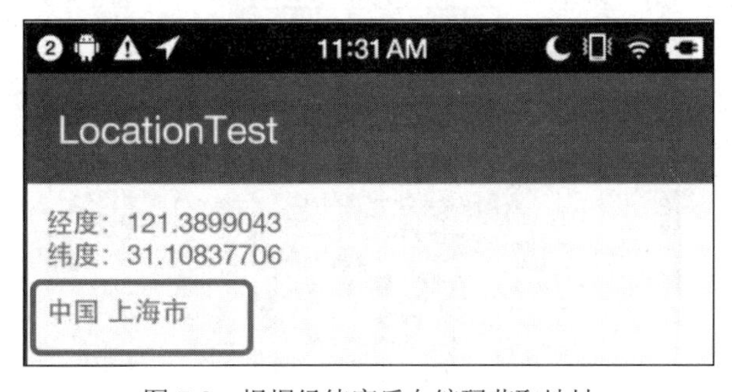

图 7-8　根据经纬度反向编码获取地址

从图 7-8 中可以看到显示了所在的国家/地区和城市名。当然，还可以调用 Address 的 getFeatureName 方法将街道地址显示出来。

经过测试，发现在国内买的锤子手机能根据经纬度反向编码出地址，但是从国外买的 Pixel（Google 与 HTC 合作）手机无法反编码出地址。

7.7　NDK 与 JNI 开发

本节学习如何在 Android 中使用 C++开发。

7.7.1　什么是 NDK

NDK（Native Development Kit）提供了一系列的工具集合，帮助开发者快速开发 C（或 C++）的动态库，并且能够自动将 So 和 Java 应用一起打包成 APK。NDK 集成了交叉编译器，可以针对不同的 CPU 编译不同的 so 文件。

使用 NDK 的优点如下：

- 运行效率高（C/C++的运行速度高于 Java）。
- 代码安全性高（Java 是半解释型语言，容易被别人反编译代码）。
- 代码复用与移植（用 C++开发的代码，不仅可以在 Android 中使用，还可以嵌入其他平台，例如 iOS）。

7.7.2 NDK 下载

在 Android Studio 中下载 NDK 很简单，具体步骤如下：

步骤01 点击标题栏 File→Project Structure，如图 7-9 所示。

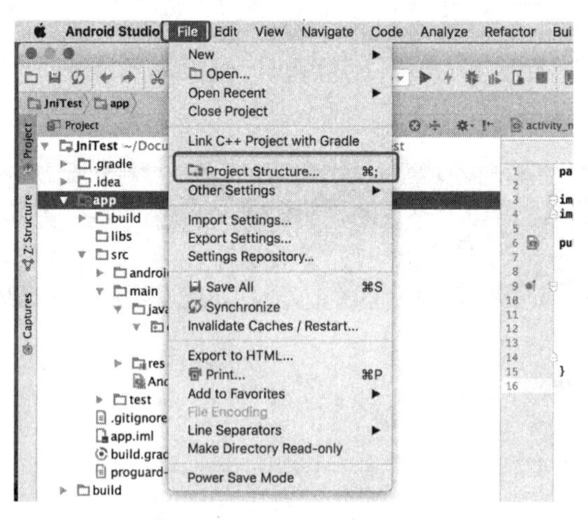

图 7-9　点击 Project Structure

步骤02 进入如图 7-10 所示的 Project Structure 对话框，如果还未下载过 NDK，则 NDK 路径输入框中什么都没有，点击 Download 链接进行下载。

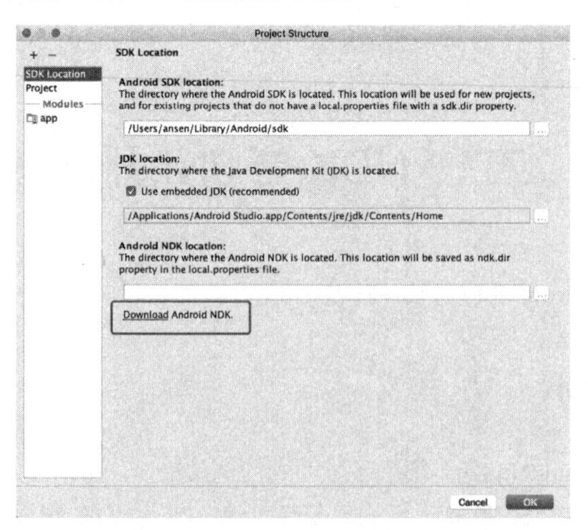

图 7-10　Project Structure 对话框

步骤03 下载过程会有点慢，大家耐心等待。下载完成之后，点击 Finish 按钮就可以看到 NDK

在本地的磁盘路径地址了。

7.7.3　在 Mac 下加入 NDK 环境变量

步骤01 打开 Terminal 终端命令窗口，使用命令切换到 home 目录下：

```
cd ~
```

步骤02 使用 touch 命令。这个命令有两个功能：

- 如果文件存在，就将已存在文件的时间标签更新为系统当前时间。
- 如果文件不存在，就创建新的空文件。

```
touch .bash_profile
```

步骤03 使用如下命令打开文件：

```
open -e .bash_profile
```

步骤04 添加环境变量。在打开的文件最后增加如下两行代码保存文件。其中，NDK_ROOT 指向的路径需要替换为自己的 NDK 路径。这个路径在下载 NDK 的页面中就能够复制。

```
export NDK_ROOT=/Users/ansen/Library/Android/sdk/ndk-bundle
export PATH=${PATH}:${NDK_ROOT}
```

步骤05 使用修改后的文件：

```
source .bash_profile
```

步骤06 验证 NDK 环境变量是否配置成功。在终端中输入“ndk-build”，如果显示下面类似的内容就表示成功了：

```
Android NDK: Could not find application project directory !
Android NDK: Please define the NDK_PROJECT_PATH variable to point to it.
/Users/ansen/Library/Android/sdk/ndk-bundle/build/core/build-local.mk:151:
*** Android NDK: Aborting    .  Stop.
```

终端操作步骤的效果如图 7-11 所示。

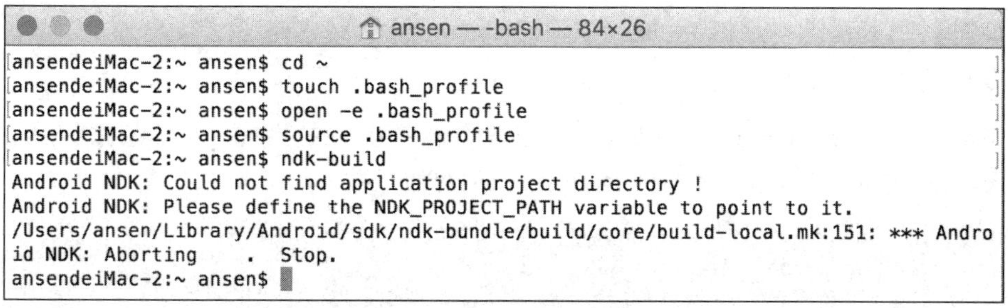

```
ansen — -bash — 84×26
ansendeiMac-2:~ ansen$ cd ~
ansendeiMac-2:~ ansen$ touch .bash_profile
ansendeiMac-2:~ ansen$ open -e .bash_profile
ansendeiMac-2:~ ansen$ source .bash_profile
ansendeiMac-2:~ ansen$ ndk-build
Android NDK: Could not find application project directory !
Android NDK: Please define the NDK_PROJECT_PATH variable to point to it.
/Users/ansen/Library/Android/sdk/ndk-bundle/build/core/build-local.mk:151: *** Andro
id NDK: Aborting    .  Stop.
ansendeiMac-2:~ ansen$
```

图 7-11　加入 NDK 环境变量

7.7.4 什么是 JNI

JNI 是 Java Native Interface 的缩写，提供了若干的 API 实现 Java 和其他语言的通信（主要是 C 和 C++）。从 Java 1.1 开始，JNI 标准成为 Java 平台的一部分，允许 Java 代码和其他语言写的代码进行交互。

JNI 一开始是为了本地已编译语言（尤其是 C 和 C++）而设计的，但是它并不妨碍你使用其他编程语言，只要调用约定受支持就可以了。

使用 Java 与本地已编译的代码交互，通常会丧失平台可移植性。但是，有些情况下这样做是可以接受的，甚至是必须的。例如，使用一些旧的库，与硬件、操作系统进行交互，或者为了提高程序的性能。JNI 标准至少要保证本地代码能工作在任何 Java 虚拟机环境。

7.7.5 NDK 与 JNI 的简单使用

1. 生成头文件

首先创建一个 JNITest 类与 MainActivity 路径同级，其中包含一个 native 方法。

```
package com.ansen.jnitest;

/**
 * @author ansen
 * @create time 2017-09-12
 */
public class JNITest {
    public native int plus(int x, int y);//这个是需用 C 语言实现的函数
}
```

打开 Android Studio 的终端（Terminal），默认是在当前项目路径，需要用 cd 命令定向到项目的 Java 目录：

```
cd app/src/main/java
```

接着用 javah 命令生成.h 文件。后面跟着的是包名.类名，命令执行完成后，在当前目录下会生成一个.h 头文件。

```
javah com.ansen.jnitest.JNITest
```

2. C 代码实现头文件

接下来，在 JniTest/app/src/main 目录下新建一个名字为 jni 的目录，然后将刚才生成的.h 文件剪切过来，并且在 jni 下新建一个文件 JNITest.c，内容如下：

```
#include <jni.h>
#include "com_ansen_jnitest_JNITest.h"
#ifdef __cplusplus //最好有这个，否则可能会被编译器改了函数名字
extern "C" {
#endif

JNIEXPORT jint JNICALL Java_com_ansen_jnitest_JNITest_plus
```

```
(JNIEnv * env, jclass cla, jint x, jint y)
{
    return x + y;
}

#ifdef __cplusplus
}
#endif
```

这个文件主要实现了生成头文件的方法，并将传入的两个参数相加。

3. 配置 NDK

打开 JniTest/app 下的 build.gradle 文件，在 defaultConfig 节点中添加以下代码：

```
ndk{
    moduleName "mylib"//生成的 so 名字,也是 libname
    ldLibs "log", "z", "m"
    abiFilters "armeabi","armeabi-v7a", "x86"
}
```

单独解释一下 abiFilters。上面的代码指定了三种 abi 体系结构下的 so 库，现在市面上 90%的手机都是 ARM 架构的 CPU，所以一般情况只需要编译 armeabi-v7a 架构即可。毕竟打包 apk 时里面多几个 so 文件也会让包变得很大。

打开 JniTest/gradle.properties 文件，尾部加上如下代码：

```
android.useDeprecatedNdk=true
```

在 Android Studio 中修改文件的截图如图 7-12 所示。

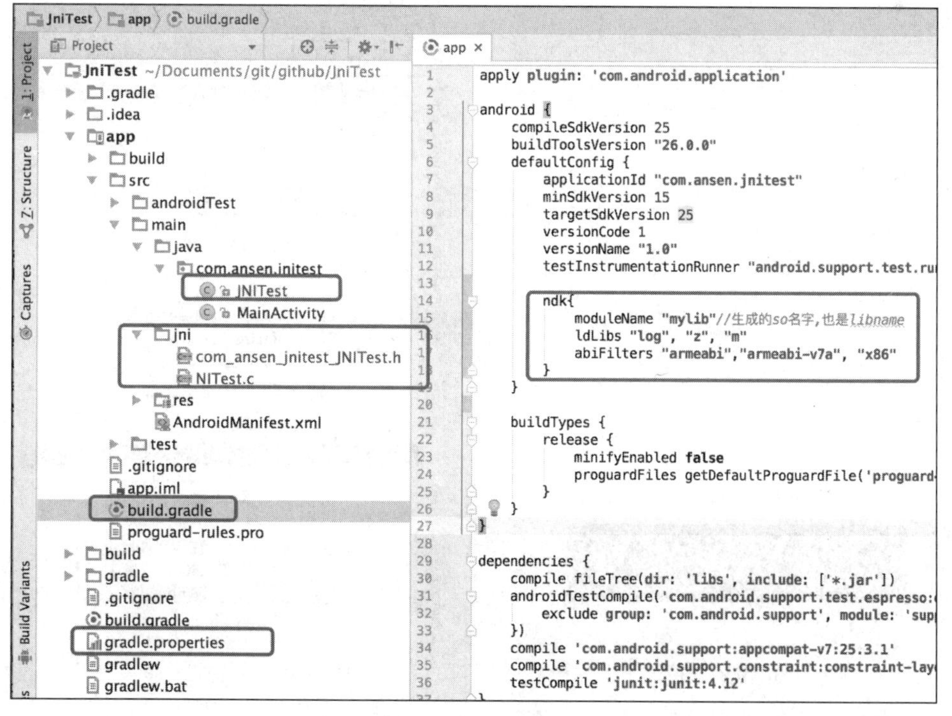

图 7-12　修改文件

4. 调用 native 方法

在 MainActivity 中使用 static 代码块的方式加载 so 库，在 oncreate 方法中创建一个 JNITest 对象，直接调用 plus 方法即可。

```
public class MainActivity extends AppCompatActivity {
    @Override
    protected void onCreate(Bundle savedInstanceState) {
        super.onCreate(savedInstanceState);
        setContentView(R.layout.activity_main);

        JNITest jniTest=new JNITest();

        TextView tvResult= (TextView) findViewById(R.id.tv_result);
        tvResult.setText("运行结果:"+jniTest.plus(100,10));
    }

    static {
        //libname 就是我们在 app/build.gradle 中 moduleName 的值
        System.loadLibrary("mylib");
    }
}
```

直接运行代码，效果如图 7-13 所示。

这里并没有编译 so 文件，为什么加载 so 库时没报错并且还能运行成功呢？那是因为 Android Studio 帮我们编译了 so 文件，并且打包到 apk 中了。在项目的 build 文件夹下可以看到编译后的 so 文件，如图 7-14 所示。

图 7-13　运行结果

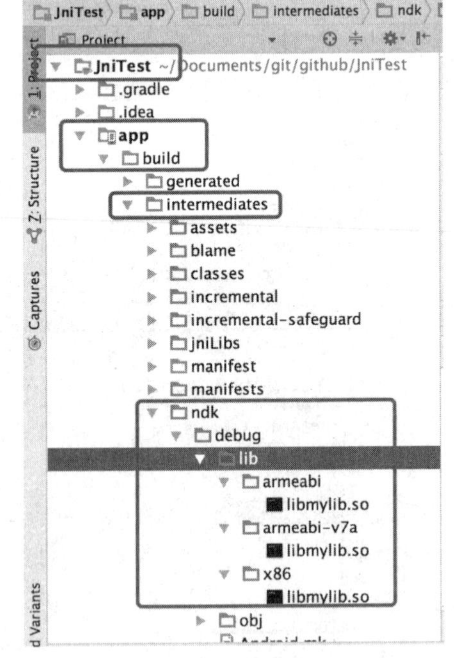

图 7-14　编译后的 so 文件

5. 直接使用编译好的 so 文件

上一个步骤虽然能直接使用，但是每次运行都会编译 so 文件（不确定性），还有一个情况就是不可能把 C/C++代码直接给别人编译运行，所以在某些情况下需要我们手动编译 so 文件。

首先在 JniTest/app/src/main 目录下新建一个名字为 jniLibs 的目录，然后将编译后的 so 文件复制过来，连同目录一起复制。当需要加载 so 文件时，默认会去 src/main/jniLibs 目录下查找。接下来，注释掉 app/build.gradle 中配置的 ndk 代码。

项目修改后的截图如图 7-15 所示。

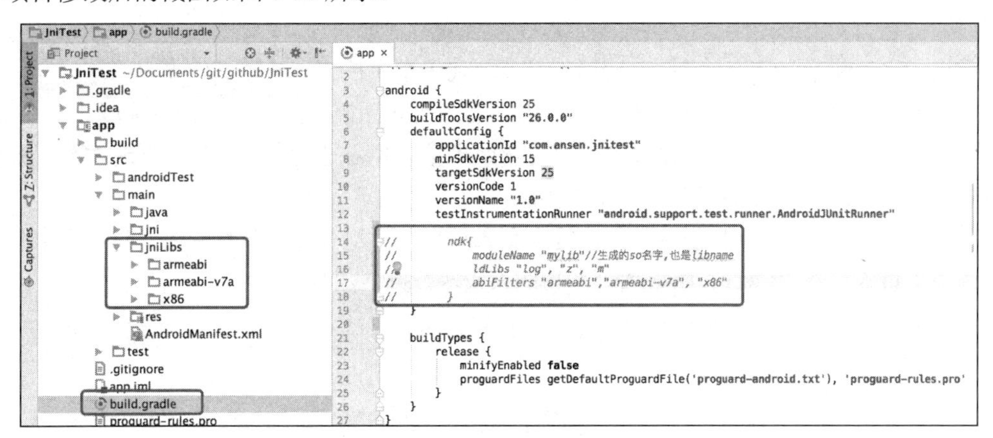

图 7-15　修改项目

运行修改后的代码，结果与之前一样。

7.8　使用 SourceTree 上传项目到 GitHub

在工作中，一般都是多人开发一个项目、维护一套代码，这时就需要一个版本控制系统来管理我们的项目。

7.8.1　什么是 Git

Git 是一个免费、开源的分布式版本控制系统（Version Control System，VCS），可以有效、高速地处理从很小到非常大的项目版本管理。Git 是 Linus Torvalds 为了帮助管理 Linux 内核开发而开发的一个开放源码的版本控制软件。

2005 年 Linux 内核开发社区正面临严峻的挑战：不能继续使用 BitKeeper（一个分布式版本控制系统），原因是当时 Bitkeeper 著作权所有者决定收回授权，内核开发团队与其协商无果，同时又没有其他的 SCM（Software Configuration Management）可以满足分布式系统的需求。

Linus Torvalds 花费一周的时间开发了 Git，最初 Git 的开发是为了辅助 Linux 内核开发。实际上内核开发团队决定开始开发和使用 Git 来作为内核开发的版本控制系统时，世界开源社群的反对声音不少，最大的理由是 Git 太艰涩难懂，从 Git 的内部工作机制来讲，的确是这样。随着开发的深入，Git 的正常使用都由一些友好的脚本命令来执行，使 Git 变得非常好用，即使是用来管理我们自己的开发项目，Git 也是一个友好、有力的工具。现在，越来越多的著名项目采用 Git 来管理开发项目。

7.8.2 什么是 GitHub

GitHub 是一个面向开源及私有软件项目的托管平台，因为只支持 Git 作为唯一的版本库格式进行托管，故名为 GitHub。

GitHub 于 2008 年 4 月 10 日正式上线，除了 Git 代码仓库托管以及基本的 Web 管理界面以外，还提供了订阅、讨论组、文本渲染、在线文件编辑器、协作图谱（报表）、代码片段分享（Gist）等功能。目前，其注册用户已经超过 900 万，托管版本数量也是非常之多，其中不乏知名开源项目 Ruby on Rails、jQuery、Python 等。

在 GitHub 上提交开源项目是免费的，提交私人项目（不想开源）是收费的。收费标准是 7 美元一个月，支持 5 个私有项目。如果项目参与的人较多，则费用也会增加。

基于 Git 的其他托管平台还有几个。

1. GitLib

GitLab 是一个用于仓库管理系统的开源项目。使用 Git 作为代码管理工具，并在此基础上搭建 Web 服务。可以通过 Web 界面访问公开的或者私有的项目。它拥有与 GitHub 类似的功能，能够浏览源代码，管理缺陷和注释，可以管理团队对仓库的访问，易于浏览提交过的版本，并且提供了一个文件历史库。团队成员可以利用内置的简单聊天程序（Wall）进行交流。它还提供了一个代码片段收集功能，可以轻松实现代码复用。下载安装地址为 https://bitnami.com/stack/gitlab/installer。

公司可以用 GitLab 来搭建自己的代码管理工具，供公司内部使用。

2. BitBucket

BitBucket 是一家源代码托管网站，采用 Mercurial 和 Git 作为分布式版本控制系统，同时提供商业计划和免费账户。

BitBucket 可以创建免费的私有项目，只能有 5 个人参与项目，如果超过 5 个人就需要收费，适合刚创业的小团队。BitBucket 的访问速度比 GitHub 要慢。

3. 码云

码云是中国人自己的代码托管网站，开源中国社区主推，也是基于 Git 的。相对于 GitHub、GitLib、BitBucket，码云具有以下几点优势：

- 访问速度快，毕竟服务器在国内。
- 更适合中国人的使用习惯，界面都是中文的。
- 不管私有项目还是开源项目都不收费。

当然，也有一些缺点：

- 如果你是一个开源项目的作者，肯定希望自己的项目被越来越多的人使用，GitHub 的曝光率更高，并且是全球性。

- 如果你是企业，用码云创建私有项目的话，你的代码就在别人的服务器上，安全性是一个问题。

因此，如果你是开源项目作者，建议将项目提交到GitHub。如果是公司，建议自己搭建GitLib。如果是小公司，为了省钱，可以选择BitBucket和码云的私有库。

7.8.3 什么是SourceTree

SourceTree是Windows和Mac OS X下免费的Git和Hg客户端管理工具，同时也是Mercurial和Subversion版本控制系统工具，支持创建、克隆、提交、push、pull和合并等操作。

SourceTree拥有一个精美简洁的界面，大大简化了开发者与代码库之间的Git操作方式，对于那些不熟悉Git命令的开发者来说非常实用。

每天我们都要使用Git提交、更新与合并代码，记住这些Git命令是一件很困难的事，用SourceTree就简单了，新手也能立马上手。

7.8.4 使用SourceTree操作GitHub

1. Git下载

GitHub是基于Git进行版本控制的，所以先要下载Git客户端。官方下载页面地址为https://git-scm.com/downloads。有四个客户端可供选择，分别是Mac、Windwos、Linux和Solaris，可根据自己的操作系统进行选择。

2. SourceTree下载

SourceTree的官网地址为https://www.sourcetreeapp.com/。在官网首页就有“Download for Mac OS X”下载按钮。（因为笔者的电脑是Mac系统，所以是for Mac OS X，各操作系统显示的名字略有区别。）

3. 安装Git与SourceTree

下载了Git与SourceTree客户端后，安装过程很简单，这里就不解释了。安装SourceTree时需要一个Atlassian账户，按照提示注册即可。需要注意的是，先安装Git再安装SourceTree。

4. 注册GitHub账号和创建项目

GitHub的官网地址为https://github.com/。用浏览器打开官网链接，显示如图7-16所示的页面。在官网首页可以直接测试，输入用户名、邮箱和密码，然后点击“Sign up for GitHub”按钮即可。GitHub会给注册邮箱发送一条激活的邮件，激活之后就算注册成功了。

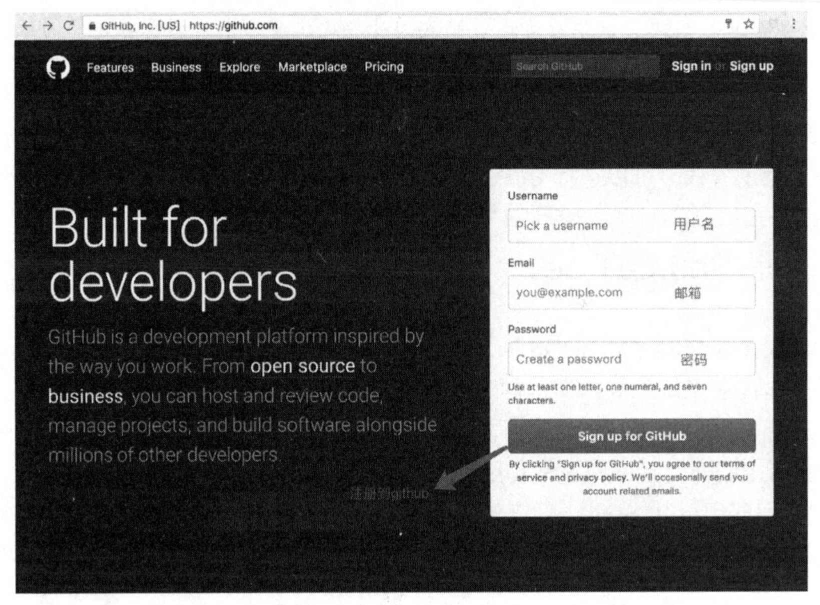

图 7-16　在网站首页测试

点击激活邮件中的 Verify email address 链接，直接进入 GitHub 首页，并且显示登录成功状态。登录成功之后的效果如图 7-17 所示。（这是笔者的账号登录之后显示的首页。）

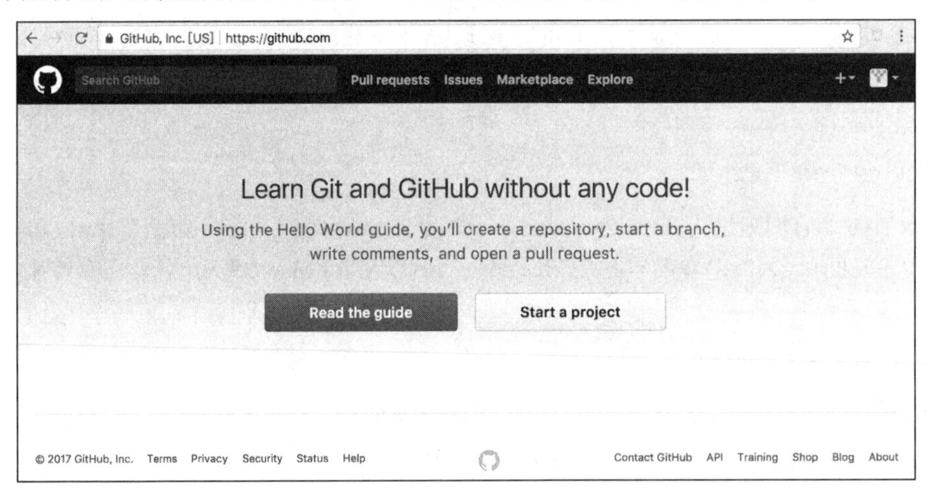

图 7-17　登录成功

现在将之前封装的 OkHttpEncapsulation 项目提交到 GitHub 上，让更多人使用封装的框架。点击 GitHub 首页的“Start a project”按钮新建一个项目，如图 7-18 所示。

图 7-18　新建项目

Repository name 输入框用于填写 OkHttpEncapsulation，和项目名称保持一致。项目描述可不写，项目类别选择“Public”，选中是否初始化阅读文档的单选按钮，毕竟一个项目没有描述文档，其他人也不知道你的项目是干什么用的。最后点击“Create repository”按钮，进入项目创建成功页面，如图 7-19 所示。

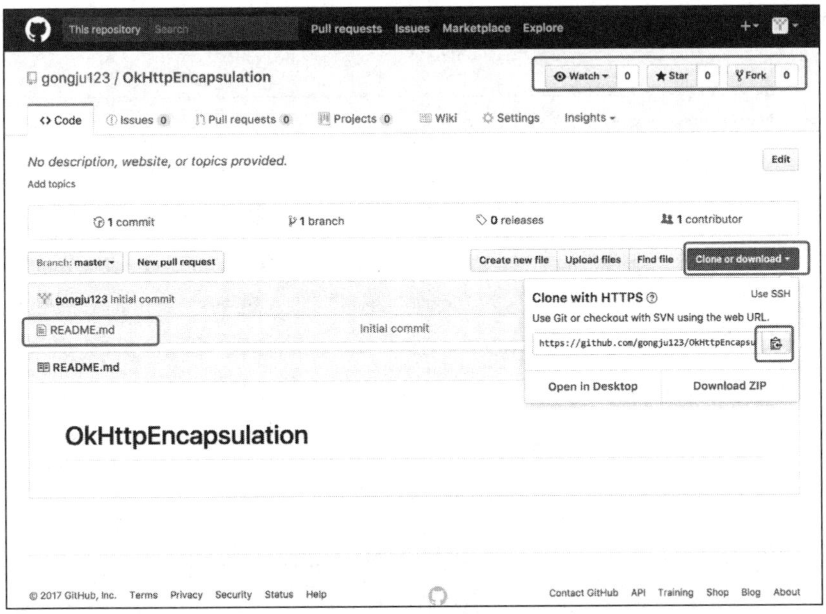

图 7-19　创建项目成功

默认显示的是 Code 列表，看到文件列表中只有一个 README.md 文件，这还是勾选了初始化阅读文件，不然一个文件都没有，然后点击“Clone or download”的下拉按钮，可以看到项目的

Git 地址，Git 地址右侧有一个复制按钮，点击此按钮即可将 Git 地址复制到剪贴板中。

右上角还有三个按钮，即 Watch、Star、Fork。

（1）Watch

Watch 即观察，点击 Watch 按钮，可以看到如图 7-20 所示的列表。

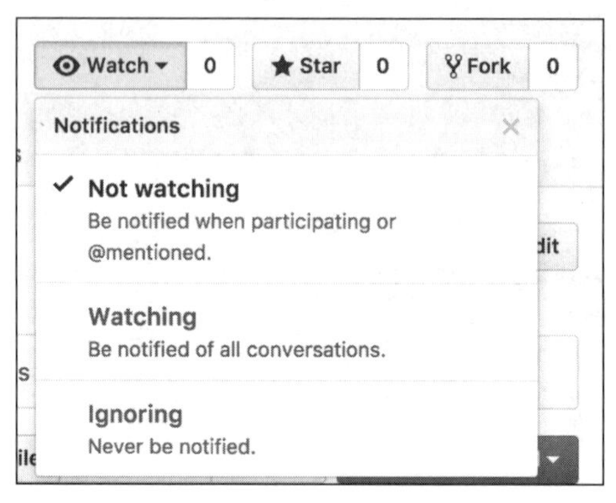

图 7-20　Watch 下拉列表

对于别人的项目，默认自己都处于 Not watching 状态。选择 Watching 后，表示以后会关注这个项目的所有动态，这个项目以后只要发生变动，如被别人提交了 Pull Request、被别人发起了问题等，都会在自己的个人通知中心收到一条通知消息。如果设置了个人邮箱，那么该邮箱也可能会收到相应的邮件。

如图 7-21 所示，Watching 了开源项目 android-cn/android-discuss 后，任何人只要在这个项目下提交了问题或者在问题下面有任何留言，通知中心就会给通知。如果配置了邮箱，还可能会因此不断地收到通知邮件。

android-cn/android-discuss	
网络请求的思考	3 hours ago
[问答]表单提交的输入验证，大家有不有啥子好的方案呢?	6 hours ago
大家有没有做过后台定时任务的? 都是怎么做的?比如我要每隔...	7 hours ago
shrinkReleaseMultiDexComponents FAILED after enabling pr...	8 hours ago
FrameLayout中有个ImageView，通过ImageView.layout(l,t,r,b...	17 days ago
[问答]如何实现微信的延时启动界面?	17 days ago
关于Activity被回收，Fragment还在的问题	20 days ago

图 7-21　查看通知

如果不想接收这个项目的所有通知，那么点击 Not watching 即可。

（2）Star

Star 解释为关注或者点赞更为合适。当你点击 Star 时，表示喜欢这个项目，或者通俗点地将其理解为朋友圈的点赞，表示对这个项目的支持。

不过，相对于朋友圈的点赞，GitHub 中有一个列表，专门收集了所有 Star 过的项目，点击

GitHub 个人头像，可以看到 Your stars 条目，点击就可以查看 Star 过的所有项目，如图 7-22 所示。

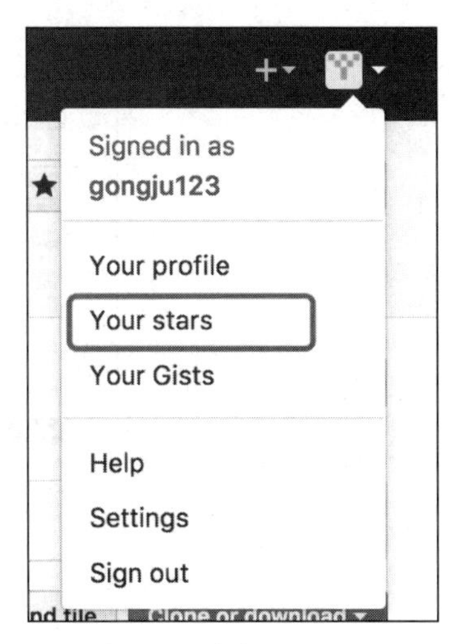

图 7-22　点击 Your stars

（3）Fork

当选择 Fork 时，相当于自己有了一份原项目的备份。当然，这个备份只是针对当时的项目文件，如果后续原项目文件发生改变，就必须通过其他的方式去同步。

一般来说，我们不需要使用 Fork 这个功能，除非有一些项目可能存在 Bug 或者可以继续优化的地方，想帮助原项目作者去完善这个项目或者单纯地想在原来项目的基础上维护一个属于自己的项目。例如，我的 AndroidWeekly 客户端，你可以复制一份，然后自己对这个项目进行修改完善。当你觉得项目没问题了，就可以尝试发起 Pull Request 给原项目作者了。接下来，静静等待他的邮件通知。

很多人都使用 Fork，将 Fork 当作类似收藏的功能，每次看到一个好的项目就 Fork。因为这样，就可以在我的仓库列表下查看 Fork 的项目了。其实完全可以使用 Star 来达到这个目的。

5. SourceTree 从 GitHub 上获取项目

首先从 GitHub 项目首页复制 Git 链接，例如刚创建的项目 Git 地址：

https://github.com/gongju123/OkHttpEncapsulation.git

打开 SourceTree 软件，此时首页什么都没有，一片空白，如图 7-23 所示。

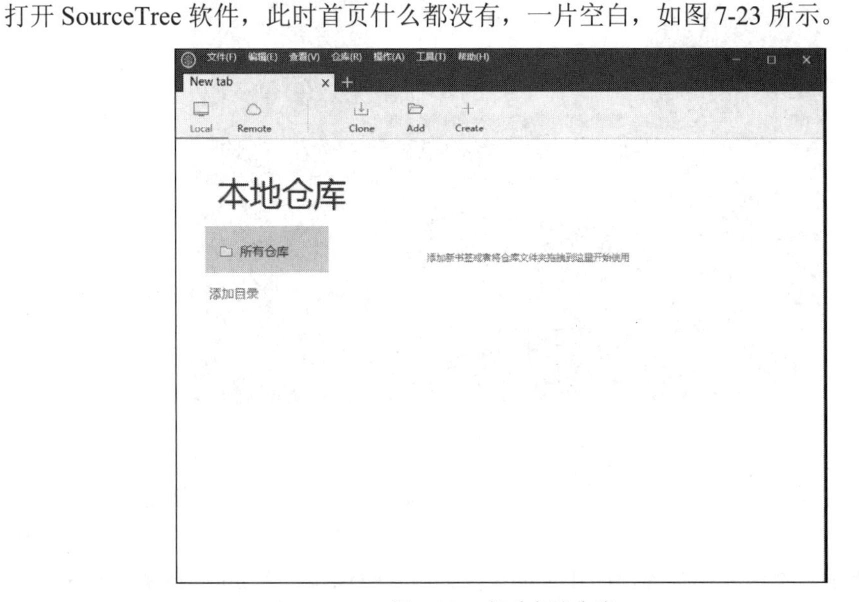

图 7-23　查看本地仓库

点击首页的 Clone 按钮，出现如图 7-24 所示的页面，然后将 Git 地址填写到第一个输入框，下面的项目保存路径、项目名称会自动生成，自己也可以根据实际需求进行修改。最后点击“克隆”按钮。

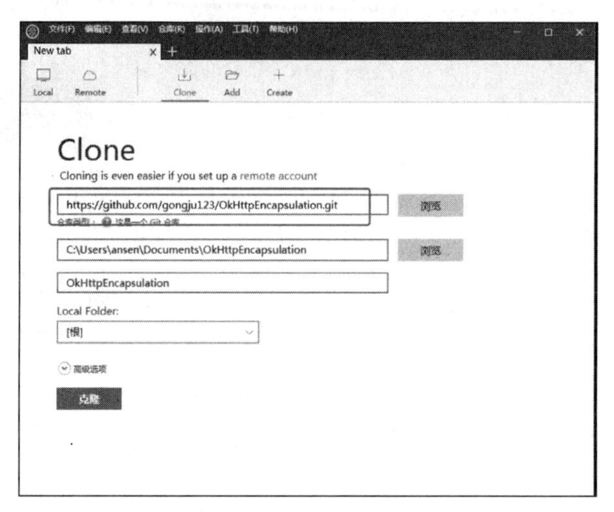

图 7-24　克隆项目

点击“克隆”按钮之后进入关联用户对话框，如图 7-25 所示。在“全名”文本框中输入自己的名字，或者用默认生成的名字。接下来，填写自己的邮箱地址，随便填写一个也可以，但是提交代码时可以看到这些信息。填写真实的信息，便于区分。点击“确定”按钮，这个项目就同步到了本地。

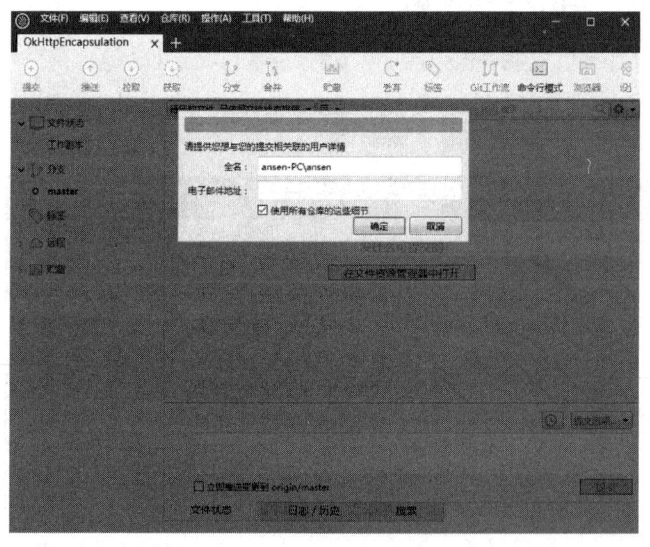

图 7-25　提交相关联的用户详情

此时，进入了可视化的项目 Git 首页，可以看到之前的提交记录，还有分支信息，如图 7-26 所示。

克隆项目时指定了本地保存路径，打开这个文件夹，其中包含一个 git 隐藏文件夹和一个初始化的 md 文件，如图 7-27 所示。

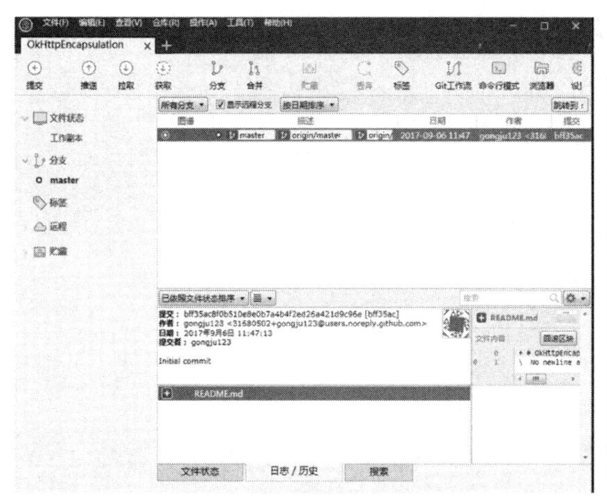

图 7-26 项目的 Git 首页

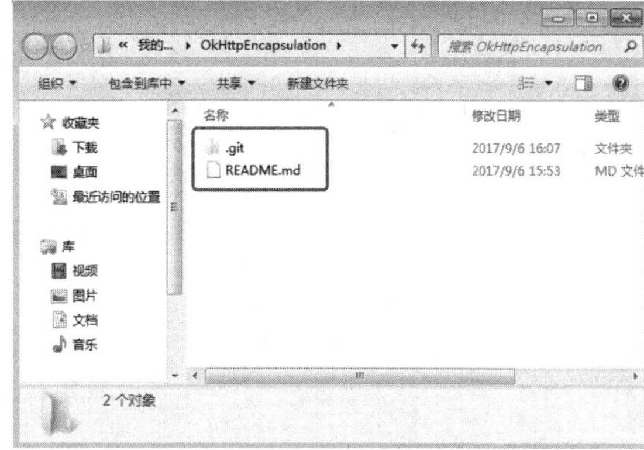

图 7-27 查看克隆的文件

6. SourceTree 提交文件到 GitHub

步骤01 将之前项目文件夹的所有文件复制到 Git 地址生成的文件夹下。复制成功之后，再次进入 SourceTree 首页，可以看到“未暂存文件”列表框中包含刚刚复制的文件，如图 7-28 所示。

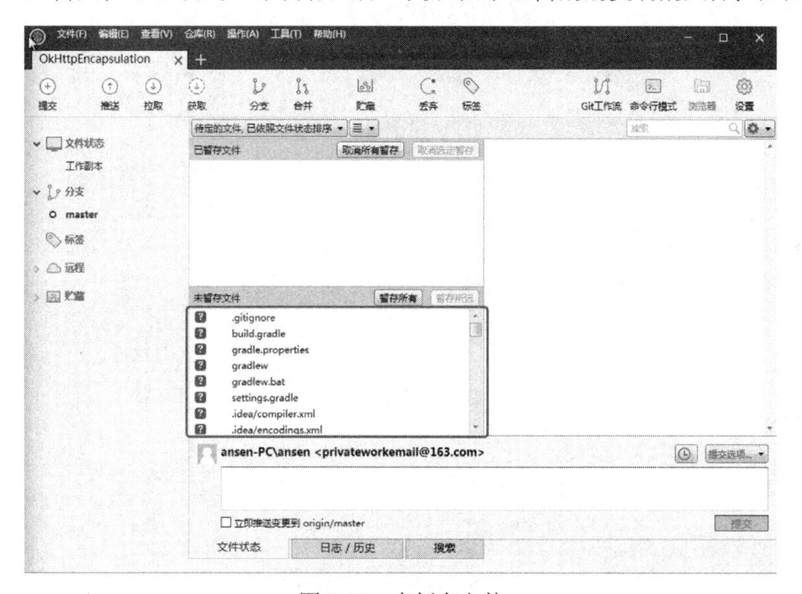

图 7-28 未暂存文件

步骤02 点击“暂存所有”按钮，然后在“提交选项”输入框中输入要提交这些内容的备注，例如第一次提交时可以填写“第一次提交”，然后选中“立即推送变更到”复选框，如图 7-29 所示。

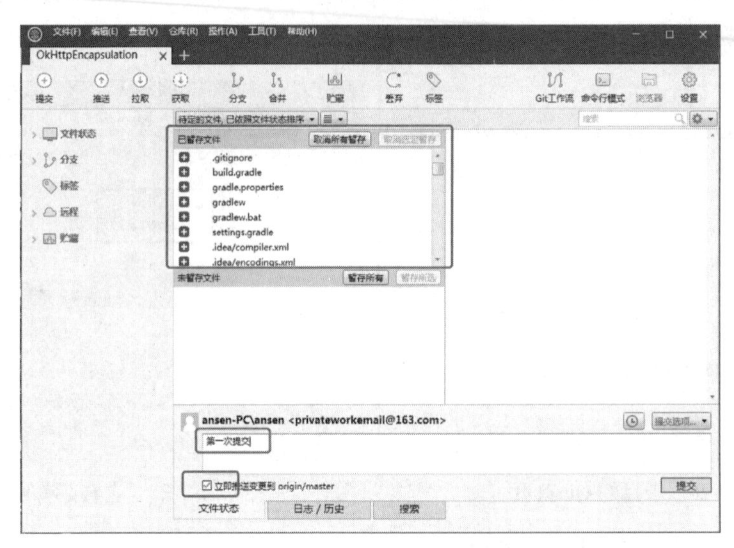

图 7-29　输入提交选项信息

步骤03 点击“提交”按钮，弹出如图 7-30 所示的 GitHub 登录窗口。因为操作的是 GitHub 网站的 Git，所以需要登录这个网站的账号来授权。

图 7-30　登录窗口

步骤04 输入用户名和密码，点击“Login”按钮，开始将代码提交到服务器，提交成功后效果如图 7-31 所示。

如何确保代码提交到了服务器呢？用浏览器打开 GitHub，登录账号，登录成功之后点击头像会弹出下拉列表，再点击下拉列表中的“Your profile”命令，如图 7-32 所示。

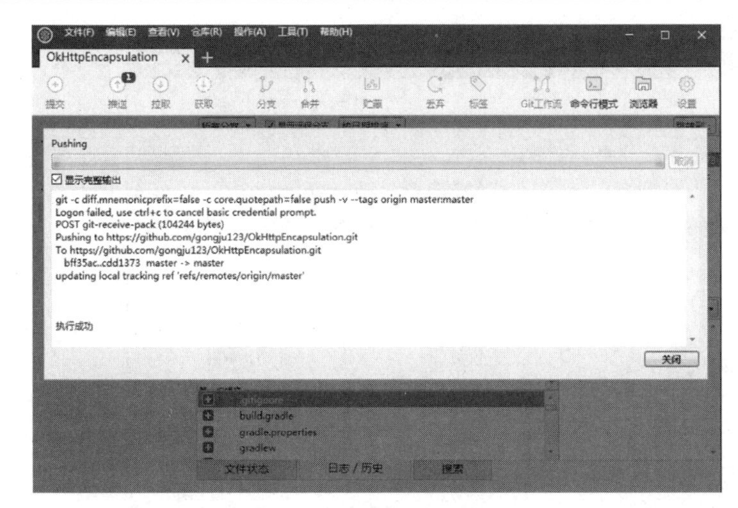

图 7-31　将代码提交到服务器

图 7-32　点击“Your profile”

接下来会进入用户首页，在首页中找到提交的那个项目 OkHttpEncapsulation，点击项目名称进入该项目首页。在项目主页可以明显地看到之前提交的文件，如图 7-33 所示。至此提交代码就算完成了。

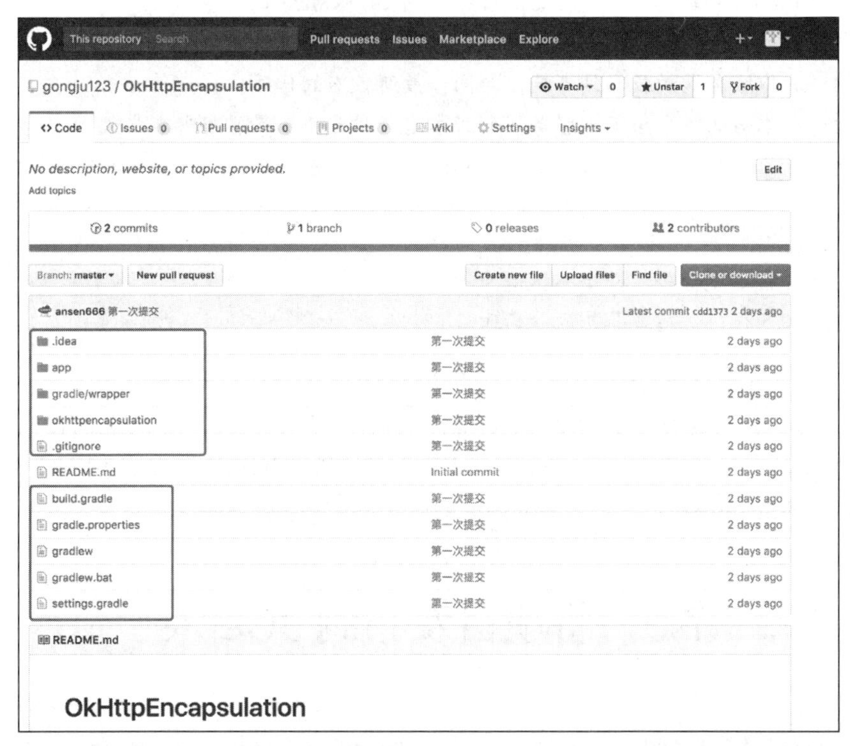

图 7-33　看到提交的文件

7. SourceTree 项目首页常见按钮的使用

在 SourceTree 项目首页中，常用的有“提交”“推送”“拉取”“获取”按钮，如图 7-34 所示。

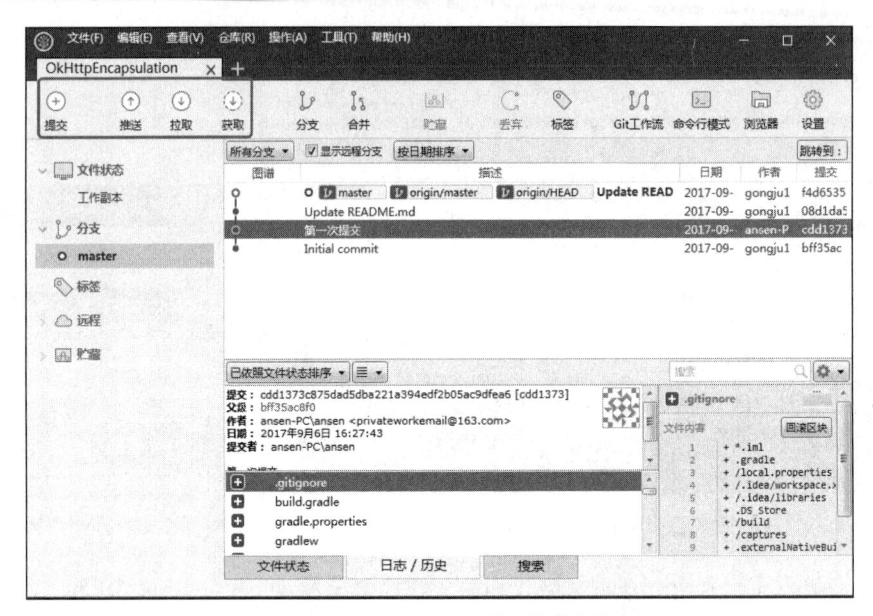

图 7-34 SourceTree 项目首页的常用按钮

- 提交：代码提交到本地，这时还没有推送到 Git 服务器上。
- 推送：代码推送到 Git 服务器上。
- 拉取：拉取服务器的版本到本地，会同步最新版本的代码。
- 获取：将本地的版本与 Git 服务器的版本进行对比。如果服务器版本大于本地，拉取的图标就会显示一个数字。这时只会提示与 Git 服务器相差几个版本，但是不会合并代码。

正常情况下如何使用呢？

- 一般提交代码流程是：提交→推送。
- 同步服务器代码流程是：获取→拉取。

如果发生冲突了怎么处理呢？

如果你和别人同时修改某个文件，恰好修改的是同一个代码块，就会发生冲突，一般是手动解决冲突。发生冲突后打开 Android Studio，查看项目中哪里有报错，因为发生冲突时会有一些非代码块，Android Studio 肯定会报错。删除那些乱码，再次提交即可。

7.9 将项目发布到 JCenter

前面使用自己封装的 OkHttp 项目时，只需要在 app/build.gradle 文件中加入一行代码就能使用项目。

```
compile 'com.ansen.http:okhttpencapsulation:1.0.1'
```

那是因为之前就把封装的模块提交到了 JCenter 服务器，所以 Android Studio 从 JCenter 服务器下载类库。

7.9.1　提交项目到 JCenter

做好一个项目或者项目中有一些好的模块想分享给别人使用时，首先将代码提交到 GitHub 上开源，但是这样还不够，别人拿你的代码作为模块依赖太麻烦了，比较酷的方法是一行代码引用。

1. 需要一个 JCenter 账号

JCenter 网站注册有点麻烦，包含企业版和个人免费版两种方式，其中个人版的项目必须要开源，个人注册地址如下：

```
https://bintray.com/signup/oss
```

一定要注意，不要注册为企业版，因为无法直接切换成个人版，必须等试用期过了重新注册。另外，注册邮箱不能是国内 163 邮箱或者 qq 邮箱，可以用 Google 等邮箱注册。

2. 上传项目到 JCenter

（1）在 JCenter 上创建 Repository

登录 JCenter 账号，进入用户首页，点击“Add New Repository”按钮，如图 7-35 所示。

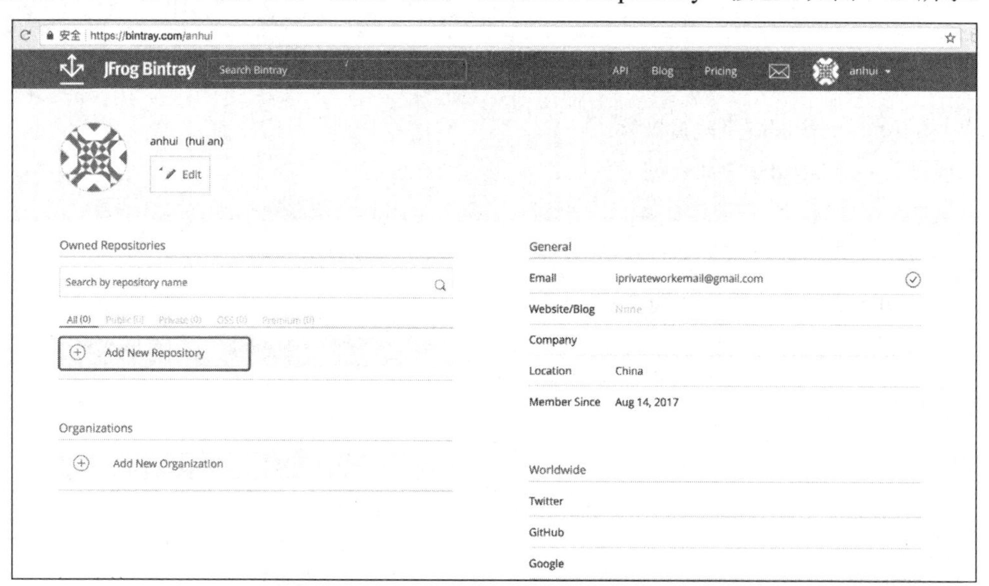

图 7-35　点击“Add New Repository”按钮

进入如图 7-36 所示的 Add New Repository 页面，在“Name”文本框中输入模块名称“okhttpencapsulation”，将“Type”设置为“Maven”，然后点击“Create”按钮，Repository 就创建成功了。

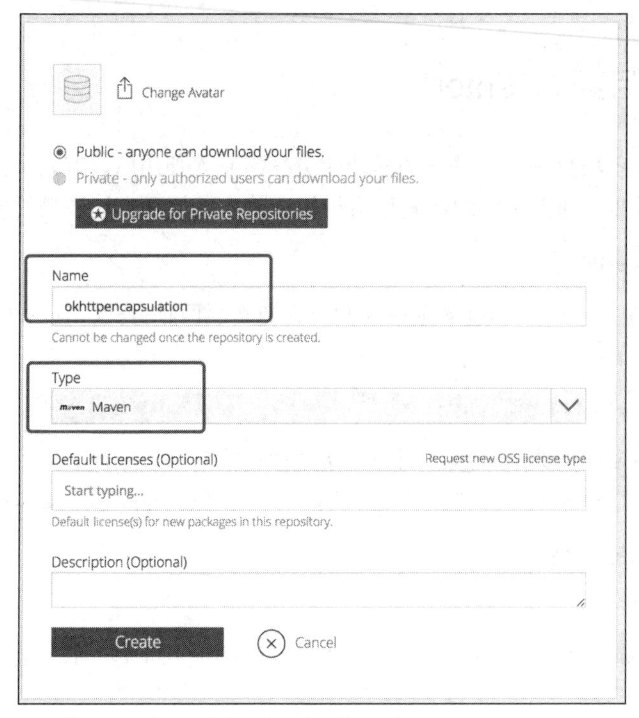

图 7-36　创建 Repository

（2）在 Android Studio 中添加代码

在项目的 build.gradle 中加入如下代码：

```
classpath 'com.github.dcendents:android-maven-gradle-plugin:1.4.1'
classpath 'com.jfrog.bintray.gradle:gradle-bintray-plugin:1.4'
```

build.gradle 修改后的效果如图 7-37 所示。

```
// Top-level build file where you can add configuration options common to all sub-projects/modules.

buildscript {
    repositories {
        jcenter()
    }
    dependencies {
        classpath 'com.android.tools.build:gradle:2.2.3'

        // NOTE: Do not place your application dependencies here; they belong
        // in the individual module build.gradle files
        // 添加下面两行代码即可
        classpath 'com.github.dcendents:android-maven-gradle-plugin:1.4.1'
        classpath 'com.jfrog.bintray.gradle:gradle-bintray-plugin:1.4'
    }
}

allprojects {
    repositories {
        jcenter()
```

图 7-37　build.gradle 修改后的效果

修改要上传的模块下的 build.gradle，例如这里的路径 OkHttpEncapsulation/okhttpencapsulation/build.gradle。在文件尾部增加如下代码：

```
// 这里添加下面两行代码
apply plugin: 'com.github.dcendents.android-maven'
apply plugin: 'com.jfrog.bintray'
```

```
    // 定义两个链接，下面会用到
    def siteUrl = 'https://github.com/ansen666/OkHttpEncapsulation' // 项目主页
    def gitUrl = 'https://github.com/ansen666/OkHttpEncapsulation.git' // Git仓库的url

    group = "com.ansen.http"
    // 唯一包名,比如compile 'com.ansen.http:okhttpencapsulation:1.0.1'中的com.ansen.http就是这里配置的
    version = "1.0.1"
    //项目引用的版本号，比如compile 'com.ansen.http:okhttpencapsulation:1.0.1'中的1.0.1就是这里配置的

    install {
        repositories.mavenInstaller {
            // 生成pom.xml和参数
            pom {
                project {
                    packaging 'aar'
                    // 项目描述，复制的话，这里需要修改。
                    name 'okhttpencapsulation'// 可选，项目名称
                    description 'okhttp project describe'// 可选，项目描述
                    url siteUrl // 项目主页，这里是引用上面定义好的

                    //软件开源协议，现在一般都是Apache License2.0，复制的话，这里不需要修改
                    licenses {
                        license {
                            name 'The Apache Software License, Version 2.0'
                            url 'http://www.apache.org/licenses/LICENSE-2.0.txt'
                        }
                    }

                    //填写开发者基本信息，复制的话，这里需要修改
                    developers {
                        developer {
                            id 'ansen' // 开发者的id
                            name 'ansen' // 开发者名字
                            email 'iprivateworkemail@gmail.com' // 开发者邮箱
                        }
                    }

                    // SCM，复制的话，这里不需要修改
                    scm {
                        connection gitUrl // Git仓库地址
                        developerConnection gitUrl // Git仓库地址
                        url siteUrl // 项目主页
                    }
                }
            }
        }
    }

   // 生成jar包的task，不需要修改
   task sourcesJar(type: Jar) {
       from android.sourceSets.main.java.srcDirs
```

```
        classifier = 'sources'
    }

    // 生成jarDoc的task，不需要修改
    task javadoc(type: Javadoc) {
        source = android.sourceSets.main.java.srcDirs
        classpath += project.files(android.getBootClasspath().join
            (File.pathSeparator))
        // destinationDir = file("../javadoc/")
        failOnError false // 忽略注释语法错误，如果用jdk1.8，注释写得不规范就不能编译
    }

    // 生成javaDoc的jar，不需要修改
    task javadocJar(type: Jar, dependsOn: javadoc) {
        classifier = 'javadoc'
        from javadoc.destinationDir
    }
    artifacts {
        archives javadocJar
        archives sourcesJar
    }

    // 这里是读取Bintray相关的信息，上传项目到GitHub上时会把gradle文件传上去，所以不要
把账号密码的信息直接写在这里，而是写在local.properties中，这里动态读取。
    Properties properties = new Properties()
    properties.load(project.rootProject.file('local.properties').newDataInputS
tream())
    bintray {
        user = properties.getProperty("bintray.user") // Bintray的用户名
        key = properties.getProperty("bintray.apikey") // Bintray刚才保存的ApiKey

        configurations = ['archives']
        pkg {
            repo = "okhttpencapsulation"//Repository名字，需要自己在bintray网站上先添加
            name = "okhttpencapsulation"// 发布到Bintray上的项目名字，这里的名字不是
compile 'com.ansen.library:circleimage:1.0.1'中的circleimage
            userOrg = 'anhui'//Bintray的组织id
            websiteUrl = siteUrl
            vcsUrl = gitUrl
            licenses = ["Apache-2.0"]
            publish = true // 是否是公开项目
        }
    }
```

需要时直接复制代码即可，然后修改一些值。在最后的 bintray 中包含从 local.properties 文件中获取用户名和 API Key。这是保密信息，我们不能暴露给别人，build.gradle 文件会提交到 Git 服务器上，但是 local.properties 文件不会提交。

打开 OkHttpEncapsulation/local.properties 文件，在尾部添加两行，这个 Key 是随意修改的，是一个错误的 Key，需要自己去替换：

```
bintray.user=anhui
bintray.apikey=ac8137c9138a8b49a18a323260041fcf1f75a6f
```

user 是我们注册的名字，apikey 需要去 JCenter 官网查看。进入修改用户界面，点击左侧“API Key”

按钮，然后输入密码就能看到了，如图 7-38 所示。将这个 Key 复制到 local.properties 中进行替换。

图 7-38　复制 API Key

（3）使用 gradle 命令上传

将项目上传到 JCenter，需要使用 gradle 命令。首先将 gradle 加入环境变量。在 Mac 系统下加入环境变量的方法如下（Windows 环境不会加入环境变量的可以自行搜索）：

```
http://blog.csdn.net/u013424496/article/details/52684213
```

在 Android Studio 底部有一个“Terminal”按钮（如图 7-39 所示），点击后进入 Terminal 页面。

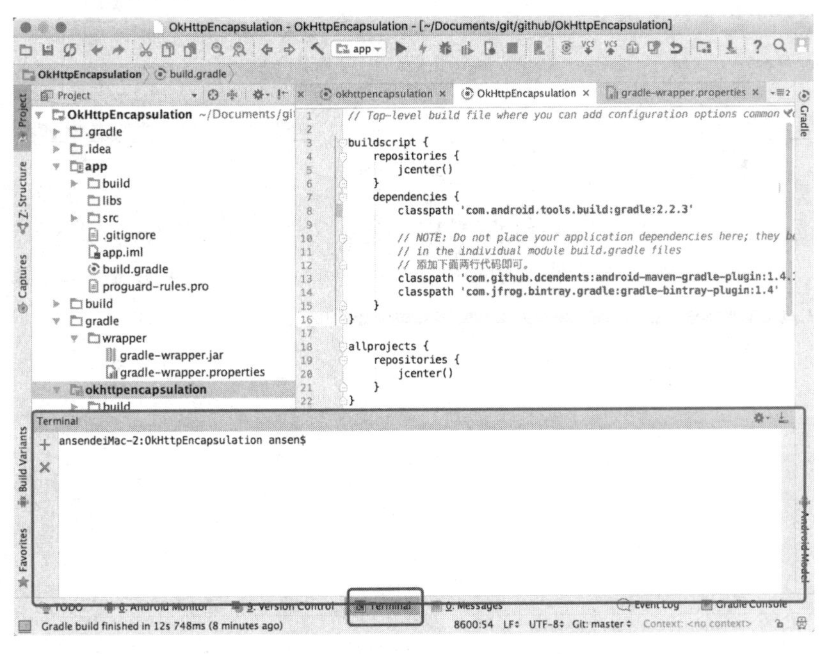

图 7-39　点击“Terminal”按钮

在 Terminal 页面中输入以下命令：

```
gradle install
```

出现 BUILD SUCCESSFUL 就表示成功了。

继续输入命令，提交项目到 bintray：

```
gradle clean build bintrayUpload
```

这个命令会提示上传进度，上传到 100%就表示成功了，最后也会出现 BUILD SUCCESSFUL。

（4）Add to JCenter

提交成功后，赶紧打开 JCenter 官网，查看是否成功。在首页中点击包名称“okhttpencapsulation”，就能够看到提交的项目，点击项目进入详细页，如图 7-40 所示。

图 7-40　项目上传成功

我们可以看到版本、gradle 在线引用的代码，还能切换到 Maven 和 lvy 方式。每个项目刚刚提交都必须审核，可以点击图 7-40 界面右边的“Add to JCenter”按钮进行审核。审核成功后仅用一行代码就能引用这个模块了。

第 8 章

Android 5.X、6.X、7.X、8.X 各版本特性

本章学习 Android 各个版本的新特性，让我们的应用程序能够兼容不同的 Android 版本。通过 5.0 版本的更新内容学习新的 Material Design 设计风格，6.0 版本的运行时添加权限可防止某些 App 在后台盗取用户隐私，7.0 版本的多窗口支持可以让用户同时处理多个任务。随着时代的发展，手机上装的 App 只会越来越多，8.0 版本新增的通知渠道功能可以更好地控制通知的显示。

作为一名软件开发人员，每时每刻学习是立足之本，不停掌握新技术才能避免被淘汰。本章内容看似比较少，但都是项目中常用的技术，希望大家好好掌握。

8.1 Android 5.X 版本新特性

Android 5.0 是谷歌公司 2014 年 10 月 15 日发布的全新 Android 系统，从 Android 5.0（Android Lollipop）开始，Android 迎来了扁平化时代，使用一种新的 Material Design 设计风格，设计了全新的通知中心，开始支持多种设备。在性能上，放弃了之前一直使用的 Dalvik 虚拟机，改用 ART 模式，程序加载时间大幅提升。增加了 Battery Saver 模式来进行省电处理，以及全新的“最近应用程序”。

提起 Android 5.0，就不得不说 Material Design，扁平化的设计理念，新的视觉语言，在基本元素的处理上，借鉴了传统的印刷设计，比如字体版式、网格系统、空间、比例、配色、图像使用等这些基础的平面设计规范。

之前需要自定义的一些效果，现在都提供了系统级的支持，用起来更加方便，而且 Android 提供的效果更加流畅。

8.1.1 悬挂式 Notification

Android 5.0 新增了悬挂式 Notification，不需要下拉通知栏就能直接在屏幕上方悬挂显示，并

且焦点仍然在用户操作的界面中。因此，不会打断用户的操作，过几秒就会自动消失。用户的 Activity 处于全屏模式中，并且使用下面的代码让 Notification 变为悬挂式 Notification：

```
builder.setFullScreenIntent(pendingIntent, true);
```

Android 5.0 中 Notification 分成了三个等级：

- VISIBILITY_PRIVATE：只有在没有锁屏时才会显示通知。
- VISIBILITY_PUBLIC：在任何情况下都会显示通知。
- VISIBILITY_SECRET：在 Pin、Password 等安全锁和没有锁的情况下显示通知。

使用时改变 setVisibility 的参数即可：

```
builder.setVisibility(Notification.VISIBILITY_PUBLIC);
```

另外，从 5.0 开始，对于通知栏图标的设计进行了修改。现在 Google 要求，所有应用程序的通知栏图标应该只使用 Alpha 图层来进行绘制，而不应该包括 RGB 图层。只使用 Alpha 图层来进行绘制，通俗点讲就是让通知栏图标不带颜色。

第 7 章我们已经学习了通知栏的使用，直接在之前的代码上进行修改，方便读者进行对比。修改后代码如下：

```
NotificationCompat.Builder mBuilder =
new NotificationCompat.Builder(this)
        .setSmallIcon(R.mipmap.ic_launcher)//小图标
        .setContentTitle("悬挂式通知-标题")
        .setContentText("悬挂式通知-内容");
Intent intent=new Intent(this,NotificationActivity.class);
PendingIntent ClickPending = PendingIntent.getActivity(this, 0, intent, 0);

mBuilder.setAutoCancel(true);//点击这条通知自动从通知栏中取消
mBuilder.setFullScreenIntent(ClickPending, true);//显示悬挂式通知
NotificationManager mNotificationManager =
(NotificationManager) getSystemService(Context.NOTIFICATION_SERVICE);
mNotificationManager.notify(id, mBuilder.build());
```

在之前的代码上只修改了一行代码，即将 setContentIntent 方法换成了 setFullScreenIntent。运行代码，效果如图 8-1 所示。

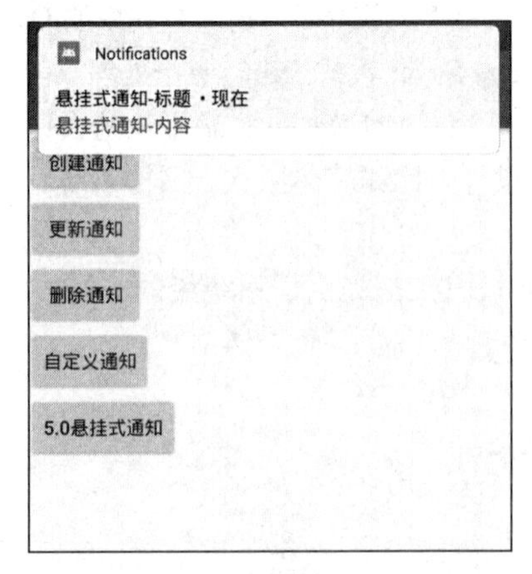

图 8-1　悬挂式通知显示

提 示

有些定制的 Android 系统可能会自动屏蔽应用的悬浮 Notification 功能，需要进行设置，在状态栏与通知中修改当前 App 允许显示通知即可。

8.1.2 利用 Drawerlayout 和 NavigationView 实现侧边栏

DrawerLayout 是 SupportLibrary 包中实现侧滑菜单效果的控件，可以说 DrawerLayout 是 Google 在第三方控件（如 MenuDrawer、slidingmenu 等）出现之后借鉴来的产物。

DrawerLayout 分为侧边菜单和主内容区两部分：侧边菜单可以根据手势展开与隐藏（DrawerLayout 自身特性），主内容区的内容可以随着菜单的点击而变化（这需要使用者自己实现）。Google 的很多 App 都使用这种侧滑菜单，例如 Google Play、Google Map、Contacts 等。因为需要用到 NavigationView，所以必须引入 design 包，这里我们使用在线依赖方式：

```
compile 'com.android.support:design:25.3.0'
```

DrawerLayout 是一个布局控件，与 LinearLayout 等控件一样，只是 DrawerLayout 带有滑动的功能。只有按照 DrawerLayout 规定的布局方式写布局，才能产生侧滑的效果。

新建项目，修改 activity_main.xml 文件：

```
<?xml version="1.0" encoding="utf-8"?>
<android.support.v4.widget.DrawerLayout
xmlns:android="http://schemas.android.com/apk/res/android"
    xmlns:app="http://schemas.android.com/apk/res-auto"
    android:id="@+id/drawer_layout"
    android:layout_width="match_parent"
    android:layout_height="match_parent">

    <!--内容布局-->
    <include
        layout="@layout/layout_sliding_menu_main"
        android:layout_width="match_parent"
        android:layout_height="match_parent" />

    <android.support.design.widget.NavigationView
        android:id="@+id/nav_view"
        android:layout_width="wrap_content"
        android:layout_height="match_parent"
        android:layout_gravity="start"
        android:fitsSystemWindows="true"
        app:headerLayout="@layout/layout_sliding_menu_header"
        app:menu="@menu/sliding_menu" />
</android.support.v4.widget.DrawerLayout>
```

从该布局中可以看出，最外层使用了 DrawerLayout，主内容区域通过 include 引入一个布局，NavigationView 为侧边抽屉栏。

注 意

主内容区的布局代码要放在侧滑菜单布局的前面，可以帮助 DrawerLayout 判断谁是侧滑菜单、谁是主内容区域。侧滑菜单部分的布局（这里是 NavigationView）可以设置 layout_gravity 属性，表示侧滑菜单位于左侧还是右侧。

NavigationView 有两个 App 属性，分别为 app:headerLayout 和 app:menu。其中，headerLayout 用于显示头部的布局（可选），menu 用于建立 MenuItem 选项的菜单。

NavigationView 控件的 app:headerLayout 属性引用了一个布局文件 layout_sliding_menu_header.xml，布局结构比较简单，最外层使用 `LinearLayout`，里面包裹一个 TextView，源码如下：

```
<?xml version="1.0" encoding="utf-8"?>
<LinearLayout xmlns:android="http://schemas.android.com/apk/res/android"
    android:layout_width="match_parent"
    android:layout_height="wrap_content"
    android:background="@color/colorAccent"
    android:orientation="horizontal"
    android:paddingBottom="80dp"
    android:paddingLeft="15dp"
    android:paddingRight="15dp"
    android:paddingTop="30dp">

    <TextView
        android:layout_width="wrap_content"
        android:layout_height="wrap_content"
        android:layout_gravity="center_vertical"
        android:layout_marginLeft="10dp"
        android:text="左侧菜单头部"
        android:textColor="#ffffffff"/>
</LinearLayout>
```

NavigationView 控件的 app:menu 属性引用了一个布局文件 sliding_menu.xml，内容如下：

```
<?xml version="1.0" encoding="utf-8"?>
<menu xmlns:android="http://schemas.android.com/apk/res/android">
    <group android:checkableBehavior="single">
        <item
            android:id="@+id/nav_camera"
            android:icon="@mipmap/ic_launcher"
            android:title="Import"/>
        <item
            android:id="@+id/nav_gallery"
            android:icon="@mipmap/ic_launcher"
            android:title="Gallery"/>
        <item
            android:id="@+id/nav_slideshow"
            android:icon="@mipmap/ic_launcher"
            android:title="Slideshow"/>
        <item
            android:id="@+id/nav_manage"
            android:icon="@mipmap/ic_launcher"
            android:title="Tools"/>
    </group>
```

```
    <item android:title="Communicate">
        <menu>
            <item
                android:id="@+id/nav_share"
                android:icon="@mipmap/ic_launcher"
                android:title="Share"/>
            <item
                android:id="@+id/nav_send"
                android:icon="@mipmap/ic_launcher"
                android:title="Send"/>
        </menu>
    </item>
</menu>
```

menu 可以用于分组，将 group 的 android:checkableBehavior 属性设置为 single 时，可以设置该组为单选。

另外，include 的首页布局文件 layout_sliding_menu_main.xml 里面也是包裹一个 TextView，内容如下：

```
<?xml version="1.0" encoding="utf-8"?>
<RelativeLayout
    xmlns:android="http://schemas.android.com/apk/res/android"
    android:layout_width="match_parent"
    android:layout_height="match_parent">
    <TextView
        android:layout_width="wrap_content"
        android:layout_height="wrap_content"
        android:layout_centerInParent="true"
        android:layout_gravity="center_vertical"
        android:layout_marginLeft="10dp"
        android:text="显示内容区域"
        android:textColor="@color/colorAccent"
        android:textSize="22sp" />
</RelativeLayout>
```

不需要修改任何 Java 代码，通过布局文件就能实现侧滑功能。Android 开发最大的好处就是 SDK 封装得很好，底层做了很多工作。运行代码，效果如图 8-2 所示。

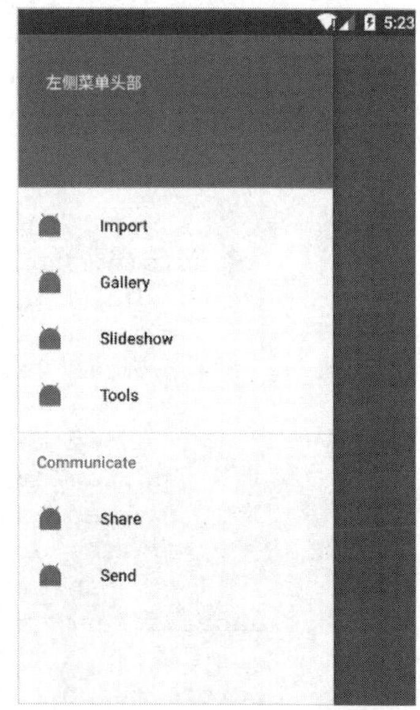

图 8-2 侧滑功能

当侧滑显示左侧菜单栏时，若用户点击菜单栏的 item，想关闭侧滑菜单栏应该怎么做？

修改 MainActivity.java 文件，在 onCreate 方法中查找 NavigationView 控件，并且设置 item 点击监听：

```
    NavigationView navigationView = (NavigationView)
findViewById(R.id.nav_view);
    navigationView.setNavigationItemSelectedListener
(this);
```

setNavigationItemSelectedListener 方法传入的是“this”，所以 MainActivity 需要实现 OnNavigationItemSelectedListener 接口，

重写 onNavigationItemSelected 方法：

```
@Override
public boolean onNavigationItemSelected(@NonNull MenuItem item) {
    DrawerLayout drawer = (DrawerLayout) findViewById(R.id.drawer_layout);
    //关闭左侧菜单栏
    drawer.closeDrawer(GravityCompat.START);
    return true;
}
```

这个方法还有一个返回参数，是一个布尔类型：值为 true 时，如果点击的是分组菜单，下次打开时就会有一个选中效果；值为 false 时，下次打开侧滑菜单没有选中效果。

8.1.3 TabLayout 和 ViewPager 结合使用

TabLayout 也是 design 包中的控件，用来替代第三方开源项目 TabPageIndicator。

新建项目，首先修改 activity_main.xml 布局文件：

```
<?xml version="1.0" encoding="utf-8"?>
<android.support.design.widget.CoordinatorLayout
xmlns:android="http://schemas.android.com/apk/res/android"
    xmlns:app="http://schemas.android.com/apk/res-auto"
    android:id="@+id/coordinatorLayout"
    android:layout_width="match_parent"
    android:layout_height="match_parent">

    <android.support.design.widget.AppBarLayout
        android:id="@+id/appBarLayout"
        android:layout_width="match_parent"
        android:layout_height="wrap_content">

        <android.support.v7.widget.Toolbar
            android:id="@+id/toolbar"
            android:layout_width="match_parent"
            android:layout_height="wrap_content"
            android:background="@color/colorAccent"
            app:layout_scrollFlags="scroll|enterAlways"/>

        <android.support.design.widget.TabLayout
            android:id="@+id/tabLayout"
            android:layout_width="match_parent"
            android:layout_height="wrap_content"
            android:layout_gravity="center_horizontal"
            android:background="#54B0E9"
            app:tabIndicatorColor="#ffffffff"
            app:tabIndicatorHeight="2dp"
            app:tabSelectedTextColor="#ffffffff"
            app:tabTextColor="#bebebe"/>
    </android.support.design.widget.AppBarLayout>

    <android.support.v4.view.ViewPager
        android:id="@+id/viewPager"
        android:layout_width="match_parent"
```

```
        android:layout_height="match_parent"
        app:layout_behavior="@string/appbar_scrolling_view_behavior"/>
</android.support.design.widget.CoordinatorLayout>
```

最外层是 CoordinatorLayout（用来调度协调子布局），里面主要分为两块：AppBarLayout 和 ViewPager。其中，AppBarLayout 中包含标题栏的 Toolbar+TabLayout，ViewPager 用来切换 Fragment 显示。

注意：为了使得 Toolbar 具有滑动效果，必须做到如下几点:

（1）CoordinatorLayout 作为布局的父布局容器。

（2）给需要滑动的组件设置 app:layout_scrollFlags= "scroll|enterAlways"属性。

（3）滑动的组件必须是 AppBarLayout 顶部组件。

（4）给滑动的组件设置 app:layout_behavior 属性。

（5）ViewPager 显示的 Fragment 中不能是 ListView，必须是 RecyclerView。

接下来查看 MainActivity 代码，和之前 ViewPager 的写法一样。之前没有学习过的就是调用 TabLayout 的 setupWithViewPager 方法将 TabLayout 和 ViewPager 关联起来。

```
public class MainActivity extends AppCompatActivity {
    @Override
    protected void onCreate(Bundle savedInstanceState) {
        super.onCreate(savedInstanceState);
        setContentView(R.layout.activity_main);

        //设置标题内容以及标题文字颜色
        Toolbar toolbar = (Toolbar) findViewById(R.id.toolbar);
        toolbar.setTitle("关注公众号[Android 开发者 666]");
        toolbar.setTitleTextColor(getResources().getColor(android.R.color.white));

        ViewPager vPager = (ViewPager) findViewById(R.id.viewPager);
        vPager.setOffscreenPageLimit(2);//设置缓存页数
        vPager.setCurrentItem(0);//设置当前显示的 item 0 表示显示第一个

        FragmentAdapter pagerAdapter = new
FragmentAdapter(getSupportFragmentManager());
        FragmentTest fragmentOne=new FragmentTest();
        FragmentTest fragmentTwo=new FragmentTest();
        FragmentTest fragmentThree=new FragmentTest();

        //把 Fragment 一个个地添加到适配器中，参数 2 是选项卡的标题
        pagerAdapter.addFragment(fragmentOne,"选项卡 1");//
        pagerAdapter.addFragment(fragmentTwo,"选项卡 2");
        pagerAdapter.addFragment(fragmentThree,"选项卡 3");

        //给 ViewPager 设置适配器
        vPager.setAdapter(pagerAdapter);

        TabLayout tabLayout = (TabLayout) findViewById(R.id.tabLayout);
        //通过 TabLayout 的 setupWithViewPager 将 ViewPager 设置进去
        tabLayout.setupWithViewPager(vPager);
    }
}
```

上面的代码用到了 FragmentAdapter 类，这个我们之前讲 ViewPager 时用到过，还有 FragmentTest 类，其中包含一个简单的 RecyclerView 列表。有不懂的读者，下载源码进行查看就明白了。

8.1.4 CoordinatorLayout、FloatingActionButton 和 Snackbar 的使用

上个知识点学习了 CoordinatorLayout（协调者布局），用于 AppBarLayout 和 ViewPager 的最外层布局，ViewPager 上下滚动时会让 AppBarLayout 隐藏。简单来说，CoordinatorLayout 是用来协调其子 View 并以触摸影响布局的形式产生动画效果的一个 Super Powered FrameLayout，其典型的子 View 包括 FloatingActionButton 和 Snackbar。注意：CoordinatorLayout 是一个顶级父 View。

1. FloatingActionButton 是什么

FloatingActionButton 实现一个默认颜色为主题中 colorAccent 的悬浮操作按钮，继承自 ImageButton，可以使用 android:src 属性设置圆形中的图标。

2. Snackbar 是什么

Snackbar 是一个操作提供轻量级、快速的反馈。Snackbar 可以在屏幕底部快速弹出消息，它和 Toast 非常相似，但是更加灵活一些。Snackbar 显示包含了文字和一个可选的操作按钮，也可以设置消失时间。

3. 实战演习

本案例是在上个案例的代码基础上进行添加。在 activity_main.xml 布局文件的 CoordinatorLayout 控件中增加如下控件：

```
<android.support.design.widget.FloatingActionButton
    android:id="@+id/fab"
    android:layout_width="wrap_content"
    android:layout_height="wrap_content"
    android:layout_gravity="end|bottom"
    android:layout_margin="15dp"
    android:src="@mipmap/ic_launcher"/>
```

在 MainActivity 中给 FloatingActionButton 设置点击事件，弹出 Snackbar，其写法和 Toast 的写法相似。

```
Snackbar.make(view, "这是内容", Snackbar.LENGTH_SHORT).setAction
        ("取消", new View.OnClickListener() {
            @Override
            public void onClick(View view) {
                Toast.makeText(MainActivity.this,
"cancel",Toast.LENGTH_SHORT).show();
            }
 }).show();
```

最终效果如图 8-3 所示。

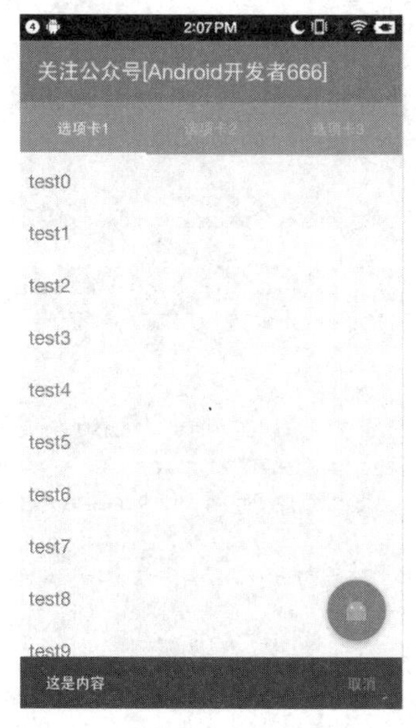

图 8-3　效果图

8.2　Android 6.X 版本新特性

Android 6.0（Android Marshmallow）最大的亮点是为用户提供了两套相互独立的解决方案，简单地说，就是为每位用户的每一个应用提供了两套数据存储方案。一套存储工作资料，另一套存储个人信息。另外，Android M 系统层面加入指纹识别，还加入了运行时权限，加入了 App Standby（应用待机）、Doze（瞌睡）Exemptions（豁免）等模式来加强电源管理。

8.2.1　6.0 运行时权限

从 Android 6.0（API 级别 23）开始，有些权限只有用户在使用软件时才向其授予权限，而不是在安装时授权。这种方法可以简化安装过程，用户在安装或者更新软件时不需要授予权限。它还能够让用户对应用的某些功能进行控制。例如，单机版的斗地主，请求访问任何权限，我都会拒绝。当然你也可以在设置界面对每个 App 的权限进行查看，以及对单个权限进行授权或者解除授权。

1. 系统权限分为两类：正常权限和危险权限

- 正常权限不会直接给用户隐私权带来风险，在 AndroidManifest.xml 中注册即可。
- 危险权限会授予应用访问用户机密数据的权限。在 AndroidManifest.xml 中注册了危险权限，同时需要用户授权你的软件才能使用这个权限。

哪些属于正常权限，哪些属于危险权限，大家可以去官网阅读。

```
https://developer.android.com/guide/topics/security/permissions.html#norma
l-dangerous
```

2. 简单的例子（6.0 手机获取危险权限）

如果 App 需要一个调用拨打电话的功能，这时就需要获取 CALL_PHONE 权限。我们来看看在 6.0 的手机上拨打电话代码需要怎么写：

```
    public class MainActivity extends AppCompatActivity implements
View.OnClickListener {
        private String[] perms = {Manifest.permission.CALL_PHONE};
        private final int PERMS_REQUEST_CODE = 200;

        @Override
        protected void onCreate(Bundle savedInstanceState) {
            super.onCreate(savedInstanceState);
            setContentView(R.layout.activity_main);

            findViewById(R.id.btn_call_phone).setOnClickListener(this);
        }

        @Override
        public void onClick(View v) {
            if (v.getId() == R.id.btn_call_phone) {//获取拨打电话的权限
                if (Build.VERSION.SDK_INT > Build.VERSION_CODES.LOLLIPOP_MR1) {
                  //Android 6.0 以上版本需要获取权限
                    requestPermissions(perms,PERMS_REQUEST_CODE);//请求权限
                } else {
                    callPhone();
                }
            }
        }

        /**
         * 获取权限回调方法
         * @param permsRequestCode
         * @param permissions
         * @param grantResults
         */
        @Override
        public void onRequestPermissionsResult(int permsRequestCode, String[]
permissions, int[] grantResults) {
            switch (permsRequestCode) {
                case PERMS_REQUEST_CODE:
                    boolean storageAccepted = grantResults[0] ==
PackageManager.PERMISSION_GRANTED;
                    if (storageAccepted) {
                        callPhone();
                    } else {
                        Log.i("MainActivity", "没有权限操作这个请求");
                    }
                    break;

            }
        }

        //拨打电话
        private void callPhone() {
```

```
            //检查拨打电话权限
            if (ActivityCompat.checkSelfPermission(this,
Manifest.permission.CALL_PHONE) == PackageManager.PERMISSION_GRANTED) {
                Intent intent = new Intent(Intent.ACTION_CALL);
                intent.setData(Uri.parse("tel:" + "10086"));
                startActivity(intent);
            }
        }
    }
```

获取权限之前先判断版本号，版本大于 6.0，就调用 requestPermissions 方法请求权限。这个方法有两个参数：参数 1 是要请求的权限数组，一次可以请求多个权限；参数 2 是一个 int 类型的请求标示。不管用户同意还是拒绝都会回调 onRequestPermissionsResult 方法。这个方法的 permsRequestCode 参数就是我们请求时的第二个参数，然后判断 permsRequestCode 数组的第一个参数。值等于 PackageManager.PERMISSION_GRANTED 时表示用户同意授权，否则用户拒绝授权这个权限。

3. 获取特殊权限

Android 特殊权限只有两种，即 SYSTEM_ALERT_WINDOW 和 WRITE_SETTINGS，与正常权限和危险权限不同。特殊权限比较敏感，因此大多数应用不应该使用它们。如果某个应用需要其中一种权限，必须在清单中声明该权限，并且发送请求用户授权的 Intent。系统将向用户显示详细管理屏幕，以响应该 Intent。

首先判断 Android 版本号是不是大于 6.0，通过 Settings.canDrawOverlays 判断有没有系统弹窗权限。如果没有权限，则发送一个 intent 请求授权。

```
    if (Build.VERSION.SDK_INT > Build.VERSION_CODES.LOLLIPOP_MR1){
        if (Settings.canDrawOverlays(this)) {
            systemAlert();
        } else {//没有系统弹窗权限,发送一个设置权限的 intent
            Intent intent = new Intent(Settings.ACTION_MANAGE_OVERLAY_PERMISSION);
            intent.setData(Uri.parse("package:" + getPackageName()));
            startActivityForResult(intent, PERMS_REQUEST_CODE);
        }
    }
```

重写 onActivityResult 方法，授权失败或者成功都会回调这个方法。在这里可以判断用户同意还是拒绝。

```
    @Override
    protected void onActivityResult(int requestCode, int resultCode, Intent data)
{
        if (requestCode == PERMS_REQUEST_CODE) {
            if (Build.VERSION.SDK_INT >= Build.VERSION_CODES.M) {
                                           //Android M 处理 Runtime Permission
                if (Settings.canDrawOverlays(this)){//有系统弹窗权限
                    Log.i("MainActivity","获取系统弹窗(特殊权限)成功");
                    systemAlert();
                } else {
                    Log.i("MainActivity","拒绝系统弹窗(特殊权限)");
                }
            }
```

```
        }
    }
```

4. 申请这么多权限岂不是很累

其实不需要每个权限都去申请，举一个例子，如果你的应用授权了读取联系人的权限，那么该应用也是被赋予了写入联系人的权限。因为读取联系人和写入联系人这两个权限都属于联系人权限分组，所以一旦组内某个权限被允许，该组的其他权限也是被允许的。

注　意
虽然在代码中请求了运行时权限，但是在 Manifest.xml 文件中也要注册权限。

```
<uses-permission android:name="android.permission.CALL_PHONE"/>
<uses-permission android:name="android.permission.SYSTEM_ALERT_WINDOW"/>
```

最后运行代码，请求打电话的权限授权弹窗如图 8-4 所示，请求特殊权限授权界面如图 8-5 所示。

图 8-4　请求打电话权限

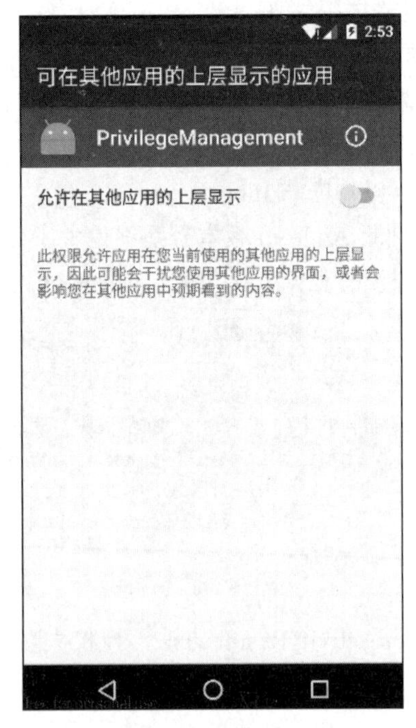

图 8-5 请求特殊权限

8.3　Android 7.X 版本新特性

Android 7.0（Android Nougat）在性能处理上有了巨大的提升，同时对文件数据加密，更加安全。添加了分屏多任务，重新设计了通知，改进 Doze 休眠机制等。总而言之，Android N 将更快、更好、更强。

8.3.1　多窗口支持

Android 7.X 可以同时显示多个应用窗口。在手机上，两个应用可以在“分屏”模式中左右并排或上下并排显示。例如，用户可以在上面窗口聊 QQ，下面窗口发送短信。

如图 8-6 所示，两个 App 在分屏模式中上下并排显示。

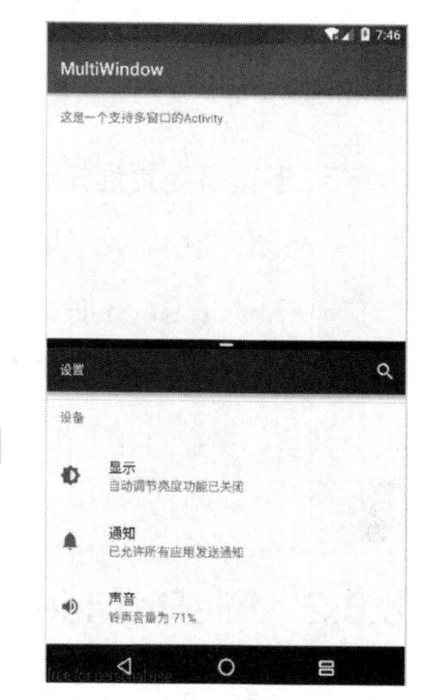

图 8-6　分屏显示

1. 让 App 支持多窗口

如果您的 App 支持 Android N，只要在 AndroidManifest.xml 文件中对<activity> 或 <application>标签设置 android:resizeableActivity 属性就能启用或者禁用多窗口显示：

```
android:resizeableActivity=["true" | "false"]
```

如果这个属性的值为 true，则 Activity 能够分屏和自由模式启动；如果这个属性的值为 false，则 Activity 不支持多窗口模式。

如果 App 支持 Android N，但是没有对该属性设置值，则该属性的默认值为 true，也就是默认支持多窗口模式。

2. 切换到多窗口模式

- 若用户打开 OverView 屏幕并长按 Activity 标题，则可以拖动该 Activity 到屏幕突出显示的区域，使 Activity 进入多窗口模式。
- 若用户长按 OverView 按钮，设备上的当前 Activity 将进入多窗口模式，同时将打开 OverView 屏幕，用户可以在该屏幕中选择要共享屏幕的另一个 Activity。

3. 多窗口的生命周期

多窗口模式不会更改 Activity 的生命周期。

在多窗口模式中，指定时间内只有最近与用户交互过的 Activity 为活动状态。该 Activity 将被视为顶级 Activity。所有其他 Activity 虽然可见，但是均处于暂停状态。不过，这些已暂停但可见的 Activity 在系统中享有比不可见 Activity 更高的优先级。如果用户与其中一个暂停的 Activity 交互，该 Activity 将恢复，而之前的顶级 Activity 将暂停。

4. 多窗口模式下的布局属性

对于 Android N，可以在 activity 标签中设置标签。标签支持以下几种属性，这些属性将影响 Activity 在多窗口模式中的效果：

- android:defaultWidth：多窗口模式下的默认宽度。
- android:defaultHeight：多窗口模式下的默认高度。
- android:gravity：多窗口模式下的初始位置。
- android:minimalHeight、android:minimalWidth：多窗口模式下的最小高度和最小宽度。如果用户在分屏模式中移动分界线，使 Activity 尺寸低于指定的最小值，系统会将 Activity 裁剪为用户请求的尺寸。

例如，以下配置如何指定在多窗口模式中显示时 Activity 的默认大小、位置和最小尺寸：

```
<activity android:name=".MyActivity">
    <layout android:defaultHeight="500dp"
          android:defaultWidth="600dp"
          android:gravity="top|end"
          android:minimalHeight="450dp"
          android:minimalWidth="300dp" />
</activity>
```

5. 多窗口变更通知和查询

Activity 类中添加了以下新方法，以支持多窗口显示。

- Activity.isInMultiWindowMode()：判断是否处于多窗口模式。
- Activity.onMultiWindowModeChanged()：进入或退出多窗口模式时会回调这个方法。

其实，多窗口在工作中应该使用的机会不大，毕竟手机屏幕比较小，分屏都看不到什么界面了。不过，作为开发者有必要去了解一下 Android 7.0 更新了哪些功能。

8.3.2 FileProvider 解决 FileUriExposedException

1. 了解 FileUriExposedException

在给 App 进行版本升级时，先从服务器下载新版本的 APK 文件到 SD 卡的路径中，然后调用安装 APK 的代码，一般写法如下：

```
private void openAPK(String fileSavePath){
    File file=new File(fileSavePath);
    Intent intent = new Intent(Intent.ACTION_VIEW);
    Uri data = Uri.fromFile(file);
    intent.setDataAndType(data, "application/vnd.android.package-archive");
    startActivity(intent);
}
```

这样的写法在 Android 7.0 之前的版本没有任何问题，只要给定一个 APK 文件路径就能打开安装，但是在 Android 7.0 版本上会报错：

```
android.os.FileUriExposedException:
file:///storage/emulated/0/Download/FileProvider.apk
exposed beyond app through Intent.getData()
```

从 Android 7.0 开始，一个应用提供自身文件给其他应用使用时，如果给出一个 file://格式的 URI 的话，应用会抛出 FileUriExposedException。这是由于谷歌认为目标 App 可能不具备文件权限，会造成潜在的问题，因此让这个行为快速失败。

2. FileProvider 方式解决

这是谷歌官方推荐的解决方案，即使用 FileProvider 来生成一个 content://格式的 URI。

（1）在 Manifest.xml 中声明一个 provider

```
<application ...>
```

```
    ...
    <provider
        android:name="android.support.v4.content.FileProvider"
        android:authorities="com.ansen.fileprovider.fileprovider"
        android:grantUriPermissions="true"
        android:exported="false">
        <meta-data
            android:name="android.support.FILE_PROVIDER_PATHS"
            android:resource="@xml/file_paths" />
    </provider>
</application>
```

android:name 值是固定的，android:authorities 随便写但是必须保证唯一性，这里用的是包名+"fileprovider"，android:grantUriPermissions 和 android:exported 是固定值。

其中包含一个 meta-data 标签，这个标签的 name 属性固定写法，android:resource 对应的是一个 xml 文件。在 res 文件夹下新建一个 xml 文件夹，再在 xml 文件夹下新建 file_paths.xml 文件。内容如下：

```
<?xml version="1.0" encoding="utf-8"?>
<paths>
    <external-path name="name" path="Download"/>
</paths>
```

name 表示生成 Uri 时的别名，path 是指相对路径。

paths 标签下的子元素一共有以下几种：

```
files-path 对应  Context.getFilesDir()
cache-path 对应  Context.getCacheDir()
external-path 对应 Environment.getExternalStorageDirectory()
external-files-path 对应  Context.getExternalFilesDir()
external-cache-path 对应  Context.getExternalCacheDir()
```

（2）修改后打开 Apk 文件的代码

首先判断一下版本号，如果手机操作系统版本号大于等于 7.0，就通过 FileProvider.getUriForFile 方法生成一个 Uri 对象。

```
private void openAPK(String fileSavePath){
    File file=new File(fileSavePath);
    Intent intent = new Intent(Intent.ACTION_VIEW);
    Uri data;
    if (Build.VERSION.SDK_INT >= Build.VERSION_CODES.N) {//判断版本大于等于 7.0
        //"com.ansen.fileprovider.fileprovider"即是在清单文件中配置的 authorities
        // 通过 FileProvider 创建一个 content 类型的 Uri
        data = FileProvider.getUriForFile(this,
"com.ansen.fileprovider.fileprovider", file);
        intent.addFlags(Intent.FLAG_GRANT_READ_URI_PERMISSION);
                // 给目标应用一个临时授权
    } else {
        data = Uri.fromFile(file);
    }
    intent.setDataAndType(data, "application/vnd.android.package-archive");
    startActivity(intent);
}
```

8.4 Android 8.X 版本新特性

Android 8.0（Android Oreo）是 Google 发布的第 14 个新系统版本，在功能、流畅、安全上都不落下风，对通知又进行了改变，增加了通知渠道。同时对静态广播的使用也做了一些限制。

8.4.1 通知栏新增通知渠道

从 Android 8.0（API 级别 26）开始，所有通知必须要分配一个渠道，对于每个渠道，可以单独设置视觉和听觉行为，然后用户可以在设置中进行修改，根据应用程序来决定哪些通知可以显示或者隐藏。

创建通知渠道之后，程序无法修改通知行为。只有用户可以修改，程序只能修改渠道名称和渠道描述。

我们可以为一个应用程序创建多个通知渠道，不同的通知类型使用不同的渠道。例如，重要通知用一个渠道，可以将这个渠道重要性设置成最高；不怎么重要的通知使用一个渠道，这个渠道重要性设置成最低。

1. 创建一个通知

要创建一个通知，有以下几个步骤：

步骤01 构造 NotificationChannel 对象，构造方法有三个参数：渠道 id、渠道名称、渠道重要性级别。

步骤02 调用 NotificationChannel.setDescription 方法可以设置渠道描述。这个描述在系统设置中可以看到。

步骤03 调用 NotificationManager.createNotificationChannel 方法创建通知渠道。

以下是注册通知渠道的代码，记得先判断版本号是不是大于或者等于 8.0，因为只有 8.0 以上才有通知渠道 API。

```
    if (Build.VERSION.SDK_INT >= Build.VERSION_CODES.O) {
        //创建通知渠道
        CharSequence name = "渠道名称 1";
        String description = "渠道描述 1";
        String channelId="channelId1";//渠道 id
        int importance = NotificationManager.IMPORTANCE_DEFAULT;//重要性级别
        NotificationChannel mChannel = new NotificationChannel(channelId, name, 
importance);
        mChannel.setDescription(description);//渠道描述
        mChannel.enableLights(true);//是否显示通知指示灯
        mChannel.enableVibration(true);//是否振动

        NotificationManager notificationManager = (NotificationManager) 
getSystemService(NOTIFICATION_SERVICE);
```

```
    notificationManager.createNotificationChannel(mChannel);//创建通知渠道
}
```

创建通知渠道不会执行任何操作，所以在启动应用程序时最好调用以上代码。

执行以上代码，就能够在“应用通知”（“设置”→“应用和通知”→“选择当前 App”→“应用通知”）界面中看到创建的通知渠道，如图 8-7 所示。可以单独开关当前的通知类别，还可以点击当前渠道，进入渠道详细界面，如图 8-8 所示。在渠道详细界面中可以设置渠道的提示音、振动、屏幕锁定时显示方式等。

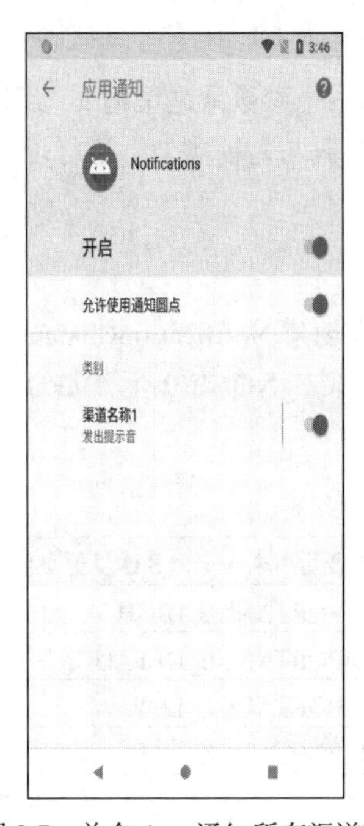

图 8-7　单个 App 通知所有渠道列表

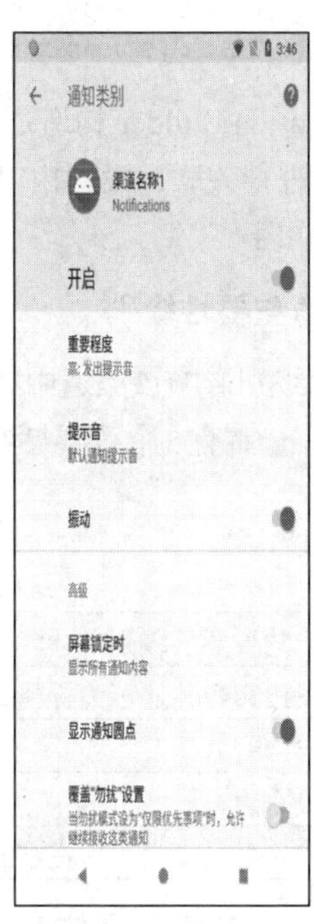

图 8-8　渠道详细界面

默认情况下，通知的振动和声音提示都是由 NotificationManagerCompat 类中重要性级别决定的，例如常量 IMPORTANCE_DEFAULT 和 IMPORTANCE_HIGH，后面会讲到这些常量的具体作用。

如果想改变该渠道的默认通知，可以调用 NotificationChannel 对象的方法进行修改。下面介绍几个常用的方法。

- enableLights(): 是否显示通知指示灯。
- setLightColor(): 设置通知灯颜色。
- setVibrationPattern(): 设置振动模式。

注　意

一旦创建了渠道，调用以上方法就会无效，只有用户在设置中才能修改。

上面那段代码只告诉了创建渠道，下面这段代码让我们在之前创建的渠道上发送一个通知。

```
//第二个参数与 channelId 对应
Notification.Builder builder = new Notification.Builder(this,channelId);
```

```
    //icon title text 必须包含，不然影响桌面图标小红点的展示
    builder.setSmallIcon(android.R.drawable.stat_notify_chat)
            .setContentTitle("通知渠道 1->标题")
            .setContentText("通知渠道 1->内容")
            .setNumber(3); //久按桌面图标时允许的此条通知的数量

    Intent intent=new Intent(this,NotificationActivity.class);
    PendingIntent ClickPending = PendingIntent.getActivity(this, 0, intent, 0);
    builder.setContentIntent(ClickPending);

    notificationManager.notify(id,builder.build());
```

使用 Notification.Builder 构造通知对象时，传入第二个参数就是渠道 id，用这个渠道 id 就能与之前创建的渠道进行关联。设置内容和发送系统通知代码与之前类似。运行以上代码，就能在 8.0 的手机上显示通知了。

2. 设置渠道重要性级别

渠道重要性级别影响该渠道中所有通知的显示，在创建 NotificationChannel 对象的构造方法中必须要指定级别，一共有 5 个重要性级别，即 IMPORTANCE_NONE(0)至 IMPORTANCE_HIGH(4)，如表 8-1 所示。

表 8-1 渠道重要性级别

用户可见的重要性级别	重要性（Android 8.0 及更高版本）
紧急（发出声音并显示为提醒通知）	IMPORTANCE_HIGH
高（发出声音）	IMPORTANCE_DEFAULT
中等（没有声音）	IMPORTANCE_LOW
低（无声音并且不会出现在状态栏中）	IMPORTANCE_MIN

3. 根据渠道 id 查看渠道信息

我们都知道渠道创建之后不能通过代码修改，但是可以先查看该渠道的振动和声音等行为，如果需要修改，可以根据返回的内容再提示用户手动去打开。

（1）调用 NotificationManager.getNotificationChannel(String channelId) 方法获取 NotificationChannel 对象，这个方法有个参数渠道 id。

（2）有了 NotificationChannel 对象就能调用 getVibrationPattern、getSound、getImportance 等方法。

示例代码如下所示：

```
    NotificationChannel
notificationChannel=notificationManager.getNotificationChannel (channelId);
    long[] vibrationPattern=notificationChannel.getVibrationPattern();//振动模式
    if(vibrationPattern!=null){
        Log.i("ansen","振动模式:"+vibrationPattern.length);
    }

    Uri uri=notificationChannel.getSound();//通知声音
    Log.i("ansen","通知声音:"+uri.toString());
```

```
    int importance=notificationChannel.getImportance();//通知等级
    Log.i("ansen","通知等级:"+importance);
```

4. 打开通知渠道设置界面

创建通知渠道之后，代码无法修改这个渠道的行为，只有用户在设置中才能进行修改（设置>应用和通知>要修改的 App）。为了让用户更快地找到当前 App 的通知渠道，可以通过代码发送 Intent 打开通知渠道的设置界面。

以下代码是打开通知渠道的设置界面：

```
    Intent channelIntent = new
Intent(Settings.ACTION_CHANNEL_NOTIFICATION_SETTINGS);
    channelIntent.putExtra(Settings.EXTRA_APP_PACKAGE, getPackageName());
    channelIntent.putExtra(Settings.EXTRA_CHANNEL_ID,channelId);
            //渠道 id 必须是我们之前注册的
    startActivity(channelIntent);
```

需要传入两个参数，应用程序的包名和渠道 id。

5. 删除通知渠道

调用 deleteNotificationChannel(String channelId)方法删除渠道，代码如下：

```
    NotificationManager mNotificationManager =
    (NotificationManager) getSystemService(Context.NOTIFICATION_SERVICE);
    mNotificationManager.deleteNotificationChannel(channelId);
```

6. 8.0 通知栏点击通知发送广播

在前面的案例中，我们点击通知栏都是跳转到指定 Activity，但是 PendingIntent 除了 getActivity 方法之外，还有 getBroadcast 跟 getService。可以点击通知发送广播，开启 Service。

于是我们修改显示通知的代码：

```
    String BROADCAST_ACTION="android.intent.action.BROADCAST_ACTION";
    if (Build.VERSION.SDK_INT >= Build.VERSION_CODES.O) {
        //第二个参数与 channelId 对应
        Notification.Builder builder = new Notification.Builder(this,channelId);
        //icon title text 必须包含，不然影响桌面图标小红点的展示
        builder.setSmallIcon(android.R.drawable.stat_notify_chat)
            .setContentTitle("通知渠道 1->标题")
            .setContentText("通知渠道 1->内容")
            .setNumber(3); //长按桌面图标时允许的此条通知的数量

        Intent intent = new Intent(BROADCAST_ACTION);
        intent.putExtra("data","12345");//带上参数
        PendingIntent pendingIntent=PendingIntent.getBroadcast(this,id,intent,
PendingIntent.FLAG_UPDATE_CURRENT);
        builder.setContentIntent(pendingIntent);
        builder.setAutoCancel(true);

        mNotificationManager.notify(id,builder.build());
    }
```

新建 Intent 对象的时候就传入了一个 action，同时给 Intent 传入一个字符串参数，然后调用 PendingIntent.getBroadcast 构造 PendingIntent 对象，最后调用 notify 方法显示通知。

新建一个类 DynamicBroadcast，继承 BroadcastReceiver，用来接收通知栏点击时发送的广播。在 onReceive 方法中打印一下我们传入的参数。

```
public class DynamicBroadcast extends BroadcastReceiver {
    @Override
    public void onReceive(Context context, Intent intent){
        String data = intent.getStringExtra("data");
        Log.i("data",data);
    }
}
```

前面的内容我们讲了广播的两种注册方式，静态注册跟动态注册，这里我们使用静态注册。在 AndroidManifest.xml 文件 application 标签中增加如下代码：

```
<receiver android:name=".DynamicBroadcast"
    android:exported="true">
    <intent-filter>
        <action android:name="android.intent.action.BROADCAST_ACTION" />
    </intent-filter>
</receiver>
```

运行之后发现在 Android 8.0 版本以下的手机正常，点击通知栏会打印广播的日志，但是在 Android 8.0 版本或以上版本点击通知栏发现没有收到广播，我以为是 8.0 增加了通知渠道的问题，后面查看官网 API，在 Broadcasts 类下面看到了这句话：

```
    Beginning with Android 8.0 (API level 26), the system imposes additional
restrictions on manifest-declared receivers. If your app targets API level 26 or
higher, you cannot use the manifest to declare a receiver for most implicit broadcasts
(broadcasts that do not target your app specifically).
```

大概意思是从 Android 8.0(API 级别 26) 开始，无法在 AndroidManifest.xml 文件中声明隐式广播，系统的隐式广播正常使用。所以，如果你的手机是 Android 8.0 以上版本的，就需要动态注册广播，这样才能接收到通知栏点击时发送的广播。

注释掉 AndroidManifest.xml 文件中声明的广播，然后在 MainActivity 的 onCreate 方法中注册广播，代码如下：

```
//动态注册广播
dynamicBroadcast=new DynamicBroadcast();
IntentFilter intentFilter=new IntentFilter(BROADCAST_ACTION);
registerReceiver(dynamicBroadcast,intentFilter);
```

静态注册的广播一定要在 onDestroy 方法中取消注册：

```
@Override
protected void onDestroy() {
    super.onDestroy();

    if(dynamicBroadcast!=null){
        unregisterReceiver(dynamicBroadcast);//取消注册广播
    }
}
```

第 9 章

常用功能模板

本章主要学习几个常用的功能模块，通过这些模块串联之前所学的知识点，让读者知道一些知识点用在哪种需求或者界面中比较合适，对之前所学的知识点进行巩固和复习。

例如，检查更新下载安装案例用到了之前学到的网络请求、文件保存 SD 卡、ProgressDialog（对话框进度条）、OkHttp 网络框架等技术。

另外，有关微信分享和百度地图的使用，让读者学习如何接入 SDK 去封装别人的 SDK。通过引导页和 Banner 广告轮播图案例，让读者更加熟练地使用 ViewPager。

9.1 启动页与首次启动的引导页

启动页与引导页是每个 App 都有的功能，一个 App 正常的启动流程如图 9-1 所示。

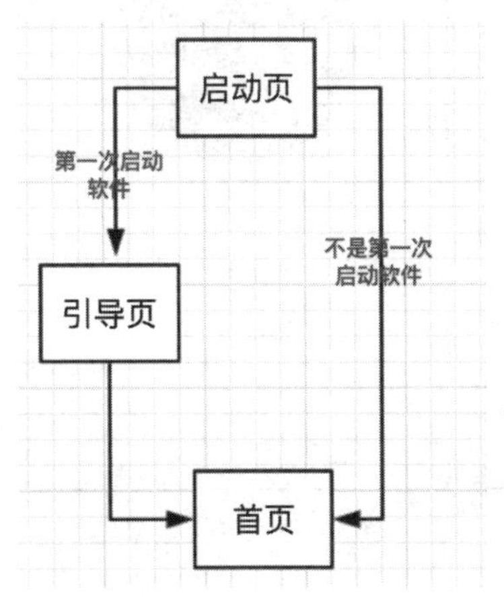

图 9-1　App 启动流程图

打开软件后，首先展示的是启动页（启动页是每次打开都会展示，一般停留 1 秒到 2 秒左右），然后判断用户是否第一次启动，如果是第一次启动就进入引导页，然后进入首页；如果不是第一次启动，就直接进入首页。

启动页比较简单，仅展示一张图片，如图 9-2 所示。

图 9-2　启动页效果（图片来源于网络）

引导页有三页，展示了三张图片，可以左右滑动进行切换，并且底部有圆点表示一共有几页，有颜色的圆点表示当前页选中，灰色圆点表示当前页未选中。效果图如图 9-3 所示。

图 9-3　引导页效果

9.1.1 需求分析

在实际开发中，当我们拿到产品需求文档跟产品模型时，第一件事情要做的就是对功能点进行分析，对模块进行拆分，这样开发更有效率。

- 启动页会停留1~2秒，需要使用Handler延迟几秒后执行。
- 启动页需要判断是否第一次启动软件，如果是第一次就进入引导页，如果不是第一次就进入首页。判断的值可以使用SharedPreferences类进行保存。
- 引导页一共有三页，可以左右滑动，也可以使用ViewPager显示。
- 引导页底部有显示圆点，可以使用LinearLayout动态添加，设置水平布局。
- 监听ViewPager的滑动来决定圆点的显示，并且滑动到最后一页就显示“立即启动”按钮；如果已处于最后一页，还向右边滑动，就直接进入首页。
- 从引导页进入首页之前，将SharedPreferences保存的boolean类型的值设置为true。

9.1.2 代码实现

新建项目，新建LauncherActivity（启动页）和FirstLauncherActivity（引导页）两个Activity类，AndroidManifest.xml中程序的入口改为LauncherActivity。修改后如下所示：

```
<?xml version="1.0" encoding="utf-8"?>
<manifest xmlns:android="http://schemas.android.com/apk/res/android"
    package="com.ansen.firstlauncher">

    <application  ......>
        <activity android:name=".LauncherActivity">
            <intent-filter>
                <action android:name="android.intent.action.MAIN" />
                <category android:name="android.intent.category.LAUNCHER" />
            </intent-filter>
        </activity>

        <activity android:name=".FirstLauncherActivity"/>
        <activity android:name=".MainActivity"/>
    </application>
</manifest>
```

9.1.3 启动页

启动页加载的布局文件为activity_launcher.xml，仅包含一个ImageView。

```
<?xml version="1.0" encoding="utf-8"?>
<LinearLayout xmlns:android="http://schemas.android.com/apk/res/android"
    android:layout_width="match_parent"
    android:layout_height="match_parent"
    android:orientation="vertical">
```

```
    <ImageView
        android:layout_width="match_parent"
        android:layout_height="match_parent"
        android:scaleType="centerCrop"
        android:src="@mipmap/bg_launcher"/>
</LinearLayout>
```

启动页 LauncherActivity.java 在 onCreate 方法中调用了 start 方法，在 start 方法中使用 Handler 延迟一秒执行 Runnable，在 Runnable 的 run 方法中调用 isFirstLauncher()来判断是否第一次启动，根据返回的 boolean 类型跳转到不同的 Activity。isFirstLauncher 方法就两行代码，首先获取 context 的 SharedPreferences 对象，调用它的 getBoolean 方法获取是否是第一次启动的 boolean 值。

```
public class LauncherActivity extends AppCompatActivity {
    public static final String  FIRST_LAUNCHER="first_launcher";//是否第一次启动
    private final long waitTime = 1000;

    @Override
    protected void onCreate(Bundle savedInstanceState) {
        super.onCreate(savedInstanceState);
        setContentView(R.layout.activity_launcher);

        start();
    }

    public void start(){
        new Handler().postDelayed(new Runnable(){
            public void run() {
                Intent intent;
                if(isFirstLauncher()){
                //为 true 时第二次启动,因为第一次启动时会把 FIRST_LAUNCHER 的值变为 true
                    intent=new Intent(LauncherActivity.this,MainActivity.class);
                }else{//第一次启动
                    intent=new Intent(LauncherActivity.this,
                            FirstLauncherActivity.class);
                }
                startActivity(intent);

                finish();
            }
        },waitTime);
    }


    /**
     * false:第一次启动，true:第二次启动，这里不能用 Activity 的 getPreferences 方法，
因为需要多个 Activity 使用一个 SharedPreference 对象，所以调用 getSharedPreferences 方法。
     * @return
     */
    public boolean isFirstLauncher(){
        SharedPreferences
sp=getSharedPreferences("ansen",Context.MODE_PRIVATE);
        return sp.getBoolean(FIRST_LAUNCHER,false);
    }
}
```

9.1.4　引导页

引导页的布局文件为 activity_first_launcher.xml，最外层是 RelativeLayout 布局，ViewPager 用于显示图片，TextView 默认是隐藏的，当滑动到最后一页时将显示“立即体验”按钮。LinearLayout 用来显示下方的圆点，告诉用户滑动到了第几页。

```
<?xml version="1.0" encoding="utf-8"?>
<RelativeLayout xmlns:android="http://schemas.android.com/apk/res/android"
    android:layout_width="match_parent"
    android:layout_height="match_parent"
    android:orientation="vertical">

    <android.support.v4.view.ViewPager
        android:id="@+id/viewPager"
        android:layout_width="match_parent"
        android:layout_height="match_parent"/>

    <TextView
        android:id="@+id/tv_goto_main"
        android:layout_above="@+id/viewGroup"
        android:layout_width="match_parent"
        android:layout_height="wrap_content"
        android:layout_marginBottom="30dp"
        android:layout_marginLeft="100dp"
        android:layout_marginRight="100dp"
        android:layout_marginTop="40dp"
        android:background="#fc0048"
        android:textColor="#ffffffff"
        android:paddingTop="10dp"
        android:paddingBottom="10dp"
        android:gravity="center_horizontal"
        android:textSize="18sp"
        android:visibility="gone"
        android:text="立即体验"/>

    <LinearLayout
        android:id="@+id/viewGroup"
        android:layout_width="match_parent"
        android:layout_height="wrap_content"
        android:layout_alignParentBottom="true"
        android:layout_marginBottom="20dp"
        android:gravity="center_horizontal"
        android:orientation="horizontal" />
</RelativeLayout>
```

新建 FirstLauncherActivity.java 类，用来展示引导页。

```
public class FirstLauncherActivity extends AppCompatActivity{
    private ViewPager viewPager;
    //图片数组
    private int[] images=new int[]{R.mipmap.bg_launcher_one,R.mipmap.bg_
                         launcher_ two,R.mipmap.bg_launcher_three};
```

```
    private List<View> mImageViews = new ArrayList<>();//显示图片View
    private List<ImageView> tips=new ArrayList<>();// 显示圆点View
    private ViewGroup group;

    private EdgeEffectCompat rightEdge;

    private TextView tvGotoMain;

    @Override
    protected void onCreate(Bundle savedInstanceState) {
        super.onCreate(savedInstanceState);
        setContentView(R.layout.activity_first_launcher);

        group = (ViewGroup) findViewById(R.id.viewGroup);
        viewPager = (ViewPager) findViewById(R.id.viewPager);
        tvGotoMain= (TextView) findViewById(R.id.tv_goto_main);

        try {
            Field rightEdgeField =
                 viewPager.getClass().getDeclaredField("mRightEdge");
            if (rightEdgeField != null){
                rightEdgeField.setAccessible(true);
                rightEdge = (EdgeEffectCompat) rightEdgeField.get(viewPager);
            }
        } catch (Exception e) {
            e.printStackTrace();
        }

        // 将图片装载到数组中
        for (int i = 0; i < images.length; i++) {
            ImageView imageView = new ImageView(this);
            imageView.setLayoutParams(new ViewGroup.LayoutParams(ViewGroup.
           LayoutParams.MATCH_PARENT, ViewGroup.LayoutParams.MATCH_PARENT));
            imageView.setScaleType(ImageView.ScaleType.CENTER_CROP);
            imageView.setImageResource(images[i]);
            mImageViews.add(imageView);
        }

        // 将圆点加入ViewGroup中
        for (int i = 0; i < images.length; i++) {
            ImageView imageView = new ImageView(this);
            imageView.setLayoutParams(new ViewGroup.LayoutParams(10, 10));
            if (i == 0) {
                imageView.setBackgroundResource(R.mipmap.icon_first_launcher_
page_select_one);
            } else {
                imageView.setBackgroundResource(R.mipmap.icon_first_launcher_
page_normal);
            }
            tips.add(imageView);
            LinearLayout.LayoutParams layoutParams = new
LinearLayout.LayoutParams(new
ViewGroup.LayoutParams(ViewGroup.LayoutParams.WRAP_CONTENT,
ViewGroup.LayoutParams. WRAP_CONTENT));
            layoutParams.leftMargin = 10;//左边距
```

```
            layoutParams.rightMargin = 10;//右边距
            group.addView(imageView,layoutParams);
        }

        viewPager.setAdapter(new PreviewImageAdapter());// 设置 Adapter
        viewPager.setOffscreenPageLimit(2);
        viewPager.addOnPageChangeListener(onPageChangeListener);
                // 设置监听，主要是设置圆点的背景

        tvGotoMain.setOnClickListener(onClickListener);
    }

    private View.OnClickListener onClickListener=new View.OnClickListener() {
        @Override
        public void onClick(View v) {
            switch (v.getId()){
                case R.id.tv_goto_main:
                    gotoMain();
                    break;
            }
        }
    };

    private ViewPager.OnPageChangeListener onPageChangeListener=new
ViewPager.OnPageChangeListener() {
        @Override
        public void onPageScrolled(int position, float positionOffset, int
positionOffsetPixels) {}

        @Override
        public void onPageSelected(int position){
            if(position==images.length-1){//最后一页
                tvGotoMain.setVisibility(View.VISIBLE);
            }else{
                tvGotoMain.setVisibility(View.INVISIBLE);
            }
            setImageBackground(position);
        }

        @Override
        public void onPageScrollStateChanged(int state) {
            if(rightEdge!=null&&!rightEdge.isFinished()){
                //到了最后一张并且继续拖动，出现蓝色限制边条
                gotoMain();
            }
        }
    };

    /**
     * 设置选中的 tip 的背景
     * @param selectItems
     */
    private void setImageBackground(int selectItems) {
        for (int i = 0; i < tips.size(); i++) {
            if (i == selectItems) {
```

```
            if(i==0){
                tips.get(i).setBackgroundResource
                (R.mipmap.icon_first_launcher_ page_select_one);
            }else if(i==1){
                tips.get(i).setBackgroundResource
                (R.mipmap.icon_first_launcher_ page_select_two);
            }else if(i==2){
                tips.get(i).setBackgroundResource
                (R.mipmap.icon_first_launcher_ page_select_three);
            }
        } else {
            tips.get(i).setBackgroundResource
                (R.mipmap.icon_first_launcher_ page_normal);
        }
    }
}

public class PreviewImageAdapter extends PagerAdapter {
    @Override
    public int getCount() {
        return mImageViews.size();
    }

    @Override
    public boolean isViewFromObject(View arg0, Object arg1) {
        return arg0 == arg1;
    }

    @Override
    public void destroyItem(View container, int position, Object object) {
        if (position < mImageViews.size()) {
            ((ViewPager) container).removeView(mImageViews.get(position));
        }
    }

    @Override
    public Object instantiateItem(View container, int position) {
        ((ViewPager) container).addView(mImageViews.get(position));
        return mImageViews.get(position);
    }
}

/**
 * 跳转到首页
 */
private void gotoMain(){
    setFirstLauncherBoolean();
    Intent intent=new Intent(FirstLauncherActivity.this,MainActivity.class);
    startActivity(intent);
    finish();
}

/**
 * 第一次启动执行完成，值设置为 true
 */
```

```
    public void setFirstLauncherBoolean(){
        SharedPreferences sp=getSharedPreferences("ansen",Context.MODE_PRIVATE);
        SharedPreferences.Editor edit=sp.edit();
        edit.putBoolean(LauncherActivity.FIRST_LAUNCHER,true);
        edit.commit();
    }
}
```

首先查看 oncreate 方法来查找三个控件，通过反射获取 ViewPager 的私有属性 mRightEdge，初始化图片数组 View，初始化圆点的 View，并且逐个加入到 LinearLayout 中。为 ViewPager 设置适配器，添加页面改变监听事件，给立即体验的 TextView 也设置点击事件。

- onClickListener 处理了立即体验的点击事件，直接调用 gotoMain 方法跳转到首页。
- onPageChangeListener 必须要重写三个方法（onPageScrolled、onPageSelected、onPageScrollStateChanged）。onPageSelected 方法是在页面选中时调用，首先判断是否处于最后一页，如果是最后一页就显示“立即体验”按钮。接下来调用 setImageBackground 方法来更新圆点的显示。onPageScrollStateChanged 方法在页面滚动状态改变时调用，如果用户滚动到最后一张图片并且还想向右滑动，则直接调用 gotoMain 方法进入首页。
- PreviewImageAdapter 继承自 PagerAdapter，是 ViewPager 控件的适配器，主要重写三个方法 getCount（获取 Item 总数）、destroyItem（删除 Item）、instantiateItem（添加 Item）。
- gotoMain 方法调用 setFirstLauncherBoolean 方法将第一次启动的布尔类型设置为 true，并且发送一个 Intent 开启首页的 Activity，关闭当前界面。
- setFirstLauncherBoolean 将是否第一次启动的布尔类型变量设置为 true，除非用户卸载软件，不然这个变量一直是 true。

9.2　检查更新并下载安装

检查更新是任何 App 都会用到的功能，任何 App 不可能在第一个版本就将所有的需求全部实现，只有通过不断地挖掘需求和完善，才能使 App 变得越来越好，这就需要软件不停地升级。检查更新自动并下载安装分为以下几个步骤：

（1）请求服务器判断是否有最新版本（通过 versionCode）。

（2）如果有最新版本，就将最新的 apk 文件下载到本地。

（3）下载完成之后，给系统发起一个安装的 Intent。

打开项目 App 的 build.gradle 文件，可以看到其中有 versionCode 和 versionName 两个属性：versionCode 是一个整型类型的值，用来更新版本；versionName 是一个浮点型的值，可以告诉用户当前的版本号。每次 App 升级时，都要对这两个值进行增加。这里使用默认值即可。

因为检查更新需要请求服务器，所以引入之前封装的 okhttp 库：

```
compile 'com.ansen.http:okhttpencapsulation:1.0.1'
```

需要访问网络和写入 SD 卡的权限，记住在 AndroidManifest.xml 中增加权限。

```
<uses-permission android:name="android.permission.INTERNET" />
<uses-permission android:name="android.permission.WRITE_EXTERNAL_STORAGE"/>
```

需要重写 Application，并且在 AndroidManifest.xml 文件中给 application 标签 name 属性指向重写的 MyApplication，在 MyApplication 中初始化 HTTPCaller。

```
public class MyApplication extends Application {
    @Override
    public void onCreate() {
        super.onCreate();

        HttpConfig httpConfig = new HttpConfig();
        httpConfig.setAgent(true);//有代理的情况能不能访问
        httpConfig.setDebug(true);//是否是 debug 模式，如果是 debug 模式就打印日志
        httpConfig.setTagName("ansen");//打印日志的 TagName

        //可以添加一些公共字段，每个接口都会带上
        httpConfig.addCommonField("pf", "android");
        httpConfig.addCommonField("version_code", "" + Utils.getVersionCode
(getApplicationContext()));

        //初始化 HTTPCaller 类
        HTTPCaller.getInstance().setHttpConfig(httpConfig);
    }
}
```

我们将 version_code 作为公共参数，通过 Utils 类的 getVersionCode 方法获取值。getVersionCode 方法需要传入一个 context 对象，通过 Context 获取包管理器，调用 PackageManager 的 getPackageInfo 方法获取包信息，调用它的公有属性 versionCode 来获取当前版本号。

```
public static int getVersionCode(Context ctx) {
    // 获取 packagemanager 的实例
    int version = 0;
    try {
        PackageManager packageManager = ctx.getPackageManager();
        //getPackageName()是当前程序的包名
        PackageInfo packInfo =
packageManager.getPackageInfo(ctx.getPackageName(), 0);
        version = packInfo.versionCode;
    } catch (Exception e) {
        e.printStackTrace();
    }
    return version;
}
```

新建项目，修改 MainActivity.java 对应的 activity_main.xml 文件，里面就两个控件，TextView 用来显示当前版本号，Button 用来“检查更新”。

```
<?xml version="1.0" encoding="utf-8"?>
<LinearLayout xmlns:android="http://schemas.android.com/apk/res/android"
    android:layout_width="match_parent"
    android:layout_height="match_parent"
    android:orientation="vertical"
    android:padding="10dp">
```

```
    <TextView
        android:id="@+id/tv_current_version_code"
        android:layout_width="match_parent"
        android:layout_height="wrap_content"
        android:gravity="center_horizontal"
        android:textSize="16sp"
        android:text="当前版本"/>

    <Button
        android:id="@+id/btn_check_update"
        android:layout_marginTop="10dp"
        android:layout_width="match_parent"
        android:layout_height="wrap_content"
        android:text="检查更新"/>
</LinearLayout>
```

接下来看布局文件对应的 MainActivity.java。为了方便学习，这里先看一部分代码，onCreate 给显示当前版本的 TextView 赋值，以及为“检查更新”按钮设置点击监听。点击按钮时发送一个 Get 请求服务器。我们想知道有没有新版本，是通过 versionCode 的值进行判断的，但是这里却没有在请求 url 后面添加参数，因为在 MyApplication 中已经将 versionCode 设置成了公共参数。

```
public class MainActivity extends AppCompatActivity implements
View.OnClickListener{
    private ProgressDialog progressDialog;

    @Override
    protected void onCreate(Bundle savedInstanceState) {
        super.onCreate(savedInstanceState);
        setContentView(R.layout.activity_main);

        TextView tvCurrentVersionCode= (TextView)
findViewById(R.id.tv_current_ version_code);
        tvCurrentVersionCode.setText("当前版本:"+ Utils.getVersionCode(this));

        findViewById(R.id.btn_check_update).setOnClickListener(this);
    }

    @Override
    public void onClick(View v) {
        switch (v.getId()){
            case R.id.btn_check_update://检查更新
                HTTPCaller.getInstance().get(CheckUpdate.class,
                "http://139.196.35.30:8080/OkHttpTest/checkUpdate.do",null,
requestDataCallback);
                break;
        }
    }

    ............
}
```

Get 请求回调监听，先判断返回的状态码。如果状态码等于 0 表示有新版本，有新版本的话下载 url 也应该赋值，将下载 url 传给 showUpdaloadDialog 方法。

```
    private RequestDataCallback<CheckUpdate> requestDataCallback=new
RequestDataCallback<CheckUpdate>(){
        @Override
        public void dataCallback(CheckUpdate obj) {
            if(obj!=null){
                if(obj.getErrorCode()==0){//有新版本
                    showUpdaloadDialog(obj.getUrl());
                }else{//没有新版本
                    Toast.makeText(MainActivity.this,obj.getErrorReason(),
Toast.LENGTH_LONG).show();
                }
            }
        }
    };
```

showUpdaloadDialog 方法弹出是否更新对话框。利用“是否更新”对话框让用户判断要不要更新，相对友好一些。某些公司的 App 在用户手机连接 WiFi 时直接从后台下载更新包，而不经过用户的同意，这种做法欠妥。这里设置一个“确认更新”对话框，如果用户点击了“确认”按钮就调用 startUpload 方法；如果点击“取消”按钮，就关闭对话框，不进行升级操作。

```
    private void showUpdaloadDialog(final String downloadUrl){
        // 这里的属性可以一直设置，因为每次设置后返回的是一个 builder 对象
        AlertDialog.Builder builder = new AlertDialog.Builder(this);
        // 设置提示框的标题
        builder.setTitle("版本升级").
                setIcon(R.mipmap.ic_launcher). // 设置提示框的图标
                setMessage("发现新版本！请及时更新").// 设置要显示的信息
                setPositiveButton("确定", new DialogInterface.OnClickListener() {
                        // 设置“确定”按钮
                    @Override
                    public void onClick(DialogInterface dialog, int which) {
                        startUpload(downloadUrl);//下载最新的版本程序
                    }
                }).setNegativeButton("取消", null);
                  //设置“取消”按钮,null 是什么都不做，并关闭对话框
        AlertDialog alertDialog = builder.create();
        // 显示对话框
        alertDialog.show();
    }
```

startUpload 方法首先创建一个进度条对话框，设置进度条样式，设置 messgage，然后调用 Utils 类的 getSaveFilePath 静态方法获取一个 sdcard 的路径，将下载的 APK 文件保存在这个路径中。接下来，调用 HTTPCaller 类的 downloadFile 方法去下载文件，包含三个参数：下载 url、文件保存路径和下载进度回调。下载进度回调是一个 ProgressUIListener 接口，用内部类方式重写。这个类有三个回调方法：onUIProgressStart 在下载开始时调用，我们将总文件长度赋值给进度条的总长度，然后显示进度条；onUIProgressChanged 方法在进度改变时调用，我们在这个方法中更新进度条；onUIProgressFinish 在下载完成时调用，在这个方法中销毁进度条，调用 openAPK 方法打开 APK 文件。

```
    private void startUpload(String downloadUrl){
        progressDialog=new ProgressDialog(this);
        progressDialog.setProgressStyle(ProgressDialog.STYLE_HORIZONTAL);
        progressDialog.setMessage("正在下载新版本");
```

```
        progressDialog.setCancelable(false);//不能手动取消下载进度对话框
        final String fileSavePath=Utils.getSaveFilePath(downloadUrl);
        HTTPCaller.getInstance().downloadFile(downloadUrl,fileSavePath,null,new
ProgressUIListener(){
            @Override
            public void onUIProgressStart(long totalBytes) {//下载开始
                progressDialog.setMax((int)totalBytes);
                progressDialog.show();
            }
            //更新进度
            @Override
            public void onUIProgressChanged(long numBytes, long totalBytes, float
percent, float speed) {
                progressDialog.setProgress((int)numBytes);
            }
            @Override
            public void onUIProgressFinish() {//下载完成
                Toast.makeText(MainActivity.this,"下载完成",Toast.LENGTH_LONG).show();
                progressDialog.dismiss();
                openAPK(fileSavePath);
            }
        });
    }
```

下载完成后给系统发送一个安装 APK 的 Intent。

```
    private void openAPK(String fileSavePath){
        Intent intent = new Intent();
        intent.addFlags(Intent.FLAG_ACTIVITY_NEW_TASK);
        intent.setAction(Intent.ACTION_VIEW);
        intent.setDataAndType(Uri.parse("file://"+fileSavePath),
         "application/vnd. android.package-archive");
        startActivity(intent);
    }
```

运行源代码，效果如图 9-4 所示。

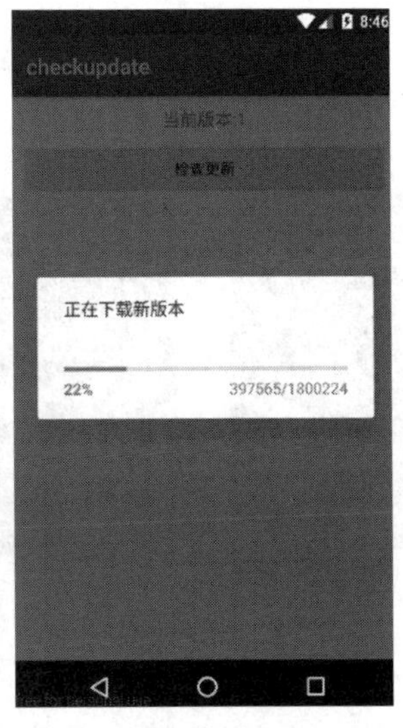

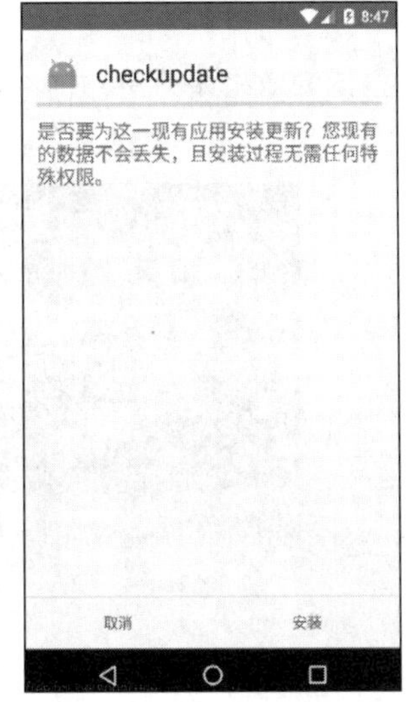

图 9-4　效果图

注 意

这里为什么有新版本？

我们查看第一张效果图，发现当前的版本是 1。其实我们预先打包了一个 versionCode 等于 2 的签名 APK 放到服务器上，所以只要给的 versionCode 参数值小于 2 就是可以升级的。

覆盖安装签名问题

通过前面的学习，我们知道调试时直接运行 App 并安装到手机上安装包是临时签名，所以在企业开发中每次提交版本到应用市场时都会用线上签名文件进行签名，保证线上（应用市场）各版本的安装包都是同一个签名，只有签名一样时才能覆盖安装。如果签名不一致，点击安装时会出现应用未安装的情况。

我们这个项目的线上签名文件在项目/jks 文件夹中，这个文件夹下还有一个已经签名的 1.0 版本 APK 文件，如果想要查看效果，可以把 1.0 版本的 APK 文件通过社交软件之类发送到手机上，点击检查更新按钮，下载安装，升级完成。当然，也可以用签名文件对 APK 签名。签名文件和密码都在 jks 文件夹下。

9.3 Banner 广告轮播图

轮播图是大部分 App 都有的效果，尤其是商品类和新闻类的 App。

轮播图的效果和第一次启动时的引导页类似，不过轮播图在引导页的基础上多了以下功能：

- 在第一页也能向左滑动，在最后一页也能向右滑动。
- 有一个计时器，每隔一段时间就滚动轮播图。

9.3.1 运行效果图

此轮播图的效果如图 9-5 所示。在图片集合中加上一个指示器，用来表示当前滚动到第几张图片。

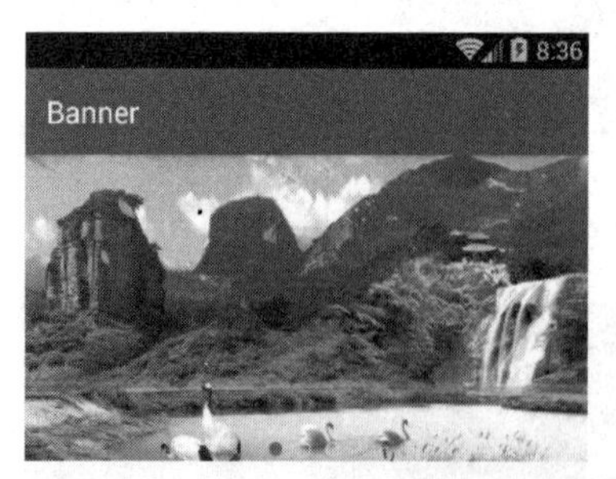

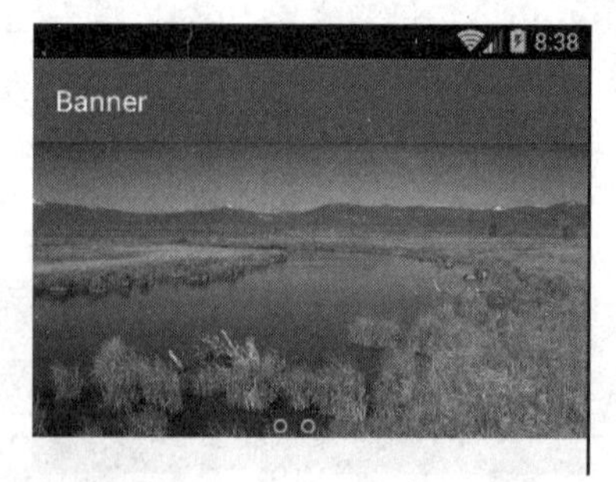

图 9-5 轮播图的效果

9.3.2　代码实现

activity_main.xml 文件的内容如下：

```
<?xml version="1.0" encoding="utf-8"?>
<RelativeLayout
    xmlns:android="http://schemas.android.com/apk/res/android"
    android:layout_width="match_parent"
    android:layout_height="200dp">

    <android.support.v4.view.ViewPager
        android:id="@+id/vp_banner"
        android:layout_width="match_parent"
        android:layout_height="match_parent"/>

    <LinearLayout
        android:id="@+id/viewGroup"
        android:layout_width="wrap_content"
        android:layout_height="wrap_content"
        android:layout_alignParentBottom="true"
        android:layout_centerHorizontal="true"
        android:layout_marginBottom="3dp"
        android:orientation="horizontal"/>
</RelativeLayout>
```

最外层是 RelativeLayout，固定高度是 200dp，其中有 ViewPager 和 LinearLayout。ViewPager 用来显示 Banner 图片，LinearLayout 显示圆点图片，表示当前 ViewPager 的选中状态。

BannerAdapter.java ViewPager 适配器继承自 PagerAdapter，重写以下四个方法。

```
public class BannerAdapter extends PagerAdapter {
    private Context context;
    private View.OnClickListener onBannerClickListener;

    //图片列表
    private int[] banners=new int[]{R.mipmap.banner_one,R.mipmap.banner_two,
     R.mipmap.banner_three};

    public BannerAdapter(Context context){
        this.context=context;
    }

    @Override
    public int getCount() {
        return Integer.MAX_VALUE;//返回 int 的最大值，可以一直滑动
    }

    @Override
    public boolean isViewFromObject(View arg0, Object arg1) {
        return arg0==arg1;
    }

    @Override
    public void destroyItem(ViewGroup container, int position, Object object){
```

```
        container.removeView((View) object);
    }

    @Override
    public Object instantiateItem(ViewGroup container, int position) {
        //position 的值范围是 0~2147483647，将这个值对图片长度求余之后，position 的
        //取值范围是 0~banners.length-1
        position %= banners.length;

        ImageView imageView = new ImageView(context);
        imageView.setScaleType(ImageView.ScaleType.CENTER_CROP);//设置图片缩放类型
        imageView.setTag(position);//将当前的下标通过 setTag 方法设置进去
        imageView.setImageResource(banners[position]);
        imageView.setOnClickListener(onClickListener);
        container.addView(imageView);
        return imageView;
    }

    private View.OnClickListener onClickListener=new View.OnClickListener() {
        @Override
        public void onClick(View v) {
            if(onBannerClickListener!=null){
                onBannerClickListener.onClick(v);
            }
        }
    };

    public void setOnBannerClickListener(View.OnClickListener
      onBannerClickListener) {
        this.onBannerClickListener = onBannerClickListener;
    }

    public int[] getBanners(){
        return banners;
    }
}
```

首先将要显示的图片资源 id 放到一个名字为 banners 的 int 数组中，这是一个实例变量，在构造方法中传入 context 对象。

- getCount：返回 int 的最大值，即使滑动到最后一页还能向后滑动。
- isViewFromObject：常规写法，参数 1==参数 2。
- destroyItem：删除一条记录。
- instantiateItem：添加一条记录，首先传入的 position 的值为 0 到 int 最大值，因为前面的 getCount 方法就返回 int 类型最大值，所以这里要对图片数量进行求余。求余之后 position 的值必定小于图片数量，初始化 ImageView 对象，设置图片缩放类型，将 position 通过 setTag 保存到 view 中，设置图片资源，并给图片设置点击事件，最后将图片添加到 ViewGroup 中。这个 ViewGroup 就是 ViewPager 自己。如果查看了源代码，就会发现 ViewPager 继承自 ViewGroup。
- onClickListener：先判断 onBannerClickListener 是否为空，如果不为空，就调用 onBannerClickListener 的 onClick 方法。

- setOnBannerClickListener：设置 banner item 点击监听，其实就是设置图片点击。
- getBanners：返回图片资源数组。

MainActivity.java 文件的内容如下：

```
public class MainActivity extends AppCompatActivity {
    public static final int CAROUSEL_TIME = 5000;//banner 滚动间隔

    private ViewPager vpBanner;//
    private ViewGroup viewGroup;//显示圆点图片，可以看到 ViewPager 当前选中状态

    private BannerAdapter bannerAdapter;//ViewPager 适配器

    private Handler handler=new Handler();
    private int currentItem = 0;//ViewPager 当前位置

    private final Runnable mTicker = new Runnable(){
        public void run() {
            long now = SystemClock.uptimeMillis();
            long next = now + (CAROUSEL_TIME - now % CAROUSEL_TIME);

            handler.postAtTime(mTicker,next);
            //延迟 5 秒再次执行 runnable，效果和计时器一样

            currentItem++;
            vpBanner.setCurrentItem(currentItem);
        }
    };

    @Override
    protected void onCreate(Bundle savedInstanceState) {
        super.onCreate(savedInstanceState);
        setContentView(R.layout.activity_main);

        vpBanner= (ViewPager) findViewById(R.id.vp_banner);
        bannerAdapter = new BannerAdapter(this);//初始化适配器
        bannerAdapter.setOnBannerClickListener(onBannerClickListener);
          //图片点击监听
        vpBanner.setOffscreenPageLimit(2);//缓存页数
        vpBanner.setAdapter(bannerAdapter);//设置适配器
        vpBanner.addOnPageChangeListener(onPageChangeListener); //页面改变监听

        viewGroup = (ViewGroup) findViewById(R.id.viewGroup);//显示圆点控件

        //将圆点加入到 ViewGroup 中
        for (int i = 0; i < bannerAdapter.getBanners().length; i++) {
            ImageView imageView = new ImageView(this);
            //设置图片宽和高
            imageView.setLayoutParams(new ViewGroup.LayoutParams(10, 10));
            if (i == 0) {
                imageView.setBackgroundResource(R.mipmap.icon_page_select);
            } else {
                imageView.setBackgroundResource(R.mipmap.icon_page_normal);
            }
            LinearLayout.LayoutParams layoutParams = new
```

```
LinearLayout.LayoutParams (new
ViewGroup.LayoutParams(ViewGroup.LayoutParams.WRAP_CONTENT,
ViewGroup.LayoutParams. WRAP_CONTENT));
                layoutParams.leftMargin = 10;//左边距
                layoutParams.rightMargin = 10;//右边距
                viewGroup.addView(imageView,layoutParams);
            }

            //给 ViewPager 设置当前页，这样刚打开软件时也能向左滑动
            currentItem = bannerAdapter.getBanners().length * 50;
            vpBanner.setCurrentItem(currentItem);

            handler.postDelayed(mTicker,CAROUSEL_TIME);//开启计时器
        }

        private ViewPager.OnPageChangeListener onPageChangeListener=new
ViewPager.OnPageChangeListener() {
            @Override
            public void onPageScrolled(int position, float positionOffset, int
positionOffsetPixels) {}

            @Override
            public void onPageSelected(int position) {
                currentItem = position;

                //改变圆点图片的选中状态
                setImageBackground(position %= bannerAdapter.getBanners().length);
            }

            @Override
            public void onPageScrollStateChanged(int state) {}
        };

        /**
         * 改变圆点的切换效果
         * @param selectItems 当前选中位置
         */
        private void setImageBackground(int selectItems){
            for (int i = 0; i < bannerAdapter.getBanners().length; i++) {
                ImageView imageView = (ImageView) viewGroup.getChildAt(i);
                imageView.setBackgroundDrawable(null);//先将背景设置为无
                if (i == selectItems){
                    imageView.setImageResource(R.mipmap.icon_page_select);
                } else {
                    imageView.setImageResource(R.mipmap.icon_page_normal);
                }
            }
        }

        private View.OnClickListener onBannerClickListener=new
View.OnClickListener() {
            @Override
            public void onClick(View view) {
                int position=(Integer) view.getTag();
                 //从 Tag 中获取当前点击的 ImageView 的位置
```

```
            Toast.makeText(MainActivity.this,"当前点击位置:"+position,
Toast.LENGTH_LONG).show();
            }
        };

        @Override
        protected void onDestroy() {
            super.onDestroy();
            handler.removeCallbacks(mTicker);//删除计时器
        }
    }
```

代码虽然比较多，但关键地方都有注释，简单解释一下：

- onCreate：查找 ViewPager 控件，设置适配器，设置图片点击监听、页面滑动监听，为 ViewGroup 添加点图片，就是滑动指示器。同时给 ViewPager 设置当前选中的 item，这个值稍微设置大一点，好让用户刚进来就能左右滑动，使 App 更加符合用户习惯。最后调用 handler.postDelayed 方法开启计时器，有两个参数：第一个参数是 Runnable，第二个参数是延迟的时间，5 秒滚动一次轮播图。
- onPageChangeListener：在 onPageSelected 方法中将当前的 position 赋值给 currentItem，调用 setImageBackground 方法改变圆点图片的选中状态，传入一个 int 类型的值（先对 position 求余之后再进行传入）。
- setImageBackground：改变指示器的选中状态。循环指示器 ViewGroup，根据显示图片的下标决定指示器选中状态。
- onBannerClickListener：图片点击监听，在这里可以处理图片点击后的操作。
- onDestroy：删除计时器。在开发过程中，如果有需要释放的内容，就一定要在 onDestroy 中添加。即时释放资源是一个程序员应养成的良好习惯。

9.4　微信登录、分享与支付

大部分 App 都有接入第三方 SDK（软件开发工具包）的需求。例如，第三方登录需要接入微信、QQ、微博等，第三方支付需要接入微信、支付宝、银联等。

在上面这些 SDK 中，需要注意的是微信不能直接调试，需要使用正式的签名进行签名才能调试。另外，这些官方的 Demo 运行不起来，缺少签名文件。

9.4.1　代码实现

现在微信的软件开发工具包已经支持 Android Studio 在线引用了，之前都是添加 JAR 的方法。本项目还需要访问微信的接口获取用户信息，所以需要将之前封装的 OkHttp 也一起在线引用。OkHttp 需要在自定义的 Application 中初始化，前面已经讲过多次，这里就不贴出代码了。新建一个项目，在 app/build.gradle 文件 dependencies 标签中加入以下两行代码，在线依赖微信 SDK 以及我们封装的 Okhttp 库：

```
compile 'com.tencent.mm.opensdk:wechat-sdk-android-without-mta:+'
compile 'com.ansen.http:okhttpencapsulation:1.0.1'
```

因为需要用到网络，所以在 AndroidManifest.xml 文件中加入网络权限:

```
<uses-permission android:name="android.permission.INTERNET" />
```

activity_main.xml 布局文件内容如下：

```
<?xml version="1.0" encoding="utf-8"?>
<LinearLayout xmlns:android="http://schemas.android.com/apk/res/android"
    android:layout_width="match_parent"
    android:layout_height="match_parent"
    android:padding="10dp"
    android:orientation="vertical">

    <TextView
        android:layout_width="wrap_content"
        android:layout_height="wrap_content"
        android:text="登录之后信息在这里显示"/>

    <RelativeLayout
        android:layout_width="wrap_content"
        android:layout_height="wrap_content">

        <TextView
            android:id="@+id/tv_nickname"
            android:layout_width="wrap_content"
            android:layout_height="wrap_content"
            android:text="昵称:"/>

        <TextView
            android:id="@+id/tv_age"
            android:layout_below="@+id/tv_nickname"
            android:layout_width="wrap_content"
            android:layout_height="wrap_content"
            android:text="年龄:"/>
    </RelativeLayout>

    <Button
        android:id="@+id/btn_login"
        android:layout_width="match_parent"
        android:layout_height="wrap_content"
        android:text="微信登录"/>

    <Button
        android:id="@+id/btn_share_friend_circle"
        android:layout_width="match_parent"
        android:layout_height="wrap_content"
        android:text="分享到朋友圈"/>

    <Button
        android:id="@+id/btn_share_friend"
        android:layout_width="match_parent"
```

```
        android:layout_height="wrap_content"
        android:text="分享给好友"/>

    <Button
        android:id="@+id/btn_pay"
        android:layout_width="match_parent"
        android:layout_height="wrap_content"
        android:text="微信支付"/>
</LinearLayout>
```

布局文件很简单，仅在 LinearLayout 中放了几个 TextView 和按钮。

WeiXin.java 用 EventBus 来传送消息。微信有一个很奇怪的地方，就是不管登录、分享还是支付之后都得用一个 Activity 来接收，所以要从接收的 Activity 中将结果信息通过 EventBus 传递给 MainActivity。虽然使用广播也能够实现，但是用 EventBus 更加简单易用。

```
public class WeiXin {
    private int type;//1:登录 2: 分享 3:微信支付
    private int errCode;//微信返回的错误码
    private String code;//登录成功才会有的代码

    public WeiXin() {
    }

    public WeiXin(int type,int errCode, String code) {
        this.type = type;
        this.errCode=errCode;
        this.code = code;
    }

    public int getType() {
        return type;
    }

    public void setType(int type) {
        this.type = type;
    }

    public String getCode() {
        return code;
    }

    public void setCode(String code) {
        this.code = code;
    }

    public int getErrCode() {
        return errCode;
    }

    public void setErrCode(int errCode) {
        this.errCode = errCode;
    }
}
```

新建 Constant.java 文件，用来定义微信 appid 和 secret 这两个常量。这样做的好处是当我们的 appid 跟 secret 变化时只需要修改常量。为了保护隐私，以下这两个值已经进行了修改。

```
public class Constant {
    public static String WECHAT_APPID="wxda6db2aec81389af";
    public static String WECHAT_SECRET="8fed5a2d510022587ef8a6194c965be3";
}
```

布局文件对应的 MainActivity.java 代码全部贴出来比较多，这里仅贴出 MainActivity 部分代码。

```
public class MainActivity extends AppCompatActivity implements
View.OnClickListener {
    private IWXAPI wxAPI;
    private TextView tvNickname,tvAge;
    public static final int IMAGE_SIZE=32768;//微信分享图片大小限制

    @Override
    protected void onCreate(Bundle savedInstanceState) {
        super.onCreate(savedInstanceState);
        setContentView(R.layout.activity_main);

        EventBus.getDefault().register(this);//注册
        wxAPI = WXAPIFactory.createWXAPI(this,Constant.WECHAT_APPID,true);
        wxAPI.registerApp(Constant.WECHAT_APPID);

        findViewById(R.id.btn_login).setOnClickListener(this);
        findViewById(R.id.btn_share_friend_circle).setOnClickListener(this);
        findViewById(R.id.btn_share_friend).setOnClickListener(this);
        findViewById(R.id.btn_pay).setOnClickListener(this);

        tvNickname= (TextView) findViewById(R.id.tv_nickname);
        tvAge=(TextView) findViewById(R.id.tv_age);
    }

    @Override
    public void onClick(View view) {
        switch (view.getId()){
            case R.id.btn_login://微信登录
                login();
                break;
            case R.id.btn_share_friend_circle://微信分享到朋友圈
                share(true);
                break;
            case R.id.btn_share_friend://微信分享给朋友
                share(false);
                break;
            case R.id.btn_pay://微信支付
             //先去服务器获取支付信息，返回一个 WeiXinPay 对象，然后调用 pay 方法
                showToast("微信支付需要服务器支持");
                break;
        }
    }

    /**
     * 这里用到了 EventBus 框架
```

```
     * @param weiXin
     */
    @Subscribe
    public void onEventMainThread(WeiXin weiXin){
        Log.i("ansen","收到eventbus请求 type:"+weiXin.getType());
        if(weiXin.getType()==1){//登录
            getAccessToken(weiXin.getCode());
        }else if(weiXin.getType()==2){//分享
            switch (weiXin.getErrCode()){
                case BaseResp.ErrCode.ERR_OK:
                    Log.i("ansen", "微信分享成功......");
                    break;
                case BaseResp.ErrCode.ERR_USER_CANCEL://分享取消
                    Log.i("ansen", "微信分享取消......");
                    break;
                case BaseResp.ErrCode.ERR_AUTH_DENIED://分享被拒绝
                    Log.i("ansen", "微信分享被拒绝......");
                    break;
            }
        }else if(weiXin.getType()==3){//微信支付
            if(weiXin.getErrCode()==BaseResp.ErrCode.ERR_OK){//成功
                Log.i("ansen", "微信支付成功......");
            }
        }
    }

    ......

    public void showToast(String message){
        Toast.makeText(this,message,Toast.LENGTH_LONG).show();
    }
    @Override
    protected void onDestroy() {
        super.onDestroy();
        EventBus.getDefault().unregister(this);//取消注册
    }
}
```

- onCreate：首先订阅 EventBus 消息，EventBus 3.0 以后版本会通过反射扫描当前类中 @Subscribe 注解的方法。接下来通过 WXAPIFactory 创建 IWXAPI 类，并且注册 appid，给四个按钮设置点击事件。查找显示名字和年龄的两个 TextView。
- onClick：点击事件监听，根据 id 来判断点击不同的按钮，跳转到相应的方法，这些方法没贴出来，后面会单独讲解。
- onEventMainThread(WeiXin weiXin)：用来接收 EventBus 消息。这个方法中有一个参数用来判断接收类型，例如我们这里是 WeiXin 对象，就是说使用 EventBus 发送事件时参数必须是 WeiXin 对象这个方法才会调用。首先判断类型，有三种：登录、分享或支付。登录成功的情况下会获取到微信的 code，这个 code 可以理解为一个微信号的唯一标示，通过 code 可以获取到微信用户信息。
- showToast：Toast 提示。
- onDestroy：取消 EventBus 注册。

9.4.2 微信登录

微信登录流程有以下三个步骤：

（1）微信授权登录。

（2）根据授权登录获取微信返回的 code，然后通过 code 获取 access_token。

（3）根据 access_token 获取微信用户资料。

当点击“登录”按钮时，调用的是 login 方法，包含在 MainActivity 中，也就是给微信发起一个登录请求，弹出一个授权界面。

```
public void login(){
    SendAuth.Req req = new SendAuth.Req();
    req.scope = "snsapi_userinfo";
    req.state = String.valueOf(System.currentTimeMillis());
    wxAPI.sendReq(req);
}
```

在包名的相应目录下新建一个 wxapi 子目录，然后在 wxapi 子目录下新增一个 WXEntryActivity 类，用来接收登录授权以及分享时微信的回调信息。这个类继承自 Activity，需要实现 IWXAPIEventHandler 接口。

```
package com.ansen.shoenet.wxapi;
public class WXEntryActivity extends Activity implements IWXAPIEventHandler {
    private IWXAPI wxAPI;

    @Override
    protected void onCreate(Bundle savedInstanceState) {
        super.onCreate(savedInstanceState);

        wxAPI = WXAPIFactory.createWXAPI(this,Constant.WECHAT_APPID,true);
        wxAPI.registerApp(Constant.WECHAT_APPID);
        wxAPI.handleIntent(getIntent(), this);
    }

    @Override
    protected void onNewIntent(Intent intent){
        super.onNewIntent(intent);
        wxAPI.handleIntent(getIntent(),this);
        Log.i("ansen","WXEntryActivity onNewIntent");
    }

    @Override
    public void onReq(BaseReq arg0) {
        Log.i("ansen","WXEntryActivity onReq:"+arg0);
    }

    @Override
    public void onResp(BaseResp resp){
        if(resp.getType()== ConstantsAPI.COMMAND_SENDMESSAGE_TO_WX){//分享
            Log.i("ansen","微信分享操作.....");
```

```
            WeiXin weiXin=new WeiXin(2,resp.errCode,"");
            EventBus.getDefault().post(weiXin);
        }else if(resp.getType()==ConstantsAPI.COMMAND_SENDAUTH){//登录
            Log.i("ansen", "微信登录操作......");
            SendAuth.Resp authResp = (SendAuth.Resp) resp;
            WeiXin weiXin=new WeiXin(1,resp.errCode,authResp.code);
            EventBus.getDefault().post(weiXin);
        }
        finish();
    }
}
```

onCreate、onNewIntent 和 onReq 这三个方法是固定写法。onResp 方法用于接收微信结果信息，首先判断类型，根据不同的类型去封装 WeiXin 对象，如果是登录操作，就将 code 传进去，然后将封装好的 WeiXin 对象通过 EventBus 发送出去。MainActivity 的 onEventMainThread 方法就会接收到这个消息，最后调用 Finish 关闭当前的 Activity。

WXEntryActivity 需要在 AndroidManifest.xml 中注册：

```
<activity
    android:exported="true"
    android:name=".wxapi.WXEntryActivity"/>
```

继续回到首页的 onEventMainThread，如果登录类型调用 getAccessToken()，并且传入 code。根据 code 获取 access_token，这个 url 是微信公开的，需要传入 appid、secret、code 三个参数。请求成功之后会返回 access_token 和 openid 等信息。

```
public void getAccessToken(String code){
    String url = "https://api.weixin.qq.com/sns/oauth2/access_token?" +
            "appid="+Constant.WECHAT_APPID+"&secret="+Constant.WECHAT_
                    SECRET+"&code="+code+"&grant_type=authorization_code";
    HTTPCaller.getInstance().get(WeiXinToken.class, url, null,
    new RequestDataCallback<WeiXinToken>() {
        @Override
        public void dataCallback(WeiXinToken obj) {
            if(obj.getErrcode()==0){//请求成功
                getWeiXinUserInfo(obj);
            }else{//请求失败
                showToast(obj.getErrmsg());
            }
        }
    });
}
```

获取到 access_token 和 openid 之后，继续调用 getWeiXinUserInfo 方法获取用户信息。这样能够取到当前微信 App 登录用户的一些信息，如昵称、年龄、头像地址、语言等基本信息。在企业开发中，到了此步就可以拿着这些信息调用自己服务器的登录接口了。当然，这里将昵称和年龄让 TextView 显示。

```
public void getWeiXinUserInfo(WeiXinToken weiXinToken){
    String url = "https://api.weixin.qq.com/sns/userinfo?access_token="+
            weiXinToken.getAccess_token()+"&openid="+weiXinToken.getOpenid();
    HTTPCaller.getInstance().get(WeiXinInfo.class, url, null,
    new RequestDataCallback<WeiXinInfo>() {
```

```
        @Override
        public void dataCallback(WeiXinInfo obj) {
            tvNickname.setText("昵称:"+obj.getNickname());
            tvAge.setText("年龄:"+obj.getAge());
            Log.i("ansen","头像地址:"+obj.getHeadimgurl());
        }
    });
}
```

WeiXinToken 和 WeiXinInfo 这两个实体类就不贴代码了。WeiXinToken 用来映射获取访问 token 接口返回的 Json，WeiXinInfo 用来映射获取用户接口返回的 Json。

9.4.3 微信分享

微信分享分为两种：分享到朋友圈和分享给好友，统一调用 share 方法。传入一个布尔类型来判断是否分享到朋友圈。

```
public void share(boolean friendsCircle){
    WXWebpageObject webpage = new WXWebpageObject();
    webpage.webpageUrl = "www.baidu.com";//分享 url
    WXMediaMessage msg = new WXMediaMessage(webpage);
    msg.title = "分享标题";
    msg.description = "分享描述";
    msg.thumbData =getThumbData();//封面图片 byte 数组

    SendMessageToWX.Req req = new SendMessageToWX.Req();
    req.transaction = String.valueOf(System.currentTimeMillis());
    req.message = msg;
    req.scene = friendsCircle ? SendMessageToWX.Req.WXSceneTimeline :
    SendMessageToWX.Req.WXSceneSession;
    wxAPI.sendReq(req);
}
```

分享内容有很多格式，如分享图片、分享视频、分享消息等。下面以分享消息为例，这也是分享比较常见的格式。首先新建一个 WXWebpageObject 对象，设置标题、内容、打开链接、封面等，然后调用 wxAPI 的 sendReq 放松一个请求。

分享和登录一样，都会回调 WXEntryActivity，然后又将分享结果发送给 MainActivity.onEventMainThread 方法。

9.4.4 微信支付

微信支付需要请求自己的服务器，从中获取支付信息。获取成功之后调用 pay 方法。

```
public void pay(WeiXinPay weiXinPay){
    PayReq req = new PayReq();
    req.appId = Constant.WECHAT_APPID;//appid
    req.nonceStr=weiXinPay.getNoncestr();//随机字符串，不长于 32 位。推荐随机数生成算法
    req.packageValue=weiXinPay.getPackage_value();//暂时填写固定值 Sign=WXPay
    req.sign=weiXinPay.getSign();//签名
    req.partnerId=weiXinPay.getPartnerid();//微信支付分配的商户号
```

```
        req.prepayId=weiXinPay.getPrepayid();//微信返回的支付交易会话 ID
        req.timeStamp=weiXinPay.getTimestamp();//时间戳

        wxAPI.registerApp(Constant.WECHAT_APPID);
        wxAPI.sendReq(req);
    }
```

正常情况下 weiXinPay 的值是从我们的服务器获取的，然后将返回信息封装到 PayReq 对象中，最后调用 wxAPI 的 sendReq 方法发起支付请求。

在 wxapi 目录下新增一个 WXPayEntryActivity 类，和 WXEntryActivity 同级，用来接收微信支付的回调信息。这个类继承自 Activity，需要实现 IWXAPIEventHandler 接口。

```
public class WXPayEntryActivity extends Activity implements IWXAPIEventHandler {
    private IWXAPI wxAPI;

    @Override
    protected void onCreate(Bundle savedInstanceState) {
        super.onCreate(savedInstanceState);
        wxAPI = WXAPIFactory.createWXAPI(this, Constant.WECHAT_APPID);
        wxAPI.handleIntent(getIntent(), this);
    }

    @Override
    protected void onNewIntent(Intent intent){
        super.onNewIntent(intent);
        setIntent(intent);
        wxAPI.handleIntent(intent, this);
    }

    @Override
    public void onReq(BaseReq baseReq) {}

    @Override
    public void onResp(BaseResp resp) {
        Log.i("ansen", "微信支付回调，返回错误码:"+resp.errCode+" 错误名称:"
              +resp.errStr);
        if (resp.getType() == ConstantsAPI.COMMAND_PAY_BY_WX){//微信支付
            WeiXin weiXin=new WeiXin(3,resp.errCode,"");
            EventBus.getDefault().post(weiXin);
        }
        finish();
    }
}
```

其他方法都是固定写法，在 onResp 中判断如果是微信登录，就封装一个 WeiXin 对象，然后发送 EventBus 请求。这样，MainActivity 的 onEventMainThread 就会接收到这个 WeiXin 对象。

WXPayEntryActivity 需要在 AndroidManifest.xml 中注册。

```
<activity
    android:exported="true"
    android:name=".wxapi.WXPayEntryActivity"/>
```

项目结构如图 9-6 所示，从中可以看到软件包名为 com.ansen.shoenet。接收微信登录支付返回的 Activity 的包名必须是 com.ansen.shoenet.wxapi。两个 Activity 的名字也是固定的写法。

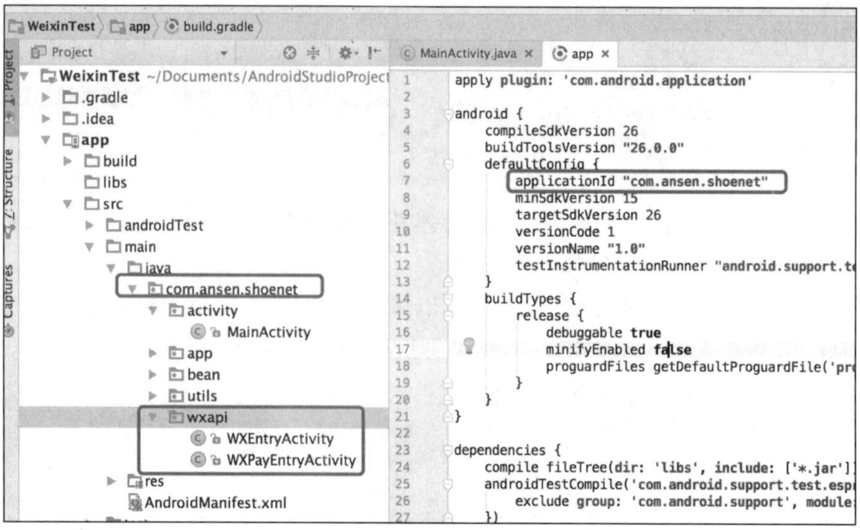

图 9-6　查看项目结构和软件包名

9.4.5 签名

微信登录分享支付都有一个签名验证，显得很麻烦，因为每次调试都需要重新签名。

首先使用 Android Studio 生成一个正式的签名文件，签名文件是以.jks 结尾的，这个签名文件是以后打包上线一直要用到的，然后用这个签名文件生成 APK（签名的整个操作步骤在前面的章节有学习）。这时 App 就有了正式签名，将正式签名的 APK 发送到手机上进行安装。

接下来下载一个 APK 文件，这是一个通过包名获取签名 MD5 的工具 APP，这个工具是微信官方开发的，官网下载地址如下：

```
https://res.wx.qq.com/open/zh_CN/htmledition/res/dev/download/sdk/Gen_Signature_Android2.apk
```

以上两个 App 都安装好之后，打开从微信下载的 App，软件名为“Gen Signature”，其中包含一个输入框，输入软件包名，点击“Get Signature”按钮，效果如图 9-7 所示。

图 9-7　获取签名

将图中那行绿色的十六进制数抄下来并保存到 txt 文本中。

9.4.6　微信开放平台官网的后台配置

在微信开放平台官网（https://open.weixin.qq.com/）的首页点击“管理中心”后，默认就是“移动应用”页面。如果还没有创建移动应用，就先进行创建；如果已有移动应用，就点击当前应用右侧的“查看”按钮，进入应用的详细页面。

在应用详细页面中一直向下滚动到最底部，看到“开发信息”。点击“修改”，进入“修改应用”页面，如图 9-8 所示。

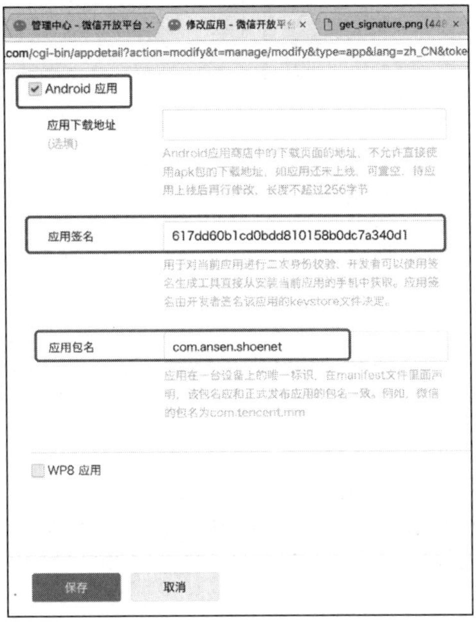

图 9-8　“修改应用”页面

首先选中“Android 应用”复选框，然后填写应用签名，这个签名就是之前让用户保存到记事本上的那个值，并输入应用包名。最后点击“保存”按钮。

9.4.7　运行软件

登录之后效果如图 9-9 所示。

点击“分享到朋友圈”按钮，其效果如图 9-10 所示。

点击“分享给好友”按钮，其效果如图 9-11 所示。

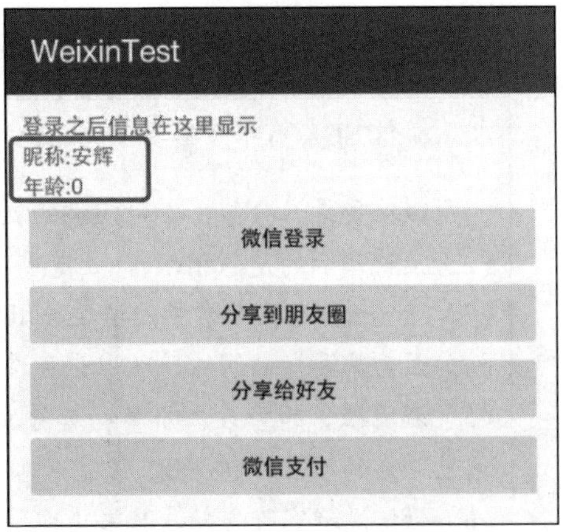

图 9-9　登录运行后的效果

图 9-10 分享到朋友圈后的效果

图 9-11 分享给好友的效果

微信支付没法测试，因为需要服务器支持。

9.4.8 微信官方开发文档

这篇文章只是针对现在微信的 SDK 版本接入，但是 SDK 是可能随时变化的，如版本变化、接口变化等。因此，建议大家还是以官方文档为主。这篇文章仅作为参考。

移动应用微信登录开发指南的网址如下：

```
    https://open.weixin.qq.com/cgi-bin/showdocument?action=dir_list&t=resource
/res_list&verify=1&id=open1419317851&token=219192a54f13e8e7011ced8e4ce5b36b699
629c4&lang=zh_CN
```

Android 微信支付开发手册的网址如下：

```
    https://open.weixin.qq.com/cgi-bin/showdocument?action=dir_list&t=resource
/res_list&verify=1&id=open1419317784&token=219192a54f13e8e7011ced8e4ce5b36b699
629c4&lang=zh_CN
```

注　意

接入微信 SDK 有很多需要注意的地方，这里进行总结。

（1）微信登录，分享，支付回调的 Activity 包名和类名一定要严格按照要求去写。

（2）接收回调的是 Activity，一定要在 AndroidManifest.xml 中注册。

（3）Constant 中两个常量的值要去微信申请并且创建应用才会有，这里需要改成申请的值。

（4）因为需要访问网络，所以记住在 AndroidManifest.xml 中添加权限。

（5）调用微信的登录、分享和支付时，安装包一定要有签名，签名信息一定要与在微信官网上配置的签名信息一致。

（6）微信没有客服支持，如果出现问题，请查看官方的 Demo 或者官方 API。

（7）微信 SDK 经常升级，开发时尽量用最新的版本。

当然，读者直接运行这里的 Demo 是不行的，因为没有 jks 文件，无法签名，并且源代码中的 appid 和 secret 也被修改过了，是不能使用的。如果只是看运行效果，笔者已经在项目下创建了一个 apk 文件夹，该文件夹下放置了一个可以调用微信登录分享的 apk 安装包。

9.5　百度地图

百度地图 Android SDK（官方网址为 http://lbsyun.baidu.com/index.php?title=androidsdk）是一套基于 Android 4.0 及以上版本设备的应用程序接口。用户可以使用该套 SDK 开发适用于 Android 系统移动设备的地图应用，通过调用地图 SDK 接口，可以轻松访问百度地图服务和数据，构建功能丰富、交互性强的地图类应用程序。

百度地图 Android SDK 提供的所有服务是免费的，接口使用无次数限制。用户申请密钥（Key）后才能使用百度地图 Android SDK。任何非营利性产品请直接使用，商业目的产品使用前请参考使用须知。

百度地图可以实现丰富的 LBS 功能：

- 地图：提供地图（2D、3D）的展示和缩放、平移、旋转、改变视角等地图操作。
- 室内图：提供展示公众建筑物室内地图的展示功能。
- Android Wear：适配 Android Wear，支持 Android 穿戴设备。
- POI 检索：可以根据关键字对 POI 数据进行周边、区域和城市内三种检索。
- 室内 POI 检索：支持设置城市和当前建筑物的室内 POI 检索。
- 地理编码：提供地理坐标和地址之间相互转换的能力。
- 线路规划：支持公交信息查询、公交换乘查询、驾车线路规划和步行路径检索。
- 覆盖物：提供多种地图覆盖物（自定义标注、几何图形、文字绘制、地形图图层、热力图图层等），满足开发者的各种需求。
- 定位：采用多种定位模式，使用定位 SDK 获取位置信息，使用地图 SDK 的“我的位置”图层进行位置展示。
- 离线地图：支持使用离线地图，节省用户流量，同时为用户带来更好的地图体验。
- 调启百度地图：利用 SDK 接口，直接在本地打开百度地图客户端或 WebApp，实现地图功能。
- 周边雷达：利用周边雷达功能，开发者可在 App 内低成本、快速实现查找周边使用相同 App 的用户位置的功能。
- LBS 云检索：支持用户检索存储在 LBS 云内的自有 POI 数据，并展示。
- 瓦片图层：支持开发者在地图上添加自有瓦片数据。
- 特色功能：提供短串分享、Place 详情检索、热力图等特色功能，帮助开发者搭建功能更加强大的应用。

9.5.1 百度定位 SDK

百度定位 SDK 是为 Android 移动端应用提供的一套简单易用的定位服务接口，专注于为广大开发者提供最好的综合定位服务。通过使用百度定位 SDK，开发者可以轻松为应用程序实现智能、精准、高效的定位功能。

1. 坐标系说明

目前国内主要有以下三种坐标系：

- WGS84：一种大地坐标系，也是目前广泛使用的 GPS（全球卫星定位系统）使用的坐标系。
- GCJ02：经过国测局加密的坐标。
- BD09：百度坐标系，其中 bd09ll 表示百度经纬度坐标，bd09mc 表示百度墨卡托米制坐标。

2. 使用百度定位 SDK 有什么产品优势

百度定位 SDK 提供 GPS、基站、WiFi、地磁、蓝牙、传感器等多种定位方式，适用于室内、室外多种定位场景，具有出色的定位性能：定位精度高、覆盖率广、网络定位请求流量小、定位速度快。

很多内容已经封装好了，不需要去考虑什么情况使用 GPS 定位、什么时候使用 WiFi 定位、如何根据基站信息从联通或者移动的 API 那里获取位置信息。

3. 百度定位 SDK 有四种开发包

百度定位 SDK 自 7.0 版本起，按照附加功能不同，向开发者提供了四种不同类型的定位开发包，可以根据不同需求，自行选择所需类型的开发包。注意，这四种开发包相互排斥，一个应用中只需集成一种定位开发包即可。

- 基础定位：开发包体积最小，但只包含基础定位能力（GPS/WiFi/基站）、基础位置描述能力。
- 离线定位：在基础定位能力基础之上，提供离线定位能力，可以在网络环境不佳时进行精准定位。
- 室内定位：在基础定位能力基础之上，提供室内高精度定位能力，精度可达 1~3 米。
- 全量定位：包含离线定位、室内高精度定位能力，同时提供更人性化的位置描述服务。

大家可以根据不同需求进行下载，一般情况下只需要基础定位即可。这个 Demo 中的就是基础定位 SDK。

```
http://lbsyun.baidu.com/sdk/download?selected=location_all
```

4. 申请密钥

无论使用定位 SDK 还是地图 SDK，都必须要申请密钥（Key）。

申请密钥之前，需要有一个百度开发者账号，如果没有就需要注册一个。在百度地图 SDK 官网首页右上角就能看到“注册”按钮，注册流程和其他网站类似。

登录注册后的账号，然后用浏览器打开申请密钥的地址：

http://lbsyun.baidu.com/apiconsole/key

进入申请密钥的首页，点击“创建应用”按钮，如图 9-12 所示。

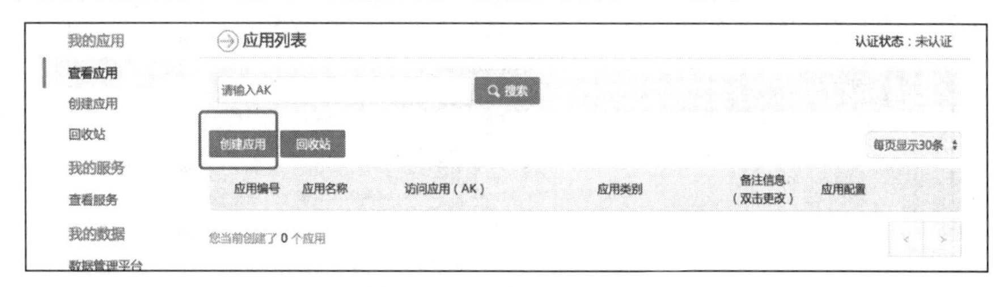

图 9-12　点击“创建应用”按钮

进入“创建应用”页面，如图 9-13 所示。

图 9-13　“创建应用”页面

在“创建应用”页面中，有四个地方需要填写值。

（1）在“应用名称”框中填写创建项目时的应用名称。

（2）在“应用类型”框中选择“Android SDK”。可以在下方的复选框中选择启用服务，这里为了测试，选择所有功能。在实际开发时，根据自己的需求进行选择。

（3）在“发布版 SHA1”框中输入证书指纹。该指纹可以通过如下方法获取：在项目的根目录下新建 jks 文件夹，用 Android Studio 创建一个.jks 签名文件，将签名文件放在 jks 文件夹下。签名文件的密码保存在密码.txt 中，也放在 jks 文件夹下。接下来，在 Android Studio 中打开终端，输入如下命令：

```
keytool -v -list -keystore jks/baidulocationtest.jks
```

就会看到证书指纹，效果如图 9-14 所示。

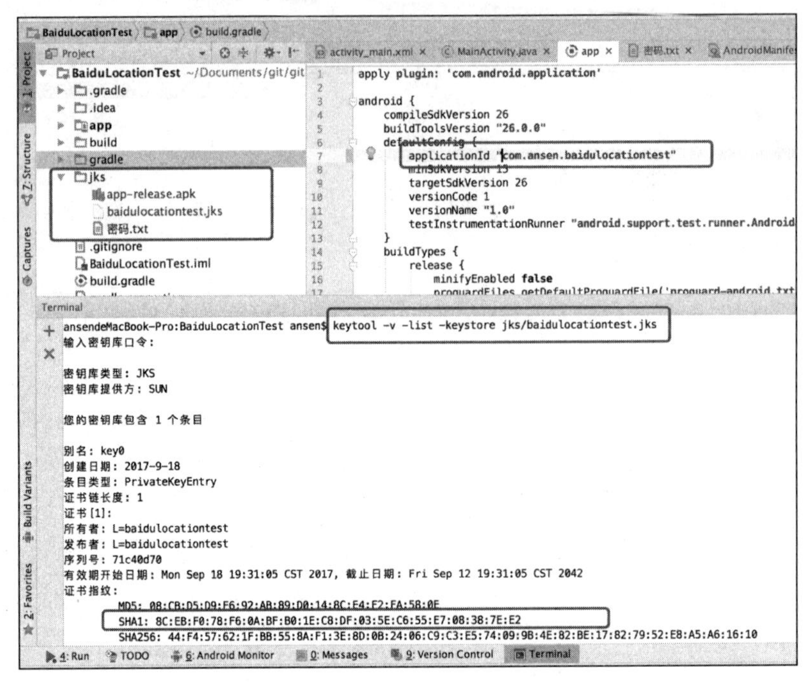

图 9-14　查看证书指纹

最后将“SHA1”后面的值复制到“SHA1”输入框中。

（4）在“包名”框中输入相应的包名，也就是 BaiduLocationTest/app/build.gradle 文件中 applicationId 对应的值。

关于这四个值的填写方法，用户可以参考官方文件：

http://lbsyun.baidu.com/index.php?title=androidsdk/guide/key

填好四个值之后，点击“提交”按钮。提交成功后会跳转到“应用列表”首页，这时就能看到刚刚创建的应用了，同时也能看到“访问应用”，如图 9-15 所示。

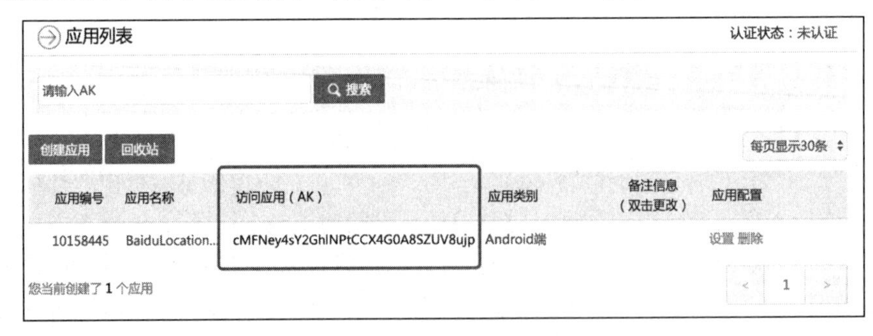

图 9-15　查看创建的应用

5. 源代码实现

新建项目，在 app/src/main 下新建 jniLibs 文件夹，将定位相关的 so 库复制过来，同时把百度定位相关的 jar 包复制到 libs 文件夹下，如图 9-16 所示。

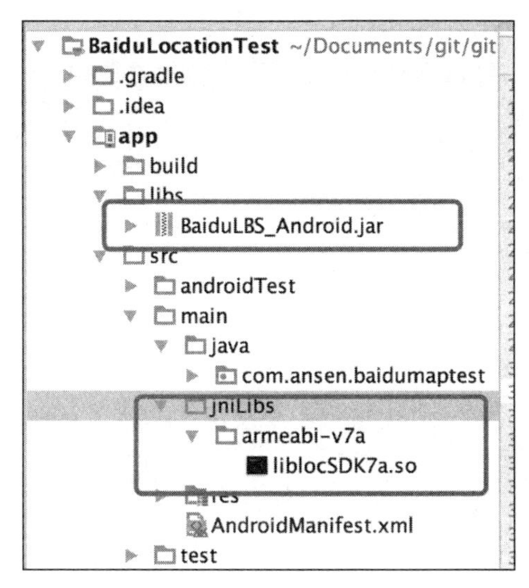

图 9-16　复制相应的文件

接下来，需要在 AndroidManifest.xml 文件中加入定位权限和访问网络的权限，以及定位的服务器，还需要一个 meta-data 标签，其中 name 是固定的。百度 SDK 会根据这个 name 来查找设置的值，这个值就是上一步申请的密钥。

```
<?xml version="1.0" encoding="utf-8"?>
<manifest xmlns:android="http://schemas.android.com/apk/res/android"
    package="com.ansen.baidumaptest">

    <uses-permission android:name="android.permission.INTERNET"/>
    <uses-permission android:name="android.permission.ACCESS_FINE_LOCATION"/>

    <application
        ...>
        ...

        <!--百度定位 SDK 需要服务-->
        <service
            android:enabled="true"
            android:process=":remote" >
            <intent-filter>
                <action android:name="com.baidu.location.service_v2.2" >
                </action>
            </intent-filter>
        </service>

        <!--从百度地图 SDK 官网申请的密钥-->
        <meta-data
            android:name="com.baidu.lbsapi.API_KEY"
            android:value="cMFNey4sY2GhlNPtCCX4G0A8SZUV8ujp"/>
    </application>
</manifest>
activity_main.xml
<?xml version="1.0" encoding="utf-8"?>
<LinearLayout xmlns:android="http://schemas.android.com/apk/res/android"
    android:layout_width="match_parent"
```

```
    android:layout_height="match_parent"
    android:padding="10dp">
    <TextView
        android:id="@+id/tv_location_result"
        android:layout_width="wrap_content"
        android:layout_height="wrap_content"
        android:text="Hello World!" />
</LinearLayout>
```

布局文件中仅有一个 TextView，用来显示定位的结果。

布局文件对应的 MainActivity.java 内容如下：

```
public class MainActivity extends AppCompatActivity {
    private LocationClient client = null;
    private TextView tvLocationResult;

    @Override
    protected void onCreate(Bundle savedInstanceState) {
        super.onCreate(savedInstanceState);
        setContentView(R.layout.activity_main);

        tvLocationResult= (TextView) findViewById(R.id.tv_location_result);

        LocationClientOption mOption=new LocationClientOption();
        //定位模式设置成高精度模式。除了高精度模式之外，还有 Battery_Saving(低功耗模式)、
        //Device_ Sensors(仅设备 Gps 模式)。
        mOption.setLocationMode(LocationClientOption.LocationMode.Hight_Accuracy);
        mOption.setCoorType("bd09ll");
         //可选，默认为 gcj02，设置返回的定位结果坐标系，如果配合百度地图使用，建议设置为
         //bd09ll mOption.setScanSpan(3000);
        //可选，默认 0，即仅定位一次，设置发起定位请求的间隔需要大于等于 1000ms 才是有效的
        mOption.setIsNeedAddress(true);//可选，设置是否需要地址信息，默认不需要
        mOption.setIsNeedLocationDescribe(true);//可选，设置是否需要地址描述
        mOption.setNeedDeviceDirect(false);//可选，设置是否需要设备方向结果
        mOption.setLocationNotify(false);
        //可选，默认为 false，设置是否当 gps 有效时按照 1S1 次频率输出 GPS 结果
        mOption.setIgnoreKillProcess(true);
        //可选，默认为 true，定位 SDK 内部是一个 SERVICE，并放到了独立进程，
        //设置是否在停止时杀死这个进程，默认不杀死
        mOption.setIsNeedLocationDescribe(true);
        //可选,默认为 false,设置是否需要位置详细信息,可以在 BDLocation.getLocationDescribe
        //中得到，结果类似于"在北京天安门附近"
        mOption.setIsNeedLocationPoiList(true);
        //可选，默认为 false，设置是否需要 POI 结果，可以在 BDLocation.getPoiList
        //中得到 mOption.SetIgnoreCacheException(false);
        //可选，默认为 false，设置是否收集 CRASH 信息，默认为收集
          mOption.setIsNeedAltitude(false);
        //可选，设置定位时是否需要海拔信息，默认为 false，即不需要，除基础定位版本外都可用

        client = new LocationClient(this);
        client.setLocOption(mOption);//设置定位参数
        client.registerLocationListener(locationListener);//注册监听

        client.start();
    }
```

```
        private BDLocationListener locationListener=new BDLocationListener() {
            @Override
            public void onReceiveLocation(BDLocation location) {
                if (null != location && location.getLocType() != BDLocation.
TypeServerError) {
                    StringBuffer sb = new StringBuffer(256);
                    sb.append("time : ");
                    /**
                     * 也可以使用systemClock.elapsedRealtime()方法,获取手机开机到现在的毫秒数,
                       手机睡眠（sleep）的时间也包括在内
                     * location.getTime()是指定位成功后的时间，如果位置不发生变化，则时间不变

                     */
                    sb.append(location.getTime());
                    sb.append("\nlocType : ");// 定位类型
                    sb.append(location.getLocType());
                    sb.append("\nlocType description : ");// *****对应的定位类型说明*****
                    sb.append(location.getLocTypeDescription());
                    sb.append("\nlatitude : ");// 纬度
                    sb.append(location.getLatitude());
                    sb.append("\nlontitude : ");// 经度
                    sb.append(location.getLongitude());
                    sb.append("\nradius : ");// 半径
                    sb.append(location.getRadius());
                    sb.append("\nCountryCode : ");
                    sb.append(location.getCountryCode());// 国家/地区代码
                    sb.append("\nCountry : ");
                    sb.append(location.getCountry());//国家/地区名称
                    sb.append("\ncitycode : ");// 城市编码
                    sb.append(location.getCityCode());
                    sb.append("\ncity : ");// 城市
                    sb.append(location.getCity());
                    sb.append("\nDistrict : ");// 区
                    sb.append(location.getDistrict());
                    sb.append("\nStreet : ");// 街道
                    sb.append(location.getStreet());
                    sb.append("\naddr : ");// 地址信息
                    sb.append(location.getAddrStr());
                    sb.append("\nUserIndoorState: ");// *****返回用户室内外判断结果*****
                    sb.append(location.getUserIndoorState());
                    sb.append("\nDirection(not all devices have value): ");
                    sb.append(location.getDirection());// 方向
                    sb.append("\nlocationdescribe: ");
                    sb.append(location.getLocationDescribe());//位置描述
                    sb.append("\nPoi: ");// POI 信息
                    if (location.getPoiList() != null && !location.getPoiList().isEmpty()){
                        for (int i = 0; i < location.getPoiList().size(); i++) {
                            Poi poi = (Poi) location.getPoiList().get(i);
                            sb.append(poi.getName() + ";");
                        }
                    }
                    if (location.getLocType() == BDLocation.TypeGpsLocation) {
                      // GPS 定位结果
                        sb.append("\nspeed : ");
```

```
                sb.append(location.getSpeed());// 速度，单位：km/h
                sb.append("\nsatellite : ");
                sb.append(location.getSatelliteNumber());// 卫星数目
                sb.append("\nheight : ");
                sb.append(location.getAltitude());// 海拔高度，单位：m
                sb.append("\ngps status : ");
                sb.append(location.getGpsAccuracyStatus());
                 // *****gps 质量判断*****
                sb.append("\ndescribe : ");
                sb.append("gps 定位成功");
            } else if (location.getLocType() ==
               BDLocation.TypeNetWorkLocation)
            {// 网络定位结果
                // 运营商信息
                if (location.hasAltitude()) {// *****如果有海拔高度*****
                    sb.append("\nheight : ");
                    sb.append(location.getAltitude());// 单位：m
                }
                sb.append("\noperationers : ");// 运营商信息
                sb.append(location.getOperators());
                sb.append("\ndescribe : ");
                sb.append("网络定位成功");
            } else if (location.getLocType() ==
              BDLocation.TypeOffLineLocation)
            {// 离线定位结果
                sb.append("\ndescribe : ");
                sb.append("离线定位成功，离线定位结果也是有效的");
            } else if (location.getLocType() == BDLocation.TypeServerError) {
                sb.append("\ndescribe : ");
                sb.append("服务端网络定位失败，可以反馈 IMEI 号和大体定位时间到 loc-
                          bugs@baidu.com，会有人追查原因");
            } else if (location.getLocType() ==
               BDLocation.TypeNetWorkException){
                sb.append("\ndescribe : ");
                sb.append("网络不同导致定位失败，请检查网络是否通畅");
            } else if
             (location.getLocType()==BDLocation.TypeCriteriaException){
                sb.append("\ndescribe : ");
                sb.append("无法获取有效定位依据导致定位失败，一般是手机的原因，处于飞
                            行模式下会造成这种结果，可以试着重启手机");
            }
            tvLocationResult.setText(sb.toString());
        }
    }
};

@Override
protected void onDestroy() {
    super.onDestroy();

    client.stop();//停止定位
}
}
```

- onCreate：查找 TextView，新建一个 LocationClientOption 对象，通过它来设置定位过滤器，

然后新建一个 LocationClient 对象，设置定位参数，注册监听，调用 start 方法开启定位。

- locationListener：定位回调接口，实现 onReceiveLocation 接口，从参数 BDLocation 对象中取出地址。
- onDestroy：Activity 销毁定位并停止。

运行软件，效果如图 9-17 所示。

图 9-17　运行效果

9.5.2　百度地图 SDK

百度地图 SDK 是一套基于 Android 2.3 及以上版本设备的应用程序接口。可以使用该套 SDK 开发适用于 Android 系统移动设备的地图应用，通过调用地图 SDK 接口，可以轻松访问百度地图服务和数据，构建功能丰富、交互性强的地图类应用程序。

1. 获取定制的百度地图 SDK

开发者可以在百度地图 SDK 的下载页面中下载到新版的地图 SDK，下载地址为：

http://developer.baidu.com/map/index.php?title=androidsdk/sdkandev-download

为了给开发者带来更优质的地图服务、满足开发者灵活使用 SDK 的需求，百度地图 SDK 自 v2.3.0 起采用可定制的形式，为用户提供开发包。百度地图 SDK 按功能可分为基础地图、检索功能、LBS 云检索、计算工具和周边雷达五个部分。开发者可以根据自身的实际需求，任意组合这五种功能，点击下载页面的“自定义下载”，即可下载相应的开发包来完成自己的应用开发。

- 基础地图：包括基本矢量地图、卫星图、实时路况图、室内图、适配 Android 穿戴设备，以及各种地图覆盖物、瓦片图层、OpenGL 绘制能力。此外，还包括各种与地图相关的操作和事件监听。
- 检索功能：包括 POI 检索（周边、区域、城市内），室内 POI 检索，Place 详情检索，公交信息查询，路线规划（驾车、步行、公交），地理编码/反地理编码，在线建议查询，短串分享等。
- LBS 云检索：包括 LBS 云检索（周边、区域、城市内、详情）。

- 计算工具：包括计算两点之间距离、计算矩形面积、坐标转换、调启百度地图客户端、判断点和圆/多边形位置关系、本地收藏夹等功能。
- 周边雷达：包含位置信息上传和检索周边相同应用的用户位置信息功能。

这里仅下载基础地图 SDK，在实际开发中，大家可以根据需求下载多功能 SDK。

2. 申请密钥

申请密钥的流程和上一节百度定位时申请密钥是一样的，这里就不重复介绍了。

3. 源代码实现

下面使用基础地图+定位来实现当前在地图上显示的位置。

新建项目，首先需要加入百度的 so 文件和 jar 包，so 文件在定位的基础上又增加了两个文件，而 jar 包数量还是一个没有变化，如图 9-18 所示。

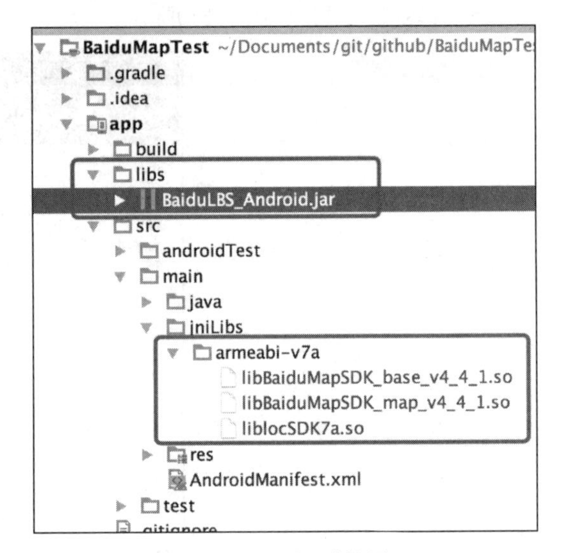

图 9-18　项目结构图

从项目结构图中看到，jar 的名字和数量与百度定位没有什么变化，但是它的内容大小是有变化的。

如果仔细对比百度定位和百度地图两个项目引用的jar包大小，就会发现定位的jar包是184KB，而有定位+地图功能的 jar 包是 3.1MB，如图 9-19 所示。

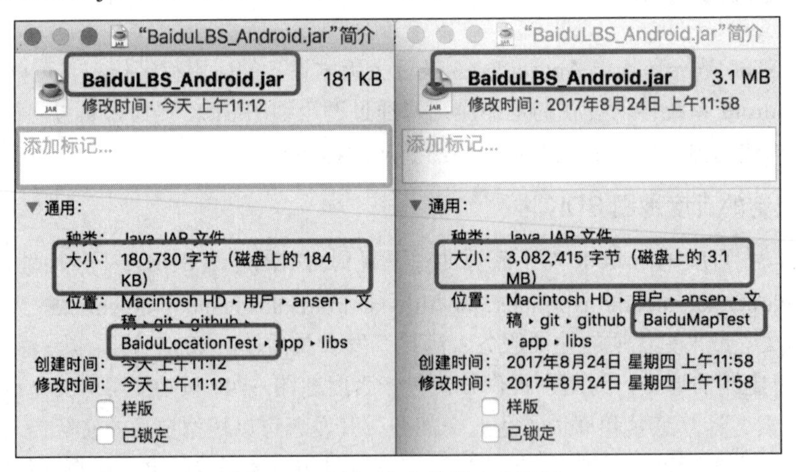

图 9-19　两个 jar 文件的大小比较

所以，当从百度地图下载 SDK 时，根据功能选择相应的 SDK，这样下载的 SDK 的 jar 包大小就能控制了，而不是选择了很多功能，然后下载一个 5MB 以上的 jar 包，结果仅使用了一个定位功能。这也是为什么写两个 Demo 的原因，读者拿到源代码复制到自己项目中时，也要选择合适的 jar 包。

引入 jar 包和 so 文件后，接下来在 AndroidManifest.xml 中加入权限，配置百度的 API Key，加入百度定位需要的服务。地图用到的权限比较多。

```
<?xml version="1.0" encoding="utf-8"?>
<manifest xmlns:android="http://schemas.android.com/apk/res/android"
    package="com.ansen.baidumaptest">

    <uses-permission android:name="com.android.launcher.permission.READ_SETTINGS" />
    <!-- 这个权限用于进行网络定位 -->
    <uses-permission android:name="android.permission.ACCESS_COARSE_LOCATION" />
    <!-- 这个权限用于访问 GPS 定位 -->
    <uses-permission android:name="android.permission.ACCESS_FINE_LOCATION" />
    <!-- 用于访问 WiFi 网络信息，WiFi 信息会用于进行网络定位 -->
    <uses-permission android:name="android.permission.ACCESS_WIFI_STATE" />
    <!-- 获取运营商信息，用于支持提供运营商信息相关的接口 -->
    <uses-permission android:name="android.permission.ACCESS_NETWORK_STATE" />
    <!-- 用于读取手机当前的状态 -->
    <uses-permission android:name="android.permission.READ_PHONE_STATE" />
    <!-- 写入扩展存储，向扩展卡写入数据，用于写入离线定位数据 -->
    <uses-permission android:name="android.permission.WRITE_EXTERNAL_STORAGE" />
    <!-- 访问网络，网络定位需要上网 -->
    <uses-permission android:name="android.permission.INTERNET" />

    <application
        ......>
        ......
        <meta-data
            android:name="com.baidu.lbsapi.API_KEY"
            android:value="HLLM33ENPr2k1Ei41Nupx5Pew0kGvLjx"/>

        <!--百度定位 SDK 需要服务-->
        <service
            android:name="com.baidu.location.f"
            android:enabled="true"
            android:process=":remote" >
            <intent-filter>
                <action android:name="com.baidu.location.service_v2.2" >
                </action>
            </intent-filter>
        </service>
    </application>
</manifest>
```

首先修改 activity_main.xml 布局文件，内容如下：

```
<?xml version="1.0" encoding="utf-8"?>
<FrameLayout xmlns:android="http://schemas.android.com/apk/res/android"
    android:layout_width="match_parent"
    android:layout_height="match_parent">

    <com.baidu.mapapi.map.MapView
        android:id="@+id/mapview"
        android:layout_width="match_parent"
        android:layout_height="match_parent"/>
</FrameLayout>
```

布局文件也比较简单，只有一个自定义的 MapView。这个控件用来显示地图，是百度 SDK 中的自定义控件。

布局文件对应的 MainActivity.java 内容如下：

```
public class MainActivity extends AppCompatActivity {
    private  MapView mapView;
    private BaiduMap baiduMap;//定义地图对象的操作方法与接口
    private boolean isFirstLoc = true;//是否首次定位

    private LocationClient locationClient = null;//定位控制类

    @Override
    protected void onCreate(Bundle savedInstanceState) {
        super.onCreate(savedInstanceState);

        //在使用 SDK 各组件之前初始化 context 信息，传入 ApplicationContext
        //注意该方法要在 setContentView 方法之前实现
        //在真实开发中，这句代码放到自定义的 Application 中比较合适
        SDKInitializer.initialize(getApplicationContext());

        setContentView(R.layout.activity_main);

        mapView= (MapView) findViewById(R.id.mapview);
        baiduMap = mapView.getMap();

        baiduMap.setMyLocationEnabled(true);//开启定位图层

        //用户自定义定位图标
        BitmapDescriptor mCurrentMarker =
        BitmapDescriptorFactory.fromResource (R.mipmap.icon_geo);

        //参数 1：有三个值
        // ·LocationMode.COMPASS 为罗盘形态，显示定位方向圈，保持定位图标在地图中心
        // ·LocationMode.FOLLOWING 为跟随形态，保持定位图标在地图中心
        // ·LocationMode.NORMAL 为普通形态，更新定位数据时不对地图做任何操作
        //参数 2：是否允许显示方向信息
        //参数 3：用户自定义定位图标
        MyLocationConfiguration config = new MyLocationConfiguration
(LocationMode.NORMAL,true,mCurrentMarker);
        baiduMap.setMyLocationConfiguration(config);//设置定位图层显示方式

        startRequestLocation();//开启定位
    }

    private void startRequestLocation() {
        LocationClientOption mOption = new LocationClientOption();
        //定位模式设置成高精度模式。除了高精度模式之外，还有 Battery_Saving(低功耗模式)、
        //Device_Sensors(仅设备 Gps 模式)。
    mOption.setLocationMode(LocationClientOption.LocationMode.Hight_Accuracy);
    mOption.setCoorType("bd09ll");//可选，默认为 gcj02，设置返回的定位结果坐标系，
                                  // 如果配合百度地图使用，建议设置为 bd09ll;
        mOption.setScanSpan(3000);//可选，默认为 0，即仅定位一次，设置发起定位请求的
                                  //间隔需要大于等于 1000ms 才是有效的
        locationClient = new LocationClient(this);
```

```
        locationClient.setLocOption(mOption);//设置定位参数
        locationClient.registerLocationListener(locationListener);//注册监听

        locationClient.start();//开启定位，目前仅支持在主线程中启动
    }

    private BDLocationListener locationListener=new BDLocationListener() {
        @Override
        public void onReceiveLocation(BDLocation location) {
            if (null != location && location.getLocType() != BDLocation.
TypeServerError){
                Log.i("ansen","getLongitude:"+location.getLongitude());

                MyLocationData locData = new MyLocationData.Builder()
                        .accuracy(location.getRadius())//设置定位数据的精度信息
                        .latitude(location.getLatitude())//
                        .longitude(location.getLongitude()).build();

                //设置定位数据，只有先允许定位图层后设置数据才会生效
                baiduMap.setMyLocationData(locData);

                if (isFirstLoc) {//第一次定位
                    isFirstLoc = false;
                    LatLng ll = new LatLng(location.getLatitude(),location.
                     getLongitude());
                    MapStatus.Builder builder = new MapStatus.Builder();
                    builder.target(ll).zoom(18.0f);//设置地图缩放级别
                    //以动画方式更新地图状态，动画耗时 300 ms
                    baiduMap.animateMapStatus(MapStatusUpdateFactory.
                    newMapStatus (builder.build()));
                }
            }
        }
    };

    @Override
    protected void onPause() {
        super.onPause();
        //在 Activity 执行 onPause 时执行 mMapView.onPause()，实现地图生命周期管理
        mapView.onPause();
    }

    @Override
    protected void onResume(){
        super.onResume();
        //在 Activity 执行 onResume 时执行 mMapView.onResume()，实现地图生命周期管理
        mapView.onResume();
    }

    @Override
    protected void onDestroy() {
        super.onDestroy();

        locationClient.stop();//停止定位
        baiduMap.setMyLocationEnabled(false);//关闭定位图层
```

```
        //在 Activity 执行 onDestroy 时执行 mapView.onDestroy()，实现地图生命周期管理
        mapView.onDestroy();
        mapView=null;
    }
}
```

- onCreate：首先初始化百度地图相关 SDK，设置首页布局文件，然后查找地图控件并开启定位图层，设置定位图层的显示方式，最后调用 startRequestLocation 方法开启定位。
- startRequestLocation：如果看过前面的内容，开启定位的代码大家应该很熟悉了，这里就不赘述了。
- locationListener：定位回调接口，在 onReceiveLocation 中处理回调 BDLocation 对象，将经纬度封装到 MyLocationData 对象中，调用 BaiduMap 的 setMyLocationData 方法更新位置数据，判断是不是第一次更新。如果是第一次，就以动画方式更新地图状态。
- onPause：实现地图生命周期管理。
- onResume：实现地图生命周期管理。
- onDestroy：停止定位，关闭定位图层，实现地图生命周期管理。

4. 签名配置

百度地图 SDK 和微信 SDK 一样，都需要签名后的 APK 才能使用。不过，如果每次直接运行 App 就会产生一个随机签名，将无法进行测试，只能在 Android Studio 中选择“Build”→“Generate Signed APK”命令来生成一个签名的 APK 文件。这样调试起来很麻烦，效率很低。

其实 gradle 支持配置签名文件，因为我们一般直接运行的是 debug 模式，所以制定 debug 模式的签名文件即可。修改后的 app/build.gradle 文件效果如图 9-20 所示。

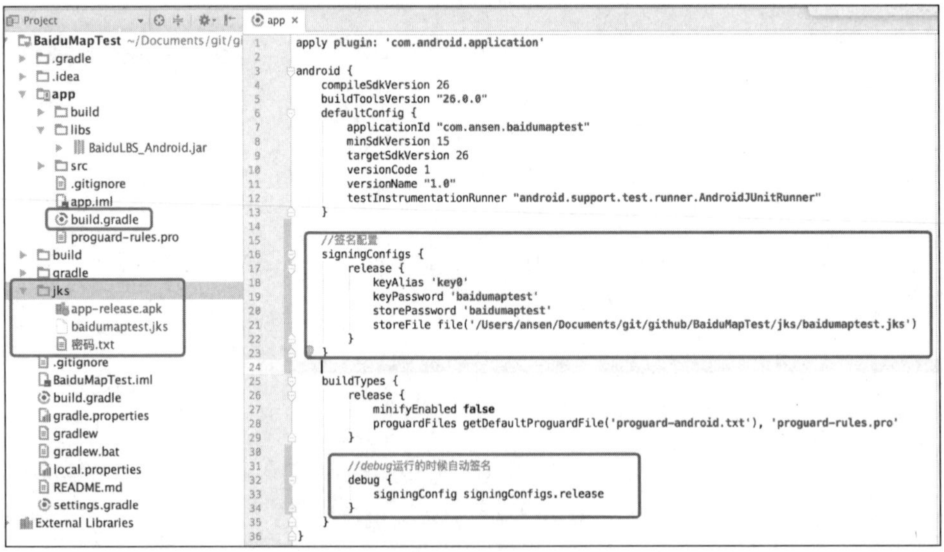

图 9-20 修改 bulid.gradle 文件

其中增加了 signingConfigs 属性，然后指定了 release 签名需要的信息，如签名文件地址、签名文件密码等，还需要在 buildTypes 下的 debug 模式中引用 signingConfigs。

现在就能够直接运行项目了，效果如图 9-21 所示。

10:05PM
BaiduMapTest
莘南花苑一街坊
高兴路
江南名邸
沪金高速
莘城公寓
入口
新城南区
宝城路
春申路
加油站
50米
春申路

图 9-21　百度地图

第 10 章

实现开发者头条

对于很多 Android 初学者而言，我们的项目经验还停留在演示的阶段，有没有一种失落的感觉？当用户真正接手大项目时，却不知道如何灵活使用自己已学的知识。本章将用一个实例引导用户如何开发项目。

开发者头条这个 App 不错，其中用到了很多 Android 中常用的技术。因此，本章以模仿开发者头条的功能，将之前学到的知识串联起来，顺便复习前面所学的知识点。

10.1 启动页实现

开始今天的正题，带领用户实现开发者头条 App 的启动页。

10.1.1 启动页的目标效果

如图 10-1 所示，从效果图中可以看出（这里列出第一个和第四个页面的效果图），使用四个页面来显示四张图片，可以左右滑动，整个滑动的界面就是使用 ViewPager 来实现的。接下来，监听 ViewPager 的滑动事件，改变页面底部四个横线小图标的切换，以及点击“开启我的头条”按钮的隐藏与显示。

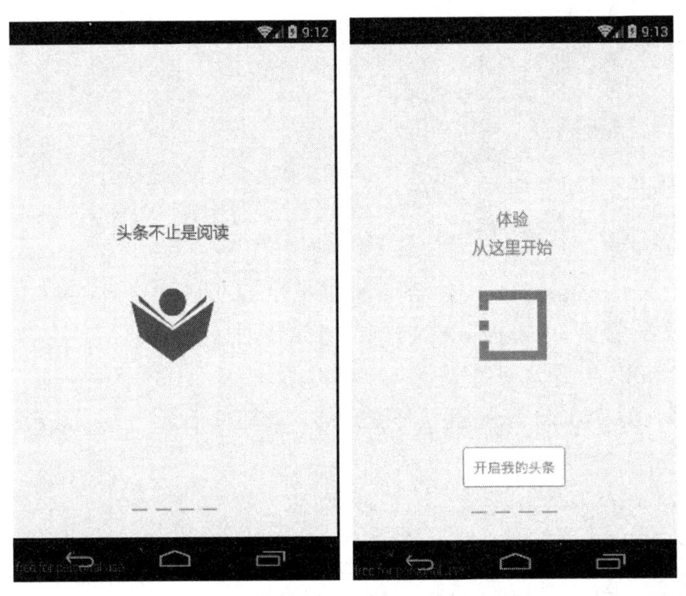

图 10-1　启动页的效果图

10.1.2　代码实现

新建项目，在 res 文件夹下新建 activity_luancher.xml 文件，内容如下：

```
<RelativeLayout xmlns:android="http://schemas.android.com/apk/res/android"
   android:layout_width="match_parent"
   android:layout_height="match_parent"
   android:background="#FCF2E4">

   <android.support.v4.view.ViewPager
      android:id="@+id/viewpager_launcher"
      android:layout_width="match_parent"
      android:layout_height="match_parent"/>

   <TextView
      android:id="@+id/tv_start_headlines"
      android:layout_alignParentBottom="true"
      android:layout_centerHorizontal="true"
      android:layout_height="wrap_content"
      android:layout_width="wrap_content"
      android:layout_marginBottom="60dp"
      android:paddingTop="12dp"
      android:paddingBottom="12dp"
      android:paddingLeft="12dp"
      android:paddingRight="12dp"
      android:background="@drawable/start_headlines_bg"
      android:textColor="@color/launcher_item_select"
      android:textSize="16sp"
      android:visibility="gone"
      android:text="开启我的头条"/>

   <LinearLayout
```

```
        android:id="@+id/viewGroup"
        android:layout_width="fill_parent"
        android:layout_height="wrap_content"
        android:layout_alignParentBottom="true"
        android:layout_marginBottom="30dp"
        android:gravity="center_horizontal"
        android:orientation="horizontal"/>
</RelativeLayout>
```

该文件的外层是 RelativeLayout，从上到下分别是 ViewPager、TextView、LinearLayout。其中，ViewPager 用来显示滑动图片；TextView 只有滑动到最后一页才会显示，点击之后将跳转到首页；LinearLayout 用来显示滑动状态。

“开启我的头条”这个 TextView 默认是隐藏的，该控件还有一个背景 drawable，设置了背景色、弧度以及边框。

```
<?xml version="1.0" encoding="utf-8"?>
<shape xmlns:android="http://schemas.android.com/apk/res/android" >
    <!-- 默认背景色 -->
    <solid android:color="@color/white_normal"/>

    <!–设置边框 -->
    <stroke
        android:width="1dp"
        android:color="@color/launcher_item_select" />

    <!-- 设置弧度 -->
    <corners android:radius="3dp"/>
</shape>
```

activity_luancher.xml 布局文件对应的 LauncherActivity.java 内容如下：

```
public class LauncherActivity extends FragmentActivity{
    private ViewPager viewPager;
    private LauncherPagerAdapter adapter;

    private ImageView[] tips;
    private TextView tvStartHeadlines;

    @Override
    protected void onCreate(Bundle savedInstanceState) {
        super.onCreate(savedInstanceState);

        setContentView(R.layout.activity_luancher);

        if(!isFirst()){//不是第一次启动，则直接进入首页
            gotoMain();
        }

        tvStartHeadlines= (TextView) findViewById(R.id.tv_start_headlines);
        tvStartHeadlines.setOnClickListener(onClickListener);

        viewPager = (ViewPager) findViewById(R.id.viewpager_launcher);
        viewPager.setOffscreenPageLimit(2);//设置缓存页数
        viewPager.setAdapter(adapter = new LauncherPagerAdapter(this));//设置适配器
```

```
        viewPager.setOnPageChangeListener(changeListener);//页面监听

        ViewGroup group=(ViewGroup)findViewById(R.id.viewGroup);
            //初始化底部显示控件
        tips = new ImageView[4];
        for (int i = 0; i < tips.length; i++) {
            ImageView imageView = new ImageView(this);
            if (i == 0) {
                imageView.setBackgroundResource(R.drawable.page_indicator_
                  focused);
            } else {
                imageView.setBackgroundResource(R.drawable.page_indicator_
                  unfocused);
            }
            tips[i] = imageView;
            LinearLayout.LayoutParams layoutParams = new
LinearLayout.LayoutParams (new ViewGroup.LayoutParams(LayoutParams.WRAP_CONTENT,
LayoutParams.WRAP_CONTENT));
            layoutParams.leftMargin = 10;// 设置小横线View的左边距
            layoutParams.rightMargin = 10;// 设置小横线View的右边距
            group.addView(imageView, layoutParams);
        }
    }

    private OnPageChangeListener changeListener = new OnPageChangeListener() {
        @Override
        public void onPageScrollStateChanged(int arg0) {}
        @Override
        public void onPageScrolled(int arg0, float arg1, int arg2) {}
        @Override
        public void onPageSelected(int index) {
            setImageBackground(index);// 改变小横线的切换效果

            if (index == tips.length - 1) {//最后一个
                tvStartHeadlines.setVisibility(View.VISIBLE);
            } else {
                tvStartHeadlines.setVisibility(View.INVISIBLE);
            }
        }
    };

    /**
     * 改变小横线的切换效果
     * @param selectItems
     */
    private void setImageBackground(int selectItems) {
        for (int i = 0; i < tips.length; i++) {
            if (i == selectItems) {
                tips[i].setBackgroundResource(R.drawable.page_indicator_focused);
            } else {
                tips[i].setBackgroundResource(R.drawable.page_indicator_unfocused);
            }
        }
    }
```

```
        private View.OnClickListener onClickListener=new View.OnClickListener() {
            @Override
            public void onClick(View view) {
                switch (view.getId()){
                    case R.id.tv_start_headlines:
                        gotoMain();//进入首页
                        break;
                }
            }
        };

        public void gotoMain() {
            setFirst();//设置为第二次启动

            Intent intent = new Intent(this, MainActivity.class);
            startActivity(intent);
            finish();
        }

        /**
         * 是否为第一次启动：true 表示第一次启动，false 表示第二次启动
         * @return
         */
        private boolean isFirst() {
            SharedPreferences setting = getSharedPreferences("headlines",
Context.MODE_PRIVATE);
            return setting.getBoolean("FIRST",true);
        }

        /**
         * 将第一次启动的值设置为 false
         */
        public void setFirst(){
            SharedPreferences setting = getSharedPreferences("headlines",
Context.MODE_PRIVATE);
            setting.edit().putBoolean("FIRST",false).commit();
        }
    }
```

- onCreate：首先判断是否为第一次启动，如果不是第一次启动，则直接进入首页，给点击跳转到首页的 TextView 设置点击事件，给 ViewPager 设置适配器，并且设置页面改动监听函数，初始化底部显示控件，就是给 LinearLayout 添加 4 个 ImageView，第一个 ImageView 用于设置选中状态的图片。
- changeListener：页面滑动会回调 onPageSelected 方法，首先改变小横线的选中状态，如果选中的是最后一页，则显示“开启我的头条”这个 TextView。
- setImageBackground：改变小横线的选中状态。
- onClickListener：点击“开启我的头条”，将进入首页。
- gotoMain：设置为第二次启动，然后跳转到首页。
- isFirst：是否为第一次启动。
- setFirst：将第一次启动的值设置为 false。

ViewPager 的适配器 LauncherPagerAdapter.java 代码如下：

```
public class LauncherPagerAdapter extends PagerAdapter{
    private List<View> views;

    //每页显示的图片
    private int[] images=new int[]{R.drawable.tutorial_1,R.drawable.tutorial_2,
            R.drawable.tutorial_3,R.drawable.tutorial_4};

    public LauncherPagerAdapter(Context context){
        views=new ArrayList<View>();

        //初始化每页显示的 View
        for(int i=0;i<images.length;i++){
            View item=LayoutInflater.from(context).inflate(R.layout.activity_
luancher_pager_item,null);
            ImageView imageview=(ImageView) item.findViewById(R.id.imageview);
            imageview.setImageResource(images[i]);
            views.add(item);
        }
    }

    @Override
    public int getCount() {
        return views == null ? 0 : views.size();
    }

    @Override
    public boolean isViewFromObject(View arg0, Object arg1) {
        return arg0==arg1;
    }

    @Override
    public void destroyItem(ViewGroup container, int position, Object object){
        container.removeView(views.get(position));
    }

    @Override
    public Object instantiateItem(ViewGroup container, int position){
        container.addView(views.get(position),0);
        return views.get(position);
    }
}
```

LauncherPagerAdapter 继承自 PagerAdapter。重写 PagerAdapter 以下四个方法，并且在 LauncherPagerAdapter 构造方法中初始化四个页面的 View。

- getCount：页面数量。
- isViewFromObject：固定写法。
- destroyItem：删除一页。
- instantiateItem：添加一页。

activity_launcher_pager_item.java 是每一页的布局文件，其内容很简单，就是在 FrameLayout

中显示图片视图（ImageView），图片视图居中显示。

```
<?xml version="1.0" encoding="utf-8"?>
<FrameLayout xmlns:android="http://schemas.android.com/apk/res/android"
    android:layout_width="match_parent"
    android:layout_height="match_parent"
    android:orientation="vertical">

    <ImageView
        android:id="@+id/imageview"
        android:layout_width="wrap_content"
        android:layout_height="wrap_content"
        android:layout_gravity="center"
        android:src="@drawable/tutorial_1" />
</FrameLayout>
```

10.2 使用 DrawerLayout 控件实现侧滑菜单栏

现在开始模仿开发者头条的侧滑菜单，是该项目的第二个知识点，相信大家已经看到很多 App 使用这种侧滑功能。下面使用 Android 自带 DrawerLayout 控件实现侧滑功能。

DrawerLayout 是 SupportLibrary 包中实现侧滑菜单效果的控件，可以说 DrawerLayout 是由于第三方控件（如 MenuDrawer 等）的出现之后，Google 借鉴而来的产物。DrawerLayout 分为侧边菜单和主内容区两部分，侧边菜单可以根据手势展开与隐藏（DrawerLayout 自身的特性），主内容区的内容可以随着菜单的点击而变化（这需要使用者自己实现）。

10.2.1 侧滑菜单的目标效果

本实例的效果如图 10-2 所示。

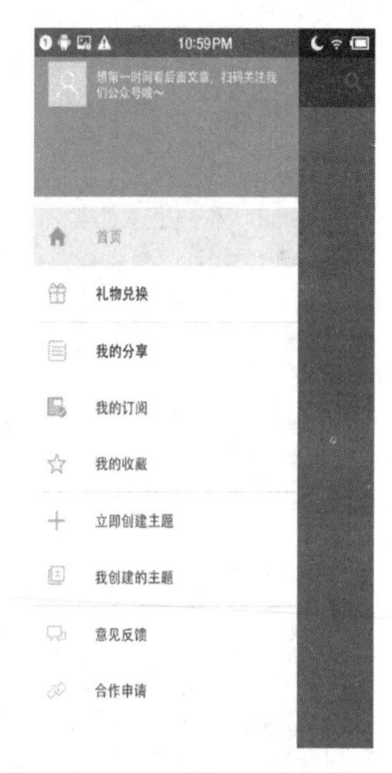

图 10-2　侧滑菜单的效果

10.2.2 代码实现

在 res 文件夹下新建 activity_main.xml 首页布局文件，内容如下：

```
<android.support.v4.widget.DrawerLayout
    xmlns:android="http://schemas.android.com/apk/res/android"
    android:id="@+id/drawer_layout"
    android:layout_width="match_parent"
    android:layout_height="match_parent" >

    <RelativeLayout
        android:layout_width="match_parent"
        android:layout_height="match_parent"
        android:clipToPadding="true"
```

```
        android:fitsSystemWindows="true" >

        <include
            android:id="@+id/rl_title"
            layout="@layout/layout_main_title" />

        <!-- The main content view -->
        <FrameLayout
            android:id="@+id/content_frame"
            android:layout_width="match_parent"
            android:layout_height="match_parent"
            android:layout_below="@+id/rl_title"
            android:background="@color/white_normal">
        </FrameLayout>
    </RelativeLayout>

    <!-- The navigation view -->
    <FrameLayout
        android:id="@+id/left_drawer"
        android:layout_width="280dp"
        android:layout_height="match_parent"
        android:layout_gravity="left">
        <!-- 左侧菜单 -->
        <include layout="@layout/layout_main_left"/>
    </FrameLayout>
</android.support.v4.widget.DrawerLayout>
```

可以看到最外层不是常用的 5 大布局之一，而是 DrawerLayout，是版本 4 包中的类，继承自 ViewGroup。其中包含 RelativeLayout 和 FrameLayout 两个子节点，RelativeLayout 显示主内容区，FrameLayout 显示左侧菜单栏。主内容区的布局代码要放在侧滑菜单布局的前面，可以帮助 DrawerLayout 判断谁是侧滑菜单、谁是主内容区。

- 主内容布局的上方是 include 标题栏布局文件，下方是一个 FrameLayout。
- 左侧菜单也是使用 include 方式加载左侧菜单布局文件。

提　示
为什么侧滑菜单栏显示在左侧呢？ 因为给 FrameLayout 设置了 android:layout_gravity 属性，这个属性的值决定侧滑出现在哪一边，如果其值为 left 就出现在左边；如果其值为 right 就出现在右边。

layout_main_title.xml 这个主内容标题布局文件如下：

```
<?xml version="1.0" encoding="utf-8"?>
<RelativeLayout xmlns:android="http://schemas.android.com/apk/res/android"
    android:layout_width="match_parent"
    android:layout_height="wrap_content"
    android:background="@color/launcher_item_select"
    android:padding="10dp"
    android:orientation="vertical">

    <ImageView
        android:id="@+id/iv_menu"
```

```
            android:layout_width="wrap_content"
            android:layout_height="wrap_content"
            android:layout_centerVertical="true"
            android:src="@drawable/ic_menu_white_24dp" />

        <TextView
            android:layout_width="wrap_content"
            android:layout_height="wrap_content"
            android:layout_marginLeft="20dp"
            android:layout_toRightOf="@+id/iv_menu"
            android:text="关注公众号[Android 开发者 666]"
            android:layout_centerVertical="true"
            android:textColor="@color/white_normal"
            android:textSize="16sp" />

        <ImageView
            android:id="@+id/iv_search"
            android:layout_width="wrap_content"
            android:layout_height="wrap_content"
            android:layout_alignParentRight="true"
            android:layout_centerVertical="true"
            android:src="@drawable/ic_search_white_24dp" />
    </RelativeLayout>
```

其中，左右两边各一张图片，中间显示文字。

layout_main_left.xml 这个侧滑菜单布局文件的内容如下：

```
<?xml version="1.0" encoding="utf-8"?>
<LinearLayout xmlns:android="http://schemas.android.com/apk/res/android"
    android:layout_width="match_parent"
    android:layout_height="match_parent"
    android:background="@color/white_normal"
    android:orientation="vertical" >

    <LinearLayout
        android:layout_width="match_parent"
        android:layout_height="wrap_content"
        android:background="@color/main_color"
        android:orientation="horizontal"
        android:paddingBottom="80dp"
        android:paddingLeft="15dp"
        android:paddingRight="15dp"
        android:paddingTop="30dp" >

        <ImageView
            android:layout_width="40dp"
            android:layout_height="40dp"
            android:src="@drawable/default_avatar" />

        <TextView
            android:layout_width="wrap_content"
            android:layout_height="wrap_content"
            android:layout_gravity="center_vertical"
            android:layout_marginLeft="10dp"
            android:text="想第一时间看后面文章，扫码关注我们公众号哦~"
```

```
            android:textColor="@color/white_normal"
            android:textSize="12sp" />
    </LinearLayout>

    <RelativeLayout
        android:id="@+id/rl_home"
        android:layout_width="match_parent"
        android:layout_height="wrap_content"
        android:layout_marginTop="10dp"
        android:background="@drawable/selector_left_menu_item"
        android:padding="@dimen/menu_left_item_padding">

        <ImageView
            android:id="@+id/iv_home"
            android:layout_width="wrap_content"
            android:layout_height="wrap_content"
            android:layout_centerVertical="true"
            android:src="@drawable/nav_icon_home"/>

        <TextView
            android:layout_width="wrap_content"
            android:layout_height="wrap_content"
            android:layout_centerVertical="true"
            android:layout_marginLeft="30dp"
            android:layout_toRightOf="@+id/iv_home"
            android:textColor="@drawable/selector_left_menu_item_text_color"
            android:text="首页" />
    </RelativeLayout>

    <View
        android:layout_width="wrap_content"
        android:layout_height="1dp"
        android:background="@color/split_line" />

    <RelativeLayout
        android:id="@+id/rl_gift"
        android:background="@drawable/selector_left_menu_item"
        android:layout_width="match_parent"
        android:layout_height="wrap_content"
        android:padding="@dimen/menu_left_item_padding">

        <ImageView
            android:id="@+id/iv_gift"
            android:layout_width="wrap_content"
            android:layout_height="wrap_content"
            android:layout_centerVertical="true"
            android:src="@drawable/nav_icon_gift" />

        <TextView
            android:layout_width="wrap_content"
            android:layout_height="wrap_content"
            android:layout_centerVertical="true"
            android:layout_marginLeft="30dp"
            android:layout_toRightOf="@+id/iv_gift"
            android:textColor="@drawable/selector_left_menu_item_text_color"
```

```
            android:text="礼物兑换" />
    </RelativeLayout>

    <View
        android:layout_width="wrap_content"
        android:layout_height="1dp"
        android:background="@color/split_line" />

    <RelativeLayout
        android:id="@+id/rl_share"
        android:background="@drawable/selector_left_menu_item"
        android:layout_width="match_parent"
        android:layout_height="wrap_content"
        android:padding="@dimen/menu_left_item_padding" >

        <ImageView
            android:id="@+id/iv_share"
            android:layout_width="wrap_content"
            android:layout_height="wrap_content"
            android:layout_centerVertical="true"
            android:src="@drawable/nav_icon_my_shares" />

        <TextView
            android:layout_width="wrap_content"
            android:layout_height="wrap_content"
            android:layout_centerVertical="true"
            android:layout_marginLeft="30dp"
            android:layout_toRightOf="@+id/iv_share"
            android:textColor="@drawable/selector_left_menu_item_text_color"
            android:text="我的分享"/>
    </RelativeLayout>
</LinearLayout>
```

基本结构都类似，因此仅贴出了上部分的布局文件。最外层使用 LinearLayout，按照垂直方向排列下来，其中每一个 RelativeLayout 代表一行。

selector_left_menu_item.xml 是左侧菜单 item 选中背景的布局文件，给当前选中的项目设置不一样的颜色，这样下次侧滑时就知道当前选中的是哪个项目。

```
<?xml version="1.0" encoding="utf-8"?>
<selector xmlns:android="http://schemas.android.com/apk/res/android">
    <item android:drawable="@color/menu_left_item_select"
android:state_selected="true"/>
    <item android:drawable="@color/white_normal"/>
</selector>
```

activity_main.xml 首页布局文件对应的 MainActivity.java 内容如下，需要注意的是 MainActivity 继承自 FragmentActivity，继承 FragmentActivity 的好处是 Fragment 可以在 3.0 以下版本的手机上使用。

```
public class MainActivity extends FragmentActivity{
    private DrawerLayout mDrawerLayout;
    private RelativeLayout rlHome, rlGift, rlShare;
    private int currentSelectItem = R.id.rl_home;//默认首页
    private ContentFragment contentFragment;
```

```
    @Override
    protected void onCreate(Bundle savedInstanceState) {
        super.onCreate(savedInstanceState);
        setContentView(R.layout.activity_main);

        mDrawerLayout = (DrawerLayout) findViewById(R.id.drawer_layout);

        findViewById(R.id.iv_menu).setOnClickListener(clickListener);

        initLeftMenu();//初始化左侧菜单

        contentFragment=new ContentFragment();
        getSupportFragmentManager().beginTransaction().add(R.id.content_frame,
         contentFragment).commit();

        setWindowStatus();
    }

    private void initLeftMenu() {
        rlHome = (RelativeLayout) findViewById(R.id.rl_home);
        rlGift = (RelativeLayout) findViewById(R.id.rl_gift);
        rlShare = (RelativeLayout) findViewById(R.id.rl_share);

        rlHome.setOnClickListener(onLeftMenuClickListener);
        rlGift.setOnClickListener(onLeftMenuClickListener);
        rlShare.setOnClickListener(onLeftMenuClickListener);

        rlHome.setSelected(true);
    }

    private OnClickListener onLeftMenuClickListener = new OnClickListener() {
        @Override
        public void onClick(View v) {
            if (currentSelectItem != v.getId()) {//防止重复点击
                currentSelectItem=v.getId();
                noItemSelect();

                switch (v.getId()) {
                case R.id.rl_home:
                    rlHome.setSelected(true);
                    contentFragment.setContent("这是首页");
                    break;
                case R.id.rl_gift:
                    rlGift.setSelected(true);
                    contentFragment.setContent("这是礼物兑换");
                    break;
                case R.id.rl_share:
                    rlShare.setSelected(true);
                    contentFragment.setContent("这是我的分享");
                    break;
                }
                mDrawerLayout.closeDrawer(Gravity.LEFT);
            }
        }
```

```
        };

        private void noItemSelect(){
            rlHome.setSelected(false);
            rlGift.setSelected(false);
            rlShare.setSelected(false);
        }

        private OnClickListener clickListener = new OnClickListener() {
            @Override
            public void onClick(View v) {
                switch (v.getId()) {
                case R.id.iv_menu:
                    mDrawerLayout.openDrawer(Gravity.LEFT);//打开左侧抽屉
                    break;
                }
            }
        };

        // 设置状态栏
        private void setWindowStatus() {
            if (Build.VERSION.SDK_INT >= Build.VERSION_CODES.KITKAT) {
                // 透明状态栏
                getWindow().addFlags(WindowManager.LayoutParams.
FLAG_TRANSLUCENT_STATUS);
                // 透明导航栏
                getWindow().addFlags(WindowManager.LayoutParams.
FLAG_TRANSLUCENT_NAVIGATION);
                // 设置状态栏颜色
                getWindow().setBackgroundDrawableResource(R.color.main_color);
            }
        }
    }
```

- oncreate：查找 DrawerLayout 控件，给标题栏左侧的按钮设置点击事件。初始化左侧菜单，主内容区域显示 ContentFragment，设置状态栏背景色。
- initLeftMenu：初始化左侧菜单，给前面三个 item（行）设置点击事件，默认选中第一个。
- onLeftMenuClickListener：左侧菜单前三个 item 点击时调用，首先判断是否重复点击，如果用户连续两次选中一个菜单，我们只让第一次点击有效，然后根据点击不同的 item 决定内容区域显示不同文字，当然最后不要忘记了关闭左侧菜单栏。
- noItemSelect：所有左侧菜单未选中。
- clickListener：点击标题栏左边图片时，开启侧滑菜单。
- setWindowStatus：Android 系统等于或者大于 4.4 版本时设置状态栏。

ContentFragment.java 文件的主内容区显示 Fragment：

```
    public class ContentFragment extends Fragment{
        private TextView tvContent;

        @Override
        public View onCreateView(LayoutInflater inflater, ViewGroup
container,Bundle savedInstanceState){
```

```
        View rootView=LayoutInflater.from(getActivity()).inflate(R.layout.
fragment_content, null);
        tvContent=(TextView) rootView.findViewById(R.id.tv_content);
        return rootView;
    }

    public void setContent(String content){
        tvContent.setText(content);
    }
}
```

仅显示一个带有 TextView 的布局文件，根据用户点击的项目来决定显示什么字符串。

DrawerLayout 与 Fragment 是什么关系？

我们看到很多使用 DrawerLayout 的代码中都同时使用了 Fragment，这会造成误解，以为使用 DrawerLayout 必须用到 Fragment，其实这是错误的。

使用 Fragment 是因为在侧滑菜单被点击时，如果主内容区的内容比较复杂，使用 Fragment 去填充会更容易；如果主内容区仅有一个简单的字符串，只想在不同菜单点击时更新一下字符串的内容，就没有必要使用 Fragment。

10.3　开发者头条首页实现

下面实现开发者头条 App 的首页，其效果如图 10-3 所示。

图 10-3　开发者头条首页

从效果图中可以看到，标题栏下面有“精选”“订阅”“发现”三个选项卡。在“精选”页面的顶部有一个轮播图，轮播图下方是文章列表。

10.3.1 源代码的实现

在 res 文件夹下新建 fragment_main.xml，用作首页 Fragment 的布局文件：

```
<?xml version="1.0" encoding="utf-8"?>
<RelativeLayout xmlns:android="http://schemas.android.com/apk/res/android"
    android:layout_width="match_parent"
    android:layout_height="match_parent"
    android:background="@color/white_normal" >

    <RelativeLayout
        android:id="@+id/ll_title"
        android:layout_width="match_parent"
        android:layout_height="44dp" >

        <LinearLayout
            android:layout_width="match_parent"
            android:layout_height="match_parent"
            android:background="@color/main_color"
            android:orientation="horizontal" >

            <TextView
                android:id="@+id/tv_selected"
                android:layout_width="0dp"
                android:layout_height="wrap_content"
                android:layout_gravity="center_vertical"
                android:layout_weight="1"
                android:gravity="center_horizontal"
                android:text="精选"
                android:textColor="@drawable/main_title_txt_sel" />

            <TextView
                android:id="@+id/tv_subscribe"
                android:layout_width="0dp"
                android:layout_height="wrap_content"
                android:layout_gravity="center_vertical"
                android:layout_weight="1"
                android:gravity="center_horizontal"
                android:text="订阅"
                android:textColor="@drawable/main_title_txt_sel" />

            <TextView
                android:id="@+id/tv_find"
                android:layout_width="0dp"
                android:layout_height="wrap_content"
                android:layout_gravity="center_vertical"
                android:layout_weight="1"
                android:gravity="center_horizontal"
                android:text="发现"
                android:textColor="@drawable/main_title_txt_sel" />
        </LinearLayout>

        <View
```

```
            android:id="@+id/view_indicator"
            android:layout_width="15dp"
            android:layout_height="2dp"
            android:layout_alignParentBottom="true"
            android:background="@color/white_normal" />
    </RelativeLayout>

    <android.support.v4.view.ViewPager
        android:id="@+id/viewpager_home"
        android:layout_width="match_parent"
        android:layout_height="match_parent"
        android:layout_below="@+id/ll_title" />

</RelativeLayout>
```

最外层是 RelativeLayout，其中有两个控件：第一个控件是 RelativeLayout，显示 Tab 选项卡；第二个控件是 ViewPager，用来显示三个选项卡对应的 Fragment。

MainFragment.java 是首页 Fragment 文件，内容如下：

```
    public class MainFragment extends Fragment {
        private int screenWidth, screenHeight;

        private List<Fragment> list = new ArrayList<Fragment>();
        private ViewPager vPager;
        private FragmentAdapter adapter;
        private TextView tvSelected,tvSubscribe,tvFind;
        private View viewIndicator;

        private int currentIndex = 0;
        private int currentSelectId;

        @Override
        public View onCreateView(LayoutInflater inflater, ViewGroup container,
Bundle savedInstanceState) {
            getScreenSize(getActivity());//获取屏幕宽高

            View rootView = inflater.inflate(R.layout.fragment_main,null);

            vPager = (ViewPager) rootView.findViewById(R.id.viewpager_home);

            SelectedFragment selectedFragment = new SelectedFragment();
            SubscribeFragment subscribeFragment = new SubscribeFragment();
            FindFragment findFragment = new FindFragment();

            list.add(selectedFragment);//添加精选 Fragment
            list.add(subscribeFragment);//添加订阅 Fragment
            list.add(findFragment);//添加发现 Fragment

            adapter=new FragmentAdapter(getActivity().getSupportFragmentManager(),
                    list);
            vPager.setAdapter(adapter);//设置适配器
            vPager.setOffscreenPageLimit(2);//缓存页数
            vPager.setOnPageChangeListener(pageChangeListener);//改变监听

            tvSelected = (TextView) rootView.findViewById(R.id.tv_selected);//精选
```

```
        tvSubscribe = (TextView) rootView.findViewById(R.id.tv_subscribe);//订阅
        tvFind = (TextView) rootView.findViewById(R.id.tv_find);//发现

        tvSelected.setOnClickListener(clickListener);
        tvSubscribe.setOnClickListener(clickListener);
        tvFind.setOnClickListener(clickListener);

        tvSelected.setSelected(true);//默认选中“精选”

        viewIndicator = rootView.findViewById(R.id.view_indicator);//指示器View
        initCursorPosition();//初始化指示器位置
        return rootView;
    }

    private OnClickListener clickListener = new OnClickListener() {
        @Override
        public void onClick(View v) {
            if (currentSelectId != v.getId()) {//防止重复点击
                switch (v.getId()) {
                    case R.id.tv_selected:
                        vPager.setCurrentItem(0);
                        break;
                    case R.id.tv_subscribe:
                        vPager.setCurrentItem(1);
                        break;
                    case R.id.tv_find:
                        vPager.setCurrentItem(2);
                        break;
                }
                currentSelectId = v.getId();
            }
        }
    };

    private void initCursorPosition() {
        LayoutParams layoutParams = viewIndicator.getLayoutParams();
        layoutParams.width = screenWidth / 3;
        viewIndicator.setLayoutParams(layoutParams);

        TranslateAnimation animation = new TranslateAnimation(-screenWidth
/3,0,0,0);
        animation.setFillAfter(true);
        viewIndicator.startAnimation(animation);
    }

    private OnPageChangeListener pageChangeListener = new OnPageChangeListener() {
        @Override
        public void onPageSelected(int index) {
            translateAnimation(index);//移动指示器
            changeTextColor(index);//改变文字颜色
            currentIndex = index;//设置当前选中
        }

        @Override
        public void onPageScrolled(int arg0, float arg1, int arg2) {
```

```
        }

        @Override
        public void onPageScrollStateChanged(int arg0) {
        }
    };

    /**
     * 改变标题栏的字体颜色
     *
     * @param index
     */
    private void changeTextColor(int index) {
        tvSelected.setSelected(false);
        tvSubscribe.setSelected(false);
        tvFind.setSelected(false);

        switch (index) {
            case 0:
                tvSelected.setSelected(true);
                break;
            case 1:
                tvSubscribe.setSelected(true);
                break;
            case 2:
                tvFind.setSelected(true);
                break;
        }
    }

    /**
     * 移动标题栏的圆点...
     *
     * @param index
     */
    private void translateAnimation(int index) {
        TranslateAnimation animation = null;
        switch (index) {
            case 0://订阅->精选
                animation = new TranslateAnimation((screenWidth/3), 0, 0, 0);
                break;
            case 1://
                if (0 == currentIndex) {//精选->订阅
                    animation = new TranslateAnimation(0, screenWidth / 3, 0, 0);
                } else if (2 == currentIndex) {//发现->订阅
                    animation = new TranslateAnimation((screenWidth/3) * 2,
screenWidth / 3, 0, 0);
                }
                break;
            case 2://订阅-> 发现
                animation = new TranslateAnimation(screenWidth / 3,
(screenWidth/3) * 2, 0, 0);
                break;
        }
        animation.setFillAfter(true);
```

```
        animation.setDuration(300);
        viewIndicator.startAnimation(animation);
    }

    //获取屏幕的宽与高
    private void getScreenSize(Activity context) {
        DisplayMetrics dm = new DisplayMetrics();
        context.getWindowManager().getDefaultDisplay().getMetrics(dm);
        screenWidth = dm.widthPixels;
        screenHeight = dm.heightPixels;
    }
}
```

- onCreateView：首先获取屏幕的宽与高，加载布局文件，然后初始化三个 Fragment，添加到列表集合，初始化适配器，给 ViewPager 设置适配器，设置缓存页数，设置页面改变监听回调接口，查找顶部 Tab 选项卡的 TextView，默认选中“精选”，最后初始化指示器位置。
- clickListener：如果不是重复点击，根据点击 Tab 选项卡来设置 ViewPager 当前选中的页面。
- initCursorPosition：初始化当前指示器位置。
- pageChangeListener：在 onPageSelected 处理页面选择事件，移动指示器，改变 Tab 选项卡的文字颜色，将当前选中的位置赋值给全局变量保存起来。
- changeTextColor：改变 Tab 选项卡的选中状态，因为给 TextView 设置了选择器，所以选中和未选中状态决定了显示哪种颜色。
- translateAnimation：移动指示器，产生一个移动动画。
- getScreenSize：获取屏幕的宽与高。

10.3.2 精选 Fragment

从整体来看，就是一个 ListView，顶部轮播图是 ListView 的头部。头部轮播也是用 ViewPager 来实现的，与开发者头条 App 启动页的实现原理相似。然后添加一个定时器，隔一段时间设置 ViewPager 的当前页面即可。

说 明

这里的图片用的是静态（本地）的，一个商业 App 的轮播图片肯定是从服务器获取的，例如真正开发者头条 App 就是从服务器获取图片 Url。

fragment_selected.java 是精选 Fragment 的布局文件，其中仅一个 ListView：

```
<?xml version="1.0" encoding="utf-8"?>
<LinearLayout xmlns:android="http://schemas.android.com/apk/res/android"
    android:layout_width="match_parent"
    android:layout_height="match_parent"
    android:orientation="vertical" >

    <ListView
        android:id="@+id/list"
        android:scrollbars="none"
        android:layout_height="wrap_content"
```

```
        android:layout_width="match_parent"/>
</LinearLayout>
```

fragment_selected_header.java 是精选头布局，用来显示 Banner：

```
<?xml version="1.0" encoding="utf-8"?>
<LinearLayout xmlns:android="http://schemas.android.com/apk/res/android"
    android:layout_width="match_parent"
    android:layout_height="match_parent"
    android:orientation="vertical" >

    <RelativeLayout
        android:layout_width="wrap_content"
        android:layout_height="200dp" >

        <android.support.v4.view.ViewPager
            android:id="@+id/viewpager"
            android:layout_width="match_parent"
            android:layout_height="match_parent" />

        <TextView
            android:id="@+id/tv_content"
            android:layout_width="wrap_content"
            android:layout_height="wrap_content"
            android:layout_alignParentBottom="true"
            android:layout_marginBottom="10dp"
            android:layout_marginLeft="5dp"
            android:text="公众号:ansen_666" />

        <RelativeLayout
            android:layout_width="fill_parent"
            android:layout_height="wrap_content"
            android:orientation="vertical" >

            <LinearLayout
                android:id="@+id/viewGroup"
                android:layout_width="fill_parent"
                android:layout_height="wrap_content"
                android:layout_alignParentBottom="true"
                android:layout_marginBottom="5dp"
                android:gravity="center_horizontal"
                android:orientation="horizontal"/>
        </RelativeLayout>
    </RelativeLayout>
</LinearLayout>
```

SelectedFragment.java 是精选 Fragment 的文件，内容如下：

```
public class SelectedFragment extends Fragment{
    private ViewPager viewPager;
    private SelectedPagerAdapter selectedPagerAdapter;

    private ImageView[] tips;//底部...
    private Timer timer;
    private final int  CAROUSEL_TIME = 3000;//滚动间隔
```

```
    private int currentIndex=0;//当前选中

    private TextView tvContent;
    private String[] carousePageStr=new String[]{"Android 开发 666","公众
号:Ansen_666","Python 的练手项目有哪些值得推荐"};

    private ListView listView;
    private SelectedAdapter selectedAdapter;

    @Override
    public View onCreateView(LayoutInflater inflater, ViewGroup container,
Bundle savedInstanceState){
        View rootView = LayoutInflater.from(getActivity()).inflate
                    (R.layout.fragment_selected, null);

        View headView = LayoutInflater.from(getActivity()).inflate(R.layout.
                    fragment_selected_header, null);

        tvContent=(TextView) headView.findViewById(R.id.tv_content);
        tvContent.setText(carousePageStr[0]);

        viewPager = (ViewPager)headView.findViewById(R.id.viewpager);
        selectedPagerAdapter=new SelectedPagerAdapter(getActivity(),
                        carousePagerSelectView);
        viewPager.setOffscreenPageLimit(2);
        viewPager.setCurrentItem(0);
        viewPager.setOnPageChangeListener(onPageChangeListener);
        viewPager.setAdapter(selectedPagerAdapter);

        ViewGroup group = (ViewGroup) headView.findViewById(R.id.viewGroup);
            // 初始化底部显示控件
        tips = new ImageView[3];
        for (int i = 0; i < tips.length; i++){
            ImageView imageView = new ImageView(getActivity());
            if (i == 0) {
                imageView.setBackgroundResource(R.drawable.page_indicator_focused);
            } else {
                imageView.setBackgroundResource(R.drawable.page_indicator_unfocused);
            }
            tips[i] = imageView;
            LinearLayout.LayoutParams layoutParams = new LinearLayout.
             LayoutParams (new ViewGroup.LayoutParams(LayoutParams.WRAP_
             CONTENT,LayoutParams.WRAP_CONTENT));
            layoutParams.leftMargin = 10;// 设置指示器 ItemView(横线)的左边距
            layoutParams.rightMargin = 10;// 设置指示器 ItemView(横线)的右边距
            group.addView(imageView, layoutParams);
        }

        timer = new Timer(true);//初始化计时器
        timer.schedule(task, 0, CAROUSEL_TIME);//延时 0ms 后执行,3000ms 执行一次

        listView=(ListView) rootView.findViewById(R.id.list);
        listView.addHeaderView(headView);
        listView.setAdapter(selectedAdapter=new SelectedAdapter(getActivity()));
        return rootView;
```

```
    }

    private ICarousePagerSelectView carousePagerSelectView=new ICarousePagerSelectView(){
        @Override
        public void carouseSelect(int index) {
            Toast.makeText(getActivity(), carousePageStr[index],
Toast.LENGTH_SHORT).show();
        }
    };

    TimerTask task = new TimerTask() {
        public void run() {
            handler.sendEmptyMessage(CAROUSEL_TIME);
        }
    };

    private Handler handler=new Handler(){
        public void handleMessage(Message msg) {
            switch (msg.what) {
            case CAROUSEL_TIME:
                if(currentIndex>=tips.length-1){//已经滚动到最后,从第一页开始
                    viewPager.setCurrentItem(0);
                }else{//开始下一页
                    viewPager.setCurrentItem(currentIndex+1);
                }
                break;
            }
        };
    };

    @Override
    public void onDestroy(){
        task.cancel();
        System.exit(0);
    }

    private OnPageChangeListener onPageChangeListener=new OnPageChangeListener() {
        @Override
        public void onPageSelected(int index) {
            tvContent.setText(carousePageStr[index]);
            setImageBackground(index);// 改变圆点的切换效果
            currentIndex=index;
        }

        @Override
        public void onPageScrolled(int arg0, float arg1, int arg2){}
        @Override
        public void onPageScrollStateChanged(int arg0) {}
    };

    /**
     * 改变圆点的切换效果
     * @param selectItems
     */
    private void setImageBackground(int selectItems) {
```

```
        for (int i = 0; i < tips.length; i++) {
            if (i == selectItems) {
                tips[i].setBackgroundResource(R.drawable.page_indicator_focused);
            } else {
                tips[i].setBackgroundResource(R.drawable.page_indicator_unfocused);
            }
        }
    }
}
```

- onCreateView：设置布局文件，初始化头文件，初始化显示广告的 ViewPager，设置适配器，给 Banner 增加滚动状态，初始化计时器，3 秒执行一次。给文章列表的 ListView 添加头文件，并且设置适配器。
- carousePagerSelectView：Banner 的项目点击回调接口。
- task：计时器回调，给 handler 发送一个消息。
- handler：更新 Banner 的选中位置，显示下一个 Banner。
- onDestroy：关闭计时器。
- onPageChangeListener：显示广告的 ViewPager 页面改变监听。
- setImageBackground：广告选中状态的变化。

有关显示广告，前面已讲过多次了，这里就不贴出 SelectedPagerAdapter 的代码了。

接下来，看看精选文章的 SelectedAdapter.java 文件：

```
public class SelectedAdapter extends BaseAdapter{
    private LayoutInflater inflater;
    private List<SelectedArticle> selectedArticles;

    public SelectedAdapter(Context context){
        inflater = LayoutInflater.from(context);
        selectedArticles=new ArrayList<SelectedArticle>();
        initData();
    }

    private void initData(){
        for(int i=0;i<50;i++){
            SelectedArticle selectedArticle=new SelectedArticle(i, "Android开
发666", i, i, "");
            selectedArticles.add(selectedArticle);
        }
    }

    @Override
    public int getCount() {
        return selectedArticles.size();
    }

    @Override
    public Object getItem(int position) {
        return selectedArticles.get(position);
    }

    @Override
```

```
    public long getItemId(int position) {
        return selectedArticles.get(position).getId();
    }

    @Override
    public View getView(int position, View convertView, ViewGroup parent){
        ViewHolder holder;
        if (convertView == null) {
            convertView = inflater.inflate(R.layout.fragment_selected_item, null);
            holder = new ViewHolder();
            holder.title = (TextView) convertView.findViewById(R.id.tv_title);
            holder.like = (TextView) convertView.findViewById(R.id.tv_like);
            holder.comment = (TextView) convertView.findViewById(R.id.tv_comment);
            convertView.setTag(holder);
        } else {
            holder = (ViewHolder) convertView.getTag();
        }
        SelectedArticle selectedArticle=selectedArticles.get(position);
        holder.title.setText(selectedArticle.getTitle());
        holder.like.setText(""+selectedArticle.getLikeNumber());
        holder.comment.setText(""+selectedArticle.getCommentNumber());
        return convertView;
    }

    private class ViewHolder {
        private TextView title;
        private TextView like;
        private TextView comment;
    }
}
```

和其他适配器没啥区别，继承 BaseAdapter，重写四个方法，前面我们已经写了很多遍了。`SelectedArticle` 类跟 Item 布局文件代码比较简单，就不逐一贴出代码了，大家下载源代码之后一看就懂。

10.4　开发者头条首页优化

Google 在 2015 年的 I/O 大会上给我们带来了更加详细的 Material Design 设计规范。同时，也带来了全新的 Android Design Support 库。在这个 Support 库中，Google 提供了更加规范的 MD 设计风格的控件。最重要的是，Android Design Support 库的兼容性更广，直接可以向下兼容到 Android 2.2。这不得不说是一个良心之作。

下面就用 Design 包重写首页，效果如图 10-4 所示。

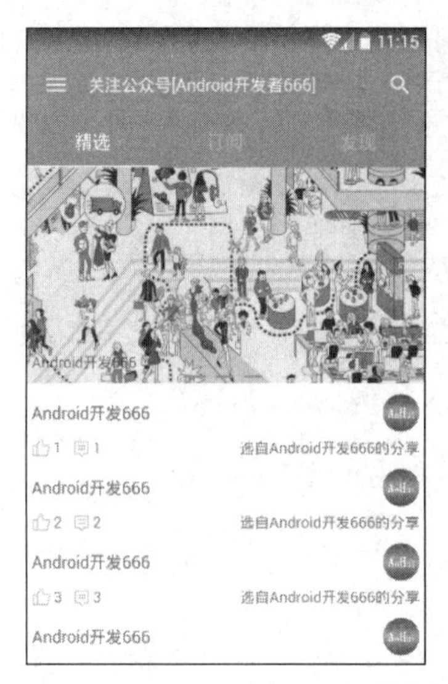

图 10-4　重写首页后的运行效果图

单从效果图上看，和前面的首页没有太大变化，唯一变化的就是列表上拉时会隐藏标题栏，其实里面的代码几乎重写了一遍，已经使用了 Android Design Support 库。

10.4.1　需要在线依赖

```
    compile 'com.android.support:design:22.2.0'
    compile 'com.android.support:recyclerview-v7:22.2.0'
```

为什么多了一个 recyclerview 依赖呢？因为在第 3 版中使用的是 ListView 来实现文章列表，而在这个版本中改用 Recyclerview 来实现文章列表。

10.4.2　标题栏和三个切换选项卡

修改 fragment_main.xml 首页 Fragment 布局文件：

```
    <?xml version="1.0" encoding="utf-8"?>
    <android.support.design.widget.CoordinatorLayout
xmlns:android="http://schemas.android.com/apk/res/android"
        xmlns:app="http://schemas.android.com/apk/res-auto"
        android:id="@+id/coordinatorLayout"
        android:layout_width="match_parent"
        android:layout_height="match_parent">

        <android.support.design.widget.AppBarLayout
            android:id="@+id/appBarLayout"
            android:layout_width="match_parent"
            android:layout_height="wrap_content">

            <android.support.v7.widget.Toolbar
                android:id="@+id/toolbar"
                android:layout_width="match_parent"
```

```
            android:layout_height="wrap_content"
            android:background="@color/launcher_item_select"
            app:layout_scrollFlags="scroll|enterAlways"
            app:titleTextAppearance="@style/ansenTextTitleAppearance"/>

        <android.support.design.widget.TabLayout
            android:id="@+id/tabLayout"
            android:layout_width="match_parent"
            android:layout_height="wrap_content"
            android:layout_gravity="center_horizontal"
            android:background="@color/main_color"
            app:tabIndicatorColor="@color/white_normal"
            app:tabIndicatorHeight="2dp"
            app:tabSelectedTextColor="@color/main_title_text_select"
            app:tabTextAppearance="@style/AnsenTabLayoutTextAppearance"
            app:tabTextColor="@color/main_title_text_normal" />
    </android.support.design.widget.AppBarLayout>

    <android.support.v4.view.ViewPager
        android:id="@+id/viewPager"
        android:layout_width="match_parent"
        android:layout_height="match_parent"
        app:layout_behavior="@string/appbar_scrolling_view_behavior" />
</android.support.design.widget.CoordinatorLayout>
```

最外层是 CoordinatorLayout，其中主要分为两块：AppBarLayout 和 ViewPager（AppBarLayout 中包含标题栏的 Toolbar+TabLayout，ViewPager 用来切换 Fragment 显示）。

有关标题栏，之前引用过一个布局文件，现在改成了 Toolbar，仅一个控件就够了。三个切换的选项卡之前用的是三个 TextView，现在改成了 TabLayout。更换之后有如下一些优点：

- 跟得上时代，规格提高了，更加规范的 MD 设计风格。
- 控件变少了，现在不管多少个 Tab 都用一个 TabLayout 控件。
- 指示器的滑动功能，只需在 xml 中添加一个属性即可。
- 隐藏显示标题栏很方便，只需要在布局文件中改动即可。

为了使 Toolbar（标题栏）产生滑动效果，必须做到如下三点：

- CoordinatorLayout 作为布局的父布局容器。
- 给需要滑动的组件设置 app:layout_scrollFlags= "scroll|enterAlways"属性。
- 滑动的组件必须是 AppBarLayout 顶部组件。
- 给滑动的组件设置 app:layout_behavior 属性。
- ViewPager 显示的 Fragment 中不能是 ListView，必须是 RecyclerView。

接下来修改 MainFragment.java 代码：

初始化 Toolbar，加载菜单布局，实现标题栏的自定义。给 NavigationIcon 设置点击事件等。下面贴出相关的代码，有关菜单的布局文件就不贴出来了，因为相对比较简单。

```
Toolbar toolbar = (Toolbar) rootView.findViewById(R.id.toolbar);
toolbar.inflateMenu(R.menu.ansen_toolbar_menu);
toolbar.setNavigationIcon(R.mipmap.ic_menu_white);//设置导航图标
toolbar.setTitle("关注公众号[Android开发者666]");
```

```
toolbar.setTitleTextColor(getResources().getColor(android.R.color.white));
toolbar.setNavigationOnClickListener(onClickListener);//导航图标点击
```

导航图标点击事件处理，当我们点击导航图标时，应该打开左侧菜单栏，但是 MainFragment 是没有 mDrawerLayout 对象的，只有 MainActivity 才有，于是我们通过接口方式调用，在 onClick 方法中调用 drawerListener.open()方法打开导航栏。

```
private View.OnClickListener onClickListener=new View.OnClickListener(){
    @Override
    public void onClick(View view) {
        if(drawerListener!=null){
            drawerListener.open();
        }
    }
};
```

drawerListener 对象是创建 MainFragment 时通过构造方法传参带过来的。MainFragment 构造方法代码如下：

```
private MainActivity.MainDrawerListener drawerListener;

public MainFragment(MainActivity.MainDrawerListener drawerListener){
    this.drawerListener=drawerListener;
}
```

接下来我们看 MainActivity 代码做了什么改动，首先增加一个接口：

```
public interface MainDrawerListener{
    public void open();//打开 Drawer
}
```

然后用内部类方式实现这个接口，当 open 方法调用时打开左侧菜单栏：

```
private MainDrawerListener drawerListener=new MainDrawerListener() {
     @Override
     public void open() {
        mDrawerLayout.openDrawer(Gravity.LEFT);
     }
};
```

还有最后一步，初始化 Fragment 时将 MainDrawerListener 对象传递过去，这样 MainFragment 持有 MainActivity 的 MainDrawerListener 对象，当在 MainFragment 调用 open 方法时，MainActivity 中 drawerListener 的 open 方法调用，然后打开左侧菜单栏。

```
mainFragment=new MainFragment(drawerListener);
```

继续修改 MainFragment 代码，在 onCreateView 方法中给 ViewPager 设置 Fragment 适配器，TabLayout 绑定 ViewPager，这样 ViewPager 滑动时或者选择选项卡时都会切换 Fragment。

```
vPager = (ViewPager) rootView.findViewById(R.id.viewPager);
vPager.setOffscreenPageLimit(2);//设置缓存页数
vPager.setCurrentItem(0);

FragmentAdapter pagerAdapter = new FragmentAdapter(getActivity().
                               getSupportFragmentManager());
```

```
    SelectedFragment selectedFragment=new SelectedFragment();
    SubscribeFragment subscribeFragment=new SubscribeFragment();
    FindFragment findFragment=new FindFragment();

    pagerAdapter.addFragment(selectedFragment,"精选");
    pagerAdapter.addFragment(subscribeFragment,"订阅");
    pagerAdapter.addFragment(findFragment,"发现");

    vPager.setAdapter(pagerAdapter);

    TabLayout tabLayout = (TabLayout) rootView.findViewById(R.id.tabLayout);
    tabLayout.setupWithViewPager(vPager);
```

10.4.3 分析 TabLayout 切换源代码

我们调用 TabLayout 的 setupWithViewPager(ViewPager viewPager)方法时，其实在内部就给 ViewPager 设置了页面滑动监听，不信你看 setupWithViewPager 方法源代码：

```
        public void setupWithViewPager(ViewPager viewPager) {
            PagerAdapter adapter = viewPager.getAdapter();
            if(adapter == null) {
                throw new IllegalArgumentException("ViewPager does not have a PagerAdapter
set");
            } else {
                this.setTabsFromPagerAdapter(adapter);
                viewPager.addOnPageChangeListener(new TabLayout.
TabLayoutOnPageChangeListener(this));
                this.setOnTabSelectedListener(new TabLayout.
ViewPagerOnTabSelectedListener(viewPager));
            }
        }
```

从上面代码中可以看到主要设置了两个监听函数。先讲述第一个监听，在 TabLayout 中有一个静态类 TabLayoutOnPageChangeListener，用来处理 ViewPager 改变状态（切换或者增加）监听。

```
     viewPager.addOnPageChangeListener(new TabLayout.
TabLayoutOnPageChangeListener(this));
```

TabLayoutOnPageChangeListener 实现了 ViewPager 的 OnPageChangeListener 接口，在 onPageSelected 方法中调用了当前选中的某个选项卡的 select 方法。

```
           public void onPageSelected(int position) {
               TabLayout tabLayout = (TabLayout)this.mTabLayoutRef.get();
               if(tabLayout != null) {
                   tabLayout.getTabAt(position).select();
               }

           }
```

然后继续跟踪 TabLayout.Tab 类的 select()查看是如何实现的。我们可以看到又调用了父类（TabLayout）的 selectTab。

```
           public void select() {
```

```
            this.mParent.selectTab(this);
        }
```

然后跟踪 selectTab 方法，这里大家可以看到参数是某个具体 Tab 对象，首先判断是不是当前的选项卡，如果不是，就设置选择当前的选项卡，并开启选项卡的滑动动画效果。

```
    void selectTab(TabLayout.Tab tab) {
        if(this.mSelectedTab == tab) {
            if(this.mSelectedTab != null) {
                if(this.mOnTabSelectedListener != null) {
                    this.mOnTabSelectedListener.onTabReselected(this.mSelectedTab);
                }

                this.animateToTab(tab.getPosition());
            }
        } else {
            int newPosition = tab != null?tab.getPosition():-1;
            this.setSelectedTabView(newPosition);
            if((this.mSelectedTab == null || this.mSelectedTab.getPosition() ==
-1) && newPosition != -1) {
                this.setScrollPosition(newPosition, 0.0F, true);
            } else {
                this.animateToTab(newPosition);
            }

            if(this.mSelectedTab != null && this.mOnTabSelectedListener != null) {
                this.mOnTabSelectedListener.onTabUnselected(this.mSelectedTab);
            }

            this.mSelectedTab = tab;
            if(this.mSelectedTab != null && this.mOnTabSelectedListener != null) {
                this.mOnTabSelectedListener.onTabSelected(this.mSelectedTab);
            }
        }

    }
```

上面的代码就不逐一解释了，直接查看最下面的两行代码，调用选项卡的选择方法：

```
    if(this.mSelectedTab != null && this.mOnTabSelectedListener != null) {
        this.mOnTabSelectedListener.onTabSelected(this.mSelectedTab);
    }
```

Tab 选择监听的接口：

```
    public interface OnTabSelectedListener {
        void onTabSelected(TabLayout.Tab var1);
        void onTabUnselected(TabLayout.Tab var1);
        void onTabReselected(TabLayout.Tab var1);
    }
```

在 TabLayout 内部实现了 OnTabSelectedListener 接口，在 onTabSelected 方法中调用了 ViewPager 的 setCurrentItem()，这个方法大家应该都熟悉吧，就不多做解释了。

```
    public static class ViewPagerOnTabSelectedListener implements
TabLayout.OnTabSelectedListener {
```

```
        private final ViewPager mViewPager;

        public ViewPagerOnTabSelectedListener(ViewPager viewPager) {
            this.mViewPager = viewPager;
        }

        public void onTabSelected(TabLayout.Tab tab) {
            this.mViewPager.setCurrentItem(tab.getPosition());
        }

        public void onTabUnselected(TabLayout.Tab tab) {
        }

        public void onTabReselected(TabLayout.Tab tab) {
        }
    }
```

上面讲的是第一个监听，也就是 ViewPager 滑动的时候如何切换项目、如何切换选项卡。现在来说第二个监听，就是点击选择选项卡时如何切换。继续回到 TabLayout 的 setupWithViewPager(ViewPager viewPager)方法。

```
    this.setOnTabSelectedListener(new TabLayout.ViewPagerOnTabSelectedListener
(viewPager));
```

看到 ViewPagerOnTabSelectedListener 类是不是很熟悉？其实就是第一种方法最后调用的那个类，因为点击某个选项卡时，选项卡切换的代码已经运行，所以这里只需要设置 ViewPager 当前选中的项目即可。

10.4.4　精选文章列表控件从 ListView 替换成 RecyclerView

RecyclerView 是 Android L 版本中新增的一个用来取代 ListView 的 SDK，灵活性与可替代性比 listview 更好。当然新的控件在使用过程中会碰到一些问题，但是相信 Google 会努力解决这些 bug。

修改 fragment_selected.xml 布局文件，将 ListView 替换成 RecyclerView。

```
    <?xml version="1.0" encoding="utf-8"?>
    <android.support.v7.widget.RecyclerView
        xmlns:android="http://schemas.android.com/apk/res/android"
        android:id="@+id/recyclerView"
        android:layout_width="match_parent"
        android:layout_height="match_parent" />
```

然后修改布局文件对应的 SelectedFragment.java 代码：

（1）在 oncreate 方法中查找 RecyclerView，设置布局管理器、适配器，设置 headerView。

```
    recyclerView= (RecyclerView) rootView.findViewById(R.id.recyclerView);
LinearLayoutManager mLayoutManager = new LinearLayoutManager(getActivity(),
LinearLayoutManager.VERTICAL, false);
recyclerView.setLayoutManager(mLayoutManager);
    selectedAdapter=new SelectedAdapter();
recyclerView.setAdapter(selectedAdapter);
    selectedAdapter.setHeaderView(headView);
```

（2）修改适配器 SelectedAdapter.java：

```
public class SelectedAdapter extends RecyclerView.Adapter {
private List selectedArticles;
public static final int TYPE_HEADER = 0;
public static final int TYPE_NORMAL = 1;
private View mHeaderView;//头文件 View

public SelectedAdapter() {
    selectedArticles = new ArrayList<SelectedArticle>();
    initData();
}

private void initData() {
    for (int i = 0; i < 50; i++) {
        SelectedArticle selectedArticle = new SelectedArticle(i, "Android 开发
666", i, i, "");
        selectedArticles.add(selectedArticle);
    }
}

public void setHeaderView(View headerView) {
    mHeaderView = headerView;
    notifyItemInserted(0);
}

public View getHeaderView() {
    return mHeaderView;
}

@Override
public int getItemViewType(int position) {
    if (mHeaderView == null)
        return TYPE_NORMAL;
    if (position == 0)
        return TYPE_HEADER;
    return TYPE_NORMAL;
}

@Override
public SelectedViewHolder onCreateViewHolder(ViewGroup parent, int viewType) {
    if (mHeaderView != null && viewType == TYPE_HEADER){//头类型
        return new SelectedViewHolder(mHeaderView);
    }
    View layout = LayoutInflater.from(parent.getContext()).inflate
                (R.layout.fragment_selected_item, parent, false);
    return new SelectedViewHolder(layout);
}

@Override
public void onBindViewHolder(SelectedViewHolder holder,int position){
    if(getItemViewType(position) == TYPE_HEADER)
        return;

    SelectedArticle selectedArticle = selectedArticles.get(position);
```

```
        holder.title.setText(selectedArticle.getTitle());
        holder.like.setText("" + selectedArticle.getLikeNumber());
        holder.comment.setText("" + selectedArticle.getCommentNumber());
    }

    @Override
    public long getItemId(int position) {
        return selectedArticles.get(position).getId();
    }

    @Override
    public int getItemCount() {
        return selectedArticles.size();
    }

    class SelectedViewHolder extends RecyclerView.ViewHolder {
        private TextView title;
        private TextView like;
        private TextView comment;

        public SelectedViewHolder(View view) {
            super(view);
            if(itemView == mHeaderView)
                return;
            title = (TextView) view.findViewById(R.id.tv_title);
            like = (TextView) view.findViewById(R.id.tv_like);
            comment = (TextView) view.findViewById(R.id.tv_comment);
        }
    }
}
```

适配器继承自 RecyclerView.Adapter，重写 5 个方法。在构造方法中初始化 50 条文章对象。

- setHeaderView：设置头布局，并且指定更新 RecyclerView 第一条记录。
- getItemViewType：如果头 View 不为空并且是第一条记录，就返回头布局的类型。如果是普通 Item，就返回普通 Item 的类型。
- onCreateViewHolder：创建每一行的 View，如果是头布局就显示传入进来的头 View；如果是普通类型，就通过 LayoutInflater.inflate 方法创建普通布局的 View。
- onBindViewHolder：给每一行的 View 绑定数据。
- getItemCount：要显示的条数。

10.5　RecyclerView 实现下拉刷新和上拉加载更多

将 RecyclerView 下拉刷新与上拉加载更多加入到开发者头条 App 中。这样首页的精选列表就能支持分页了。在实际工作中，大部分列表页都需要支持分页加载。

效果如图 10-5 所示。

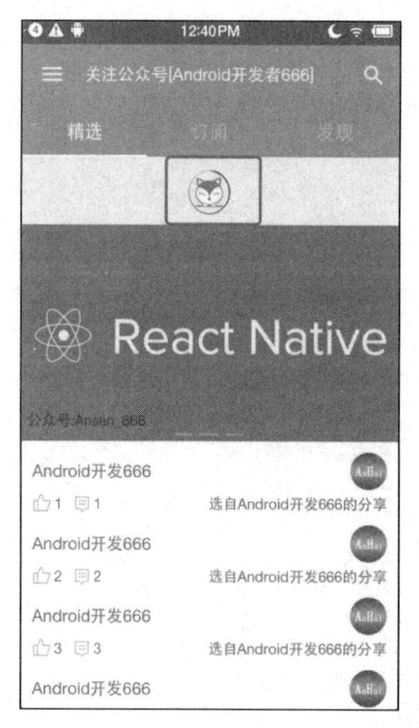

图 10-5　实现下拉刷新和上拉加载更多的精选列表

10.5.1　实现步骤

步骤01 查找一个带有下拉刷新和上拉加载更多的 RecyclerView 开源库。

步骤02 下载后运行一下，然后看看是不是我们需要的功能，觉得不错就把模块依赖进来，整合主项目。

步骤03 整合进来之后，肯定需要进行适当修改，例如本例就产生滑动冲突，有多个 headView 等问题。

10.5.2　实现详解

1. 查找 RecyclerView 下拉刷新和上拉加载的开源库

查找开源项目的首选是 Github，搜索一下，会发现有一大堆。如果效果图是想要的功能，再查找排名靠前、收藏比较多的项目。这里查找的项目是 CommonPullToRefresh，支持 ListView、RecyclerView、GridView、SwipeRefreshLayout 等常用控件。运行了一下演示文件，没有发现什么问题，挺好用的。

刷新库的 Github 地址如下：

https://github.com/Chanven/CommonPullToRefresh

2. 将库添加到项目中

步骤01 将.module 导入进来，然后主项目依赖一下。有不会的读者可查看本书第 1 章 Android Studio 如何导入模块、添加项目依赖。

步骤02 修改 SelectedFragment。首先查看布局文件的变化，在 RecyclerView 外面包裹了自定义的一个类 PtrClassicFrameLayout，内部实现了下拉刷新和上拉加载。还可以设置自定义属性，这些属性的功能就不逐一解释了，有兴趣的读者可以点击 Github 上的链接，讲解得很详细。

修改 fragment_selected.xml 文件：

```
<?xml version="1.0" encoding="utf-8"?>
<LinearLayout xmlns:android="http://schemas.android.com/apk/res/android"
   android:layout_width="match_parent"
   android:layout_height="match_parent"
   android:orientation="vertical">

   <com.chanven.lib.cptr.PtrClassicFrameLayout xmlns:cube_ptr=
     "http://schemas.android.com/apk/res-auto"
      android:id="@+id/test_recycler_view_frame"
      android:layout_width="match_parent"
      android:layout_height="match_parent"
      android:background="#f0f0f0"
      cube_ptr:ptr_duration_to_close="200"
      cube_ptr:ptr_duration_to_close_header="700"
      cube_ptr:ptr_keep_header_when_refresh="true"
      cube_ptr:ptr_pull_to_fresh="false"
      cube_ptr:ptr_ratio_of_header_height_to_refresh="1.2"
      cube_ptr:ptr_resistance="1.8">

      <android.support.v7.widget.RecyclerView
         android:id="@+id/test_recycler_view"
         android:layout_width="match_parent"
         android:layout_height="match_parent"
         android:background="@android:color/white"/>
   </com.chanven.lib.cptr.PtrClassicFrameLayout>
</LinearLayout>
```

接下来查看 SelectedFragment.java。在 onCreateView 中查找 PtrClassicFrameLayout 控件，然后调用 init 方法。

```
    @Override
    public View onCreateView(LayoutInflater inflater, ViewGroup container,Bundle
savedInstanceState){
        View rootView = LayoutInflater.from(getActivity()).inflate(R.layout.
                        fragment_selected, null);

        ptrClassicFrameLayout = (PtrClassicFrameLayout)
rootView.findViewById(R.id.test_recycler_view_frame);
        mRecyclerView = (RecyclerView)
rootView.findViewById(R.id.test_recycler_view);
        mRecyclerView.setLayoutManager(new LinearLayoutManager(getActivity()));

        init();
        return rootView;
    }
```

接下来看 init 方法如何实现。首先初始化适配器，然后用 RecyclerAdapterWithHF 类对适配器做包装，这个包装类中封装了添加头部的方法，因为 RecyclerView 不支持添加 Header，需要自己手动去实现。最后设置了下拉刷新和上拉加载的监听。

```
    private void init() {
        //初始化适配器
        selectedAdapter = new SelectedRecyclerAdapter(getActivity());
        //对适配器进行封装
        mAdapter = new RecyclerAdapterWithHF(selectedAdapter);
        //将滚动 Banner 加入头部
        mAdapter.addCarouse(initCarouselHead());
        mRecyclerView.setAdapter(mAdapter);
        ptrClassicFrameLayout.setPtrHandler(ptrDefaultHandler);//设置下拉监听
        ptrClassicFrameLayout.setOnLoadMoreListener(onLoadMoreListener);
          //设置上拉监听
        ptrClassicFrameLayout.setLoadMoreEnable(true);//设置可以加载更多
    }
```

在 init 方法中还调用了 initCarouselHead 方法，这个方法初始化了 Banner，把 Banner 布局 View 返回，用来作为 RecyclerView 的头部。

```
    private View initCarouselHead(){//初始化
        View headView = LayoutInflater.from(getActivity()).inflate
(R.layout.fragment_selected_header,mRecyclerView,false);

        tvContent=(TextView) headView.findViewById(R.id.tv_content);
        tvContent.setText(carousePageStr[0]);

        viewPager = (ViewPager)headView.findViewById(R.id.viewpager);
        selectedPagerAdapter=new SelectedPagerAdapter(getActivity(),
carousePagerSelectView);
        viewPager.setOffscreenPageLimit(2);
        viewPager.setCurrentItem(0);
        viewPager.addOnPageChangeListener(onPageChangeListener);
        viewPager.setAdapter(selectedPagerAdapter);

        ViewGroup group = (ViewGroup) headView.findViewById(R.id.viewGroup);
          // 初始化底部显示控件
        tips = new ImageView[3];
        for (int i = 0; i < tips.length; i++){
            ImageView imageView = new ImageView(getActivity());
            if (i == 0) {
                imageView.setBackgroundResource(R.mipmap.page_
                 indicator_focused);
            } else {
                imageView.setBackgroundResource(R.mipmap.page_
                 indicator_unfocused);
            }

            tips[i] = imageView;
            LinearLayout.LayoutParams layoutParams = new LinearLayout.
             LayoutParams(new LayoutParams(LayoutParams.WRAP_CONTENT,
             LayoutParams.WRAP_CONTENT));
            layoutParams.leftMargin = 10;// 设置圆点 View 的左边距
            layoutParams.rightMargin = 10;// 设置圆点 View 的右边距
            group.addView(imageView, layoutParams);
```

```
        }

        timer = new Timer(true);//初始化计时器
        timer.schedule(task, 0, CAROUSEL_TIME);//延时 0ms 后执行,3000ms 执行一次

        return headView;
    }
```

接下来查看 RecyclerView 适配器 SelectedRecyclerAdapter.java 代码。适配器前面已经学习了很多遍了，不做解释：

```
    public class SelectedRecyclerAdapter extends RecyclerView.Adapter
    <RecyclerView.ViewHolder> {
        private List<SelectedArticle> selectedArticles;
        private LayoutInflater inflater;

        public SelectedRecyclerAdapter(Context context) {
            super();
            inflater = LayoutInflater.from(context);

            selectedArticles = new ArrayList<SelectedArticle>();
            initData();
        }

        private void initData() {
            for (int i = 0; i < 10; i++) {
                SelectedArticle selectedArticle = new SelectedArticle(i, "Android
开发 666", i, i, "");
                selectedArticles.add(selectedArticle);
            }
        }

        public void loadMore(int page) {
            for (int i = 0; i < 5; i++) {
                SelectedArticle selectedArticle = new SelectedArticle(i, "第" + page
+ "页数据", i, i, "");
                selectedArticles.add(selectedArticle);
            }
        }

        public void getFirst() {
            selectedArticles.clear();
            initData();
        }

        @Override
        public int getItemCount() {
            return selectedArticles.size();
        }

        @Override
        public void onBindViewHolder(RecyclerView.ViewHolder viewHolder, int
position) {
```

```
            SelectedRecyclerHolder holder = (SelectedRecyclerHolder) viewHolder;

            SelectedArticle selectedArticle = selectedArticles.get(position);
            holder.title.setText(selectedArticle.getTitle());
            holder.like.setText("" + selectedArticle.getLikeNumber());
            holder.comment.setText("" + selectedArticle.getCommentNumber());
        }

        @Override
        public RecyclerView.ViewHolder onCreateViewHolder(ViewGroup viewHolder,
int position) {
            View view = inflater.inflate(R.layout.fragment_selected_item, null);
            return new SelectedRecyclerHolder(view);
        }

        public class SelectedRecyclerHolder extends RecyclerView.ViewHolder {
            private TextView title;//标题
            private TextView like;//喜欢数量
            private TextView comment;//评论数量

            public SelectedRecyclerHolder(View view) {
                super(view);
                title = (TextView) view.findViewById(R.id.tv_title);
                like = (TextView) view.findViewById(R.id.tv_like);
                comment = (TextView) view.findViewById(R.id.tv_comment);
            }
        }
    }
```

步骤03 解决整合进来的问题。

滑动冲突：当上拉到顶部将标题栏挤出屏幕外时，再进行下拉会触发 RecyclerView 的下拉事件，正确的情况应该是显示标题栏。

- RecyclerView 下拉刷新时先判断标题栏是否显示。如果标题栏没有显示，则不处理。
- AppBarLayout 有一个 addOnOffsetChangedListener 方法，在 AppBarLayout 的布局偏移量发生改变时被调用。

在 MainFragment 中进行监听：

```
    appBarLayout= (AppBarLayout) rootView.findViewById(R.id.appBarLayout);
    appBarLayout.addOnOffsetChangedListener(onOffsetChangedListener);
```

然后在回调函数中将值赋给 SelectedFragment：

```
    private AppBarLayout.OnOffsetChangedListener onOffsetChangedListener=new
AppBarLayout.OnOffsetChangedListener() {
        @Override
        public void onOffsetChanged(AppBarLayout appBarLayout, int i){
           //i>=0  标题栏完全显示
            selectedFragment.setPullRefresh(i>=0);
            System.out.println("i 值:"+i);
        }
    };
```

在 SelectedFragment 中，继续将值传给 PtrFrameLayout：

```
public void setPullRefresh(boolean pullRefresh) {
   ptrClassicFrameLayout.setPullRefresh(pullRefresh);
}
```

在 PtrFrameLayout 中用一个实例变量接收这个值：

```
private boolean pullRefresh=true;

public void setPullRefresh(boolean pullRefresh) {
   this.pullRefresh = pullRefresh;
}
```

找到 PtrFrameLayout 类的 dispatchTouchEvent 事件，这个方法是处理屏幕触摸事件的。

```
@Override
public boolean dispatchTouchEvent(MotionEvent e) {
   if (!isEnabled() || mContent == null || mHeaderView == null) {
      System.out.println("都是空的...");
      return dispatchTouchEventSupper(e);
   }
   int action = e.getAction();
   switch (action) {
      case MotionEvent.ACTION_UP:
         System.out.println("弹起...");
      case MotionEvent.ACTION_CANCEL:
          System.out.println("取消...");
//             if(pullRefresh){
            mPtrIndicator.onRelease();
            if (mPtrIndicator.hasLeftStartPosition()) {
               if (DEBUG) {
                  PtrCLog.d(LOG_TAG, "call onRelease when user release");
               }
               System.out.println("call onRelease when user release");
               onRelease(false);
               if (mPtrIndicator.hasMovedAfterPressedDown()) {
                  sendCancelEvent();
                  return true;
               }
            }
            return dispatchTouchEventSupper(e);
//             }
      case MotionEvent.ACTION_DOWN:
         System.out.println("按下...");
         mHasSendCancelEvent = false;
         mPtrIndicator.onPressDown(e.getX(), e.getY());

         mScrollChecker.abortIfWorking();

         mPreventForHorizontal = false;
         // The cancel event will be sent once the position is moved.
         // So let the event pass to children.
         // fix #93, #102
         return dispatchTouchEventSupper(e);
      case MotionEvent.ACTION_MOVE:
```

```
            System.out.println("移动...");
            if(pullRefresh){//Toolbar 显示
               mLastMoveEvent = e;
               mPtrIndicator.onMove(e.getX(), e.getY());
               float offsetX = mPtrIndicator.getOffsetX();
               float offsetY = mPtrIndicator.getOffsetY();

               if (mDisableWhenHorizontalMove && !mPreventForHorizontal &&
               (Math.abs(offsetX) > mPagingTouchSlop && Math.abs(offsetX) >
               Math.abs(offsetY))) {
                  if (mPtrIndicator.isInStartPosition()) {
                     mPreventForHorizontal = true;
                  }
               }
               if (mPreventForHorizontal) {
                  return dispatchTouchEventSupper(e);
               }

               boolean moveDown = offsetY > 0;
               boolean moveUp = !moveDown;
               boolean canMoveUp = mPtrIndicator.hasLeftStartPosition();

               if (DEBUG) {
                  boolean canMoveDown = mPtrHandler != null && mPtrHandler.
                  checkCanDoRefresh(this, mContent, mHeaderView);
                  PtrCLog.v(LOG_TAG, "ACTION_MOVE: offsetY:%s, currentPos: %s,
                   moveUp: %s, canMoveUp: %s, moveDown: %s: canMoveDown: %s",
                   offsetY, mPtrIndicator. getCurrentPosY(), moveUp,
                   canMoveUp, moveDown, canMoveDown);
               }

               // disable move when header not reach top
               if (moveDown && mPtrHandler != null && !mPtrHandler.
                   checkCanDoRefresh(this, mContent, mHeaderView)) {
                  return dispatchTouchEventSupper(e);
               }

               if ((moveUp && canMoveUp) || moveDown) {
               // System.out.println("是否下拉刷新:
                                          "+pullRefresh+"偏移量是多少:"+offsetY);
                  movePos(offsetY);
                  return true;
               }
            }
      }
      return dispatchTouchEventSupper(e);
}
```

只修改了一行代码，当 action==MotionEvent.ACTION_MOVE 时，先判断传入的 pullRefresh 是否为 true。

3. 顶部添加轮播

RecyclerView 的头部和底部加入 View，前面已经介绍过了，都是用适配器的封装类

RecyclerAdapterWithHF 进行控制。从效果图可以看出，轮播的 View 是加入到头部的，找到 RecyclerAdapterWithHF 类，查看源代码即可。

步骤01 需要一个保存 View 的集合，其实使用一个变量也行，因为只有一个轮播 View。

```
    private List<View> mCarouse = new ArrayList<View>();//保存轮播 View
    //可以添加轮播 View
    public void addCarouse(View view){
        mCarouse.add(view);
    }
```

步骤02 定义一个常量，用于判断类型：

```
    public static final int TYPE_CAROUSE = 7900;
```

步骤03 在 getItemViewType 中添加轮播的类型：

```
    @Override
    public final int getItemViewType(int position) {
        // check what type our position is, based on the assumption that the
        // order is headers > items > footers
        if (isHeader(position)) {
            return TYPE_HEADER;
        } else if (mCarouse.size()>0&&mHeaders.size()==position){
        //判断集合个数&&position==0，这时 mHeaders 中还没有值
            return TYPE_CAROUSE;
        }else if (isFooter(position)) {
            return TYPE_FOOTER;
        }
        int type = getItemViewTypeHF(getRealPosition(position));
        if (type == TYPE_HEADER || type == TYPE_FOOTER|| type == TYPE_CAROUSE) {
            throw new IllegalArgumentException("Item type cannot equal " +
TYPE_HEADER + " or " + TYPE_FOOTER);
        }
        return type;
    }
```

步骤04 onCreateViewHolder 也要修改一下，就是在 if 中多加了一个&&。无论是头部或底部轮播的 View，都添加到 FrameLayout 中。

```
    @Override
    public final RecyclerView.ViewHolder onCreateViewHolder(ViewGroup viewGroup,
int type) {
        // if our position is one of our items (this comes from
        // getItemViewType(int position) below)
        if (type != TYPE_HEADER && type != TYPE_FOOTER && type != TYPE_CAROUSE) {
            ViewHolder vh = onCreateViewHolderHF(viewGroup, type);
            return vh;
            // else we have a header/footer
        } else {
            // create a new framelayout, or inflate from a resource
            FrameLayout frameLayout = new FrameLayout(viewGroup.getContext());
            // make sure it fills the space
            frameLayout.setLayoutParams(new ViewGroup.LayoutParams(ViewGroup.
             LayoutParams.MATCH_PARENT, ViewGroup.LayoutParams.MATCH_PARENT));
```

```
        return new HeaderFooterViewHolder(frameLayout);
    }
}
```

步骤05 onBindViewHolder 为项目绑定数据，其实就是第四步返回的 ItemView 绑定数据。

```
    @Override
    public final void onBindViewHolder(final RecyclerView.ViewHolder vh, int
position){
        // check what type of view our position is
        if (isHeader(position)) {
            View v = mHeaders.get(position);
            // add our view to a header view and display it
            prepareHeaderFooter((HeaderFooterViewHolder) vh, v);
        }else if(mCarouse.size()>0&&position==mHeaders.size()){
            //这时 mHeaders.size()值为 0
    //System.out.println("有多少个头 View:"+mHeaders.size()+"值等于多少:"+(mHeaders.
        size()-1));
            View v = mCarouse.get(mHeaders.size());//取出轮播的 View
            prepareHeaderFooter((HeaderFooterViewHolder) vh, v);
        } else if (isFooter(position)) {
            View v = mFooters.get(position - getItemCountHF() - mHeaders.size());
            // add our view to a footer view and display it
            prepareHeaderFooter((HeaderFooterViewHolder) vh, v);
        } else {
            vh.itemView.setOnClickListener(new MyOnClickListener(vh));
            vh.itemView.setOnLongClickListener(new MyOnLongClickListener(vh));
            // it's one of our items, display as required
            onBindViewHolderHF(vh, getRealPosition(position));
        }
    }
```

步骤06 从第五步看到头部或底部轮播的 View，最终都会调用 prepareHeaderFooter 方法。查看这个方法的源代码，其实就是将类型对应的 View 添加到 Item 中。

```
    private void prepareHeaderFooter(HeaderFooterViewHolder vh, View view) {
        // if it's a staggered grid, span the whole layout
        if (mManagerType == TYPE_MANAGER_STAGGERED_GRID) {
            StaggeredGridLayoutManager.LayoutParams layoutParams = new
             StaggeredGridLayoutManager.LayoutParams
                    (ViewGroup.LayoutParams.MATCH_PARENT,ViewGroup.LayoutParams.
                     WRAP_CONTENT);
            layoutParams.setFullSpan(true);
            vh.itemView.setLayoutParams(layoutParams);
        }

        // if the view already belongs to another layout, remove it
        if (view.getParent() != null) {
            ((ViewGroup) view.getParent()).removeView(view);
        }

        // empty out our FrameLayout and replace with our header/footer
        vh.base.removeAllViews();
        vh.base.addView(view);
    }}}}}}
```